Technical Writing:
A Practical Approach

Technical Writing:
A Practical Approach

Maxine T. Turner

Georgia Institute of Technology

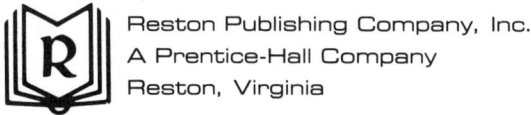

Reston Publishing Company, Inc.
A Prentice-Hall Company
Reston, Virginia

Library of Congress Cataloging in Publication Data

Turner, Maxine.
  Technical writing.

  1. Technical writing. I. Title.
T11.T786   1983   808'.0666   83-8681
ISBN 0-8359-7546-0

Interior design and production by Jeanne-Marie Peterson

©1984 by Reston Publishing Company
A Prentice-Hall Company
Reston, Virgina 22090

All rights reserved. No part of this book may be reproduced in any way, or by any means, without permission in writing from the publisher.

10   9   8   7   6   5   4   3   2   1

Printed in the United States of America

Late on a Sunday afternoon as a team of engineers worked to meet a deadline for a technical recommendations report, the project manager threw his pen across the room, buried his face in his hands, and sobbed, "@#%&! I hate to write *sooo BAD!*"

This book is lovingly dedicated to technical writers who have shared that feeling.

Contents

Preface ... xvii

Part 1 An Introduction to Technical Writing ... **1**

Chapter 1 Technical Writing Functions ... 5

Chapter 2 Technical Writing Systems ... 10

Chapter 3 How Technical Professionals Write ... 15

    Invention ... 15
    Arrangment ... 16
        Definition, 17; Description, 17; Comparison and Contrast, 17; Classification, 18; Division, 18; Process, 18; Cause and Effect, 18
    Style ... 19
    Grammar ... 19
    Mechanics ... 20

Chapter 4 How Technical Writers Use Form ... 21

		The Elements of Form Form, 21; Design, 22; Conventions, 22;	21
		Examples Memorandums, 22; Memorandum: Example 4.1, 24; Letters, 24; Letter: Example 4.2, 26	22
Chapter 5		How Technical Professionals Write for an Audience	28
		Audience Analysis Memorandums, 31; Letters, 31; Proposal, 32; Technical Report, 33; Technical Report, 33	31
Chapter 6		Technical Writing and Professional Ethics	38
		Why a Work Ethic Is Important to a Technical Writer	38
		A Definition of the Ethics of Technical Writing	39
		The Function of the Work Ethic in Technical Writing	40
		The Standard Code of Ethics in Technical Writing	42

Part 2   Short-Term Writing Projects   45

Chapter 7	Memorandums	47
	Components of the Memorandum Function, 47; Design, 48; Convention, 48; Technique, 48; Organization, 48; Basic Writing Skill, 49; Style, 49	47
	Evaluating Memos Original Assignment, 50; Progress Report Assignment, 52; Student Memo: Exercise 7.1, 53; Student Memo: Exercise 7.2, 54; Student Memo: Exercise 7.3, 56; Student Memo: Exercise 7.4, 56; Student Memo: Exercise 7.5, 57; Summary Evaluation, 58	49
	Major Types of Memos Summary Example 7.0, 58; Informative Memorandum: Example 7.A.1, 60; Informative Memorandum: Example 7.A.2, 60; Directive Memorandum: Example 7.B.1, 61; Directive Memorandum: Example 7.B.2, 62; Meeting Memorandum: Example 7.C.1, 64; Meeting Memorandum: Example 7.C.2, 65; Administrative Memorandum: Example 7.D.1, 66; Administrative Memorandum: Example 7.D.2, 67; Administrative Memorandum: Example 7.D.3, 68	58

Contents

Chapter 8	Letters	69

Components of the Letter ... 69
    Function, 69; Design, 70; Convention, 70; Techniques, 74; How to Begin a Letter: Routine Requests, Reports, Responses, 75; How to End a Letter, 76; Basic Writing Skills, 78; Situation, 78

Summary Examples of Types of Letters ... 79
    Routine Business Letter: Summary Example 8.1, 79; Routine Technical Letter: Summary Example 8.2, 80; Formal Letter Re Technical Project: Summary Example 8.3, 81; Letters, 83; Letter Which Requests Information: Design, 85

Letter Which Requests Information ... 86
    Construction, 86; Situation, 86; Letter Which Gives Information (Technical Design), 87

Letter Which Gives Information: Technical ... 88
    Construction, 88; Convention, 88; Signature Line, 88; Situation, 88; Letter Which Gives Information: Professional (Design), 89

Letter Which Gives Information: Professional ... 91
    Construction, 91; Situation, 91; Letter Which Gives Information (Sales—Public Relations): Design, 92

Letter Which Gives Information: Sales—Public Relations ... 92
    Construction, 92; Convention, 93; Enclosure, 93; Situation, 93; Letter Which Requests Action: Design, 94

Letter Which Requests Action ... 94
    Construction, 94; Situation, 95; Letter Which Reports Action: Design, 95

Letter Which Reports Action ... 96
    Construction, 96; Situation, 96; Letter Which Directs Action: Design, 97

Letter Which Directs Action ... 98
    Construction, 98; Convention, 98; Reference Line, 98; Situation, 98; Letter Which Sells: Design, 99

Letter Which Sells ... 100
    Construction, 100; Situation, 100

Conclusion
    Exercise, 101; Assignment Letter, 102; Student Letter, 103; Student Letter, 104; Student Letter, 105; Student Letter, 107; Student Letter, 109

Chapter 9	Informal Reports: Combining the Letter and Memorandum Forms	111

	Length	111
	Situation	112
	The Long Technical Letter, 112; The Technical Letter to a Personal Contact, 112; The Technical Problem or Business Disagreement, 113; Original Letter, 114; Revision: Cover Letter, 115; Memorandum, 115	
	Summary Examples	117
	Cover Letter for Memorandum Report: Summary Example 9.0, 117; Memorandum Report: Summary Example 9.1, 118; Memorandum Report: Summary Example 9.2, 119; Memorandum Report: Summary Example 9.3, 120	
Chapter 10	Correspondence Systems	121
	The Employment Project	122
	Components of Employment Correspondence, 122; Examples, 123; Letter of Application: Example 10.A.1, 124; Letter of Application: Example 10.A.2, 125; Letter of Application: Example 10.A.3, 126	
	Letter of Application	127
	Construction, 127; Resume: Summary Example, 128; Letter Requesting a Recommendation: Example 10.B.1, 129; Letter Requesting a Recommendation: Example 10.B.2, 130	
	Letter Requesting a Recommendation	131
	Construction, 131; Thanks for a Plant Trip: Example 10.C.1, 132; Thank-You Letter: Example 10.C.2, 133; Thank-You Letter: Example 10.C.3, 134	
	Thank-You Letter	134
	Construction, 134; Letter Accepting a Job: Example 10.D, 135	
	Letter Accepting a Job	135
	Construction, 135; Letter Refusing a Job: Exercises 10.1 and 10.2, 136	
	Refusing a Job	137
	Construction, 137; Exercise: Employment Letters, 137; Letter of Application: Exercise 10.3, 138; Letter of Application: Exercise 10.4, 139; Letter of Application: Exercise 10.5, 140; Letter of Application: Exercise 10.6, 141; Letter Requesting a Recommendation: Exercise 10.7, 142; Letter Accepting a Job: Exercise 10.8, 142; Letter Accepting a Job: Exercise 10.9, 143; Letter Accepting a Job: Exercise 10.10, 144	

Contents   xi

         Correspondence System for a Large Technical Project   144
            Stage 1: Initial Contact and Proposal, 145; Stage 2:
            Planning, 147; Stage 3: Actual Performance, 148;
            Stage 4: Completing a Project, 152; Conclusion, 152

Part 3   Long-Term Writing Projects   155

Chapter 11   Short Technical Reports   163

         Short Technical Report: Summary Example 11.1, 164;
         Short Report to a Small Non-Technical Audience:
         Summary Example 11.2, 170; Short Technical Report with
         Transmittal Letter: Exercise 11.1—Cover Letter, 171;
         Report, 172; Short Technical Report: Exercise 11.2, 177

Chapter 12   Case Studies   183

         Case Study: Summary Example   184
            Format, 184; Case Study: I. Summary of Analysis, 187;
            Construction, 188; II. Major Issues, 188; Construction, 188; III. Analysis, 189; Construction, 190; Construction, 192; Construction, 194; Construction, 195;
            Construction, 197

Chapter 13   Laboratory Reports   198

         Components of the Laboratory Report   198
            Subject, 198; Purpose, 198; Audience, 199; Design, 199; Style, 199; Lab Report Assignment: Example 13.1, 200; Annotated Lab Report: Example 13.2, 210; Construction, 211; Summary, 211; Construction, 212; Introduction, 212; Construction, 212; Apparatus, 213; Construction, 214; Procedure, 215; Construction, 218; Results and Discussion, 219; Construction, 219; Conclusion, 220; Construction, 220; References, 221; Graphics, 221; Appendix, 222

Chapter 14   Proposals   229

         Summary Example 14.0, 231
         General Qualities of a Proposal   234
            Proposal: Example 14.1, 235; Proposal: Example
            14.2, 236; Letter of Transmittal: Proposal to a Large

Technical Audience, 237; Letter of Transmittal: Example 14.3, 239; Proposal: Exercise 14.1, 240; Proposal: Exercise 14.2, 243; Proposal: Exercise 14.3, 247

Chapter 15   Technical Recommendations Reports   251

Guide for Writing the First Draft of a Technical Recommendations Report   253
    Who, 253; Structuring the Discussion, 254; Appendices, 258; Conclusion (The Final WHY), 259; Recommendations (The Final HOW), 259; Introduction, 259; Executive Summary, 260
Guide for Constructing the Finished Draft for a Technical Recommendations Report   261
    Table of Contents, 261; Construction, 262; 1. Executive Summary, 263; Construction, 263; 2. Introduction, 264; Construction, 265; 3. Beginning a Chapter, 267; Construction, 268; Conclusions and Recommendations, 269; Construction, 270
Reading and Evaluating Technical Reports   271
    Technical Recommendations Report: Exercise 15.1, 272

Chapter 16   Technical Presentations   312

Factors to Be Managed   313
    Acknowledging the Fear of Speaking, 313; Managing the Fear of Speaking, 314; Managing the Physical Setting, 315; Managing Speech Notes as a Physical Factor, 316; Managing A/V Equipment and Graphics as a Physical Factor, 316; Managing the Material for an Oral Presentation, 317; Managing Audience Attention, 318; Managing Time, 319
Group Presentations   320
    Design Conference, 320; Presentation, 320

Chapter 17   Professional Articles   322

Publishing a General-Interest Article   325
Reporting the Results of Research   325
    Subject: *WHAT?* 325; Audience: *WHO?* 326; Purpose: *HOW?* 327; Publication: *HOW?* 328; Publication: *WHERE?* 328

Evaluating Professional Articles 329
   Example 17.1, 331; Example 17.2, 338; Example 17.3, 344

Part 4 A Technical Writer's Handbook 353

Chapter 18 Words 355

Standard English 355
   Colloquial Usage, 356; Unidiomatic Usage, 357; Correct Usage, 358
Current Business Usage 359
   How *Not* to Begin a Letter, 359; How *Not* to End a Letter, 360
Precise Technical Usage 361
   Numbers, 361; Words Related to Quantities, 361; Words Related to Cause, 362; Verbs, 363
Exercises 364
   Listening, 364; Reading, 366

Chapter 19 Sentences 369

Basic Sentence Structure 370
Adding Components to a Sentence 370
   At the Beginning, 370; In the Middle, 370; At the End, 371
Multiple Components in Sentence Structure 371
   Coordinating Conjunctions, 371; Correlative Conjunctions, 372; Conjunctive Adverbs with Semicolons, 372
Learning to Write Sentences 372
   Exercise, 372
Punctuation Guide 374
   Commas, 374; Semicolons, 375; Colons, 375; Dashes, 375
Informal Procedure for Testing Sentence Structure 376
Common Design Flaws in Sentences 376
   Faulty Use of Coordinating Conjunctions, 377; The Loose Adjective Clause or "*Which* Afterthought," 377; Faulty Parallelism, 377; Dangling Modifier Followed by Passive Voice, 377; Run-on Sentence, 377;

Awkward Structure, 378; Unnecessary Comma, 378; Misuse of Semicolon, 378; Absence of Semicolon, 378

Chapter 20 Paragraphs 379

Classification and Division 380
  Basis, 380; Division of Objects: Example 20.1, 381; Division of Groups: Example 20.2, 382; Division of Ideas: Example 20.3; 383
Definition 383
  Technical Definition: Example 20.4, 384; Non-Technical Definition: Example 20.5, 385
Description 386
  Building: Example 20.6, 387; Site: Example 20.7, 388; Space and the Reader, 389; Space and the Writer, 389
Process 389
  Process: Example 20.8, 389
Cause and Effect 391
  Technical: Example 20.9, 391; Persuasive: Example 20.10, 392
Exercises 393
  Reading Exercise, 393; Writing Exercise, 393

Chapter 21 Graphics in Technical Writing 394

Partition, 395; Classification and Division, 398; Comparison and Contrast, 399; Process, 402; Figure in Text: Example 21.1, 405; Appended Figures: Example 21.2, 407; Page Design: Example 21.3, 408; Manuscript Specifications, 409

Chapter 22 Outlines 410

The Thinking Process 410
Techniques of Arranging Data 411
  Exercise 22.1, 411
Relationship of Parts 412
Linear Forms 412
  Historical Sequence, 412; Causal Sequence, 413; Exercise 22.2, 413
Vertical Forms 415
  Comparison and Contrast, 415; Classification and Division, 416; Exercise 22.3, 417; Exercise 22.4, 418; Selecting an Organizational Pattern, 418; Exercise 22.5, 419; Outline 22.5.1, 420; Outline 22.5.2, 421;

	Outline 22.5.3, 422; Outline 22.5.4, 423; Outline 22.5.5, 424; Outline 22.5.6, 425; Outline 22.5.7, 426	
	Intensive Reading	427
	Reading Exercise 22.1, 428; Reading Exercise 22.2, 429; Reading Exercise 22.3, 429	
Chapter 23	Technical Persuasion	430
	Ways to Persuade	431
	Argument by Exposition, 431; Argument Based on Authority, 432	
	Ways to Refute Persuasive Techniques	433
	Flaws in Cause-and-Effect Statements, 434; Flaws in Language, 436; Exercise, 436	
Chapter 24	Researching and Writing a Library Paper	437
	Sustained Writing	437
	The First Step in Sustained Writing, 438; What You Already Know, 438; How the Scope of Sustained Writing Differs, 438; How Your Relationship to a Subject Differs, 439; How Your Treatment of Subject Differs, 439; How Your Relationship to a Reader Differs, 440; How the Purpose Differs, 440	
	Guide for Selecting a Topic	440
	What to Write About, 441; A Comparative Study of Three Sub-Compact Cars, 442; Nutritional Quality of Commercial Baby Foods, 1973–1977, 446; Alternative Commuter Routes from College Park to Georgia Tech, 448; Remarks, 450	
	Guide for Collecting Data	450
	What You Should Know About Reference Sources, 450; How to Manage Library Sources, 451; Conclusion, 454	
	End Notes and Bibliography	455
	Exercises, 457	

Part 5   Appendices   461

Appendix A	Manuscript Specifications	463
	Memorandums	464
	Design, 464; Specifications, 465; Design, 466	
	Letters	468
	Overall Page Design, 468; Specifications, 469;	

Overall Page Design, 470; Spacing on Page 2 of a Letter, 471; INI/tials, 472; CC, 472; Enclosure, 472

Typist's Checklist for Producing a Long-Term Writing Project ... 475

Before Beginning to Type, 475; Beginning the Text, 476; Ending Page 1 of the Text, 476; Beginning Page 2 of the Text, 476; Break Each Page, 476; Allow Space to Add Graphics, 476; Beginning a New Section in a Continuous Text, Example, 477; Beginning a New Chapter in a Longer Text, 477; Citing Reference Sources in a Text, 477; Listing End Notes, 477; Listing Bibliography, 478; Adding Appendices, 478; Writing the Table of Contents for a Manuscript with Chapters, 479; Typing Page 2 of the Table of Contents, 479; Listing Appendices or Lists of Figures, 480; Adding Graphics to the Typed Draft, 480; Typing the Title page, 480; Binding the Paper in a Thesis Binder, 480; Adding a Cover Label, 481

One Final Note ... 481
The Title Page ... 482
Example A.1, 482; Example A.2, 483
Report Page Specifications ... 484
Memorandum Draft ... 485
Letter Draft ... 486
Report Draft ... 487

Appendix B Using Evaluation Checklists ... 488

Memorandum Checklist ... 489
Letters Checklist ... 490
Evaluation Checklist: Short Technical Reports ... 491
Evaluation Checklist: Lab Reports ... 492
Proposal Evaluation Checklist ... 494
Long Report Evaluation ... 495

Index ... 497

Preface

Technical Writing: A Practical Approach is a title which describes how every section of this book was first written: as an immediate, practical solution for a student or practicing professional with a writing problem.

After class one day an engineering student said, "It helps me when you draw a letter on the board; engineers like to *see* what the finished design should look like." That incident accounts for the emphasis in this text upon examples, summary examples, and directions for designing a document on the page.

A senior engineer in a consulting firm said in an almost pleading tone, "Give me *something* that I can put in the hands of my young people that will *show* them how to write a report." A part of the chapter on the technical recommendations report was the result. Several weeks later the same engineer said, "What you wrote didn't go far enough; we need clearer directions and more examples," and an expanded draft of the chapter on technical recommendations reports appeared in the first of many drafts.

The Society for Marketing Professional Services wanted a presentation on how to make engineering and architectural writing more sales effective, the faculty of the School of Allied Health at Georgia State wanted to learn more about publishing professional articles, the Student Section of the Society of Women Engineers wanted a presentation on how to make a technical presentation, a major retail sales corporation needed more up-to-date style in letter writing, and a middle management seminar for the Metropolitan Atlanta Transit

xvii

Authority called for a talk on overcoming barriers to good writing—and each of these situations created a part of what is now this text book. Many such situations and almost daily responses to the questions of individual technical writers at work have produced the material for this text.

The major substance of this text is contained in Parts Two and Three which describe and discuss the principal short and long forms written by technical professionals. Some texts in technical writing are divided on the basis of the types of writing all writers must perform—description, narration, process, and the like. Other texts are divided on the assumption that the long report will be assigned first as the major work of the course and therefore the long forms appear first in order of presentation.

Long-term consulting associations with an engineering firm and a design firm gave me the opportunity to observe how entry-level professionals first write in-house memorandums and then progress through the forms to simple letters, letters about more complex situations, short reports, parts of longer reports, and finally to proposals and technical presentations. Based on that experience as a consultant, I have arranged the forms on the basis of increasing complexity of content, form, and especially audience in a given situation. I believe also that a student can learn more about basic writing skill by writing the forms as they are ordered here.

Another feature of Parts Two and Three also derives from my experience as a consultant. I have organized this book on the basis of technical writing forms because I have so often seen the principle of form help a writer respond quickly and effectively in a letter writing situation by knowing what the basic form requires, and I have seen form sustain many writers through managing the long, weary process of writing an important proposal or a long technical recommendations report. Writing assignments are usually made according to form: "Write a memorandum for the files," "We need a letter to X client about the deliveries," or "The Navy has issued a Request for Proposals." Knowing the parts, the arrangement of parts, the format, and the conventions of a given form can thus carry a writer through the first, most difficult steps of writing. Such emphasis is designed to free rather than restrict a writer, in somewhat the same way that poet Robert Frost spoke of "moving easy in harness."

The central parts of this text are supported on either side by a series of introductory chapters in Part One and a Technical Writer's Handbook at the end of the text. Practical application as it is required to write the forms also requires an understanding of the task and skill to perform the task. The first and last sections of this text are designed to enhance that understanding and skill.

The introductory chapters address a real need I have found among my students to know more about technical writing in general terms. Many have read letters, of course, but few have given a close critical reading to letters; and few have read proposals, reports, and the other long forms in technical writing. Therefore, it seemed appropriate to dramatize the widespread presence of technical writing as a factor in all technical work in our society. I felt it important also to define technical writing as a specific kind of writing for special purposes. Unlike poems, essays, short stories, and scholarly articles, technical writing is most often done within a management system; as a result of interaction on a project team, with a client, or as a result of direct work experience.

The introductory chapters also reflect another point of view which I consider necessary to approach the task of technical writing confidently. Traditional rhetoric as it has been taught in freshman composition courses is perhaps the most useful tool for technical writers. Hence the "translation" of rhetorical patterns into technical writing terms in this introductory section, and an index ordered on the basis of arrangement and rhetorical patterns as we traditionally define them.

Finally, I wanted to reflect the enormous respect I have gained for the serious and professional attitude most engineers and technologists take toward their work. To relate technical writing and the work ethic, and to treat technical writing as the exercise of a professional skill of a very high order, seemed the best way to accomplish that purpose. Supplied with some larger understanding of the work of the technical professional as technical writer, I hope that students who use this text will consider technical writing not only important as a useful skill, but also as a professional discipline worthy of their best efforts.

Part Four of the text, A Technical Writer's Handbook, is a section that I did not initially set out to write. After all, there are many handbooks on the market—and I have taught dozens of them—to answer a writer's questions about diction, sentences, outlines, and the like. But it always seemed that, when I returned a set of graded memos or letters to a class, I would include lengthy memos about verbs having to do with time and motion, conventions of writing numbers, sentence variety as technical writing demands good sentence structure, and so on until I had the text of the handbook included here in a stack of ditto sheets addressed to my classes.

Many handbooks will deal in more specific terms with matters of grammar, spelling, and punctuation, but I have often wished that they would go one step further to address the specific need of technical writers: for appropriate diction in letters, for examples from technical writing to illustrate the traditional patterns of rhetoric, for a discussion of how persuading a client to buy a service or a product is different from more traditional forms of argument and persuasion. Thus the inclusion of this handbook to meet the specific needs of writers using English for a special purpose.

For the highly motivated individual who wishes to improve writing skills—as one works individually at golf, tennis, the violin, or another individual lifetime activity—there is a guide in the writer's handbook for independent work on writing. This program, however, is like most other parts of this book: it will work only for the person who is willing to work hard at it.

Appended in Part Five of the text are manuscript specifications and checklists. Like most such specifications, they are specific and sometimes arbitrary. The aim in formulating standards and specs has been to settle questions of grading standards *before* the fact. And it is especially my intention to free technical writers for the more important work of communicating the important knowledge and expertise they command.

As I first wrote parts of this book for individuals or for my classes, many anecdotes worked their way into the visions and revisions of the manuscript. My favorite occurred one day when a senior in engineering stopped me after class to say, "*Hey*, this is *useful!* I was afraid you would try to make Milton out of me." While I think we may be assured that this text will make no Miltons out of student writers, if I could justify the way this text works, it is that many writers among my students and clients have already found this approach useful. My wish for students and teachers who use this book is that they also may say, "This is *useful!*"

part 1

An Introduction to Technical Writing

1. Technical Writing Functions
2. Technical Writing Systems
3. How Technical Professionals Write
4. How Technical Writers Use Form
5. How Technical Professionals Write for an Audience
6. Technical Writing and Professional Ethics

These six short chapters describe the role which technical writing plays in the engineering and technological (E/T) professional fields. The first two chapters establish that good technical writing is part of the goal at which E/T students aim: applying what they learn to real technical problems. Working successfully requires being able to write well on the job.

The third and fourth chapters explain how technical writing differs from writing in general; it is writing for special purposes, most often in a prescribed form. The chapter on form "walks a writer through" basic writing assignments an entry-level E/T might perform on a job.

The fifth chapter introduces one of the strongest determinants in technical writing—the audience. The transmission of sound involves three essentials—the vibrating source, the transmitting medium, and the tuned receiver; similarly, technical communication does not happen unless an audience receives the message clearly.

The sixth chapter's discussion of the professional ethics of technical writing completes the introduction to technical writing: a specific skill used for a special and practical purpose, done with other workers, intended for a specific audience, and practiced as a part of a profession governed by ethical standards.

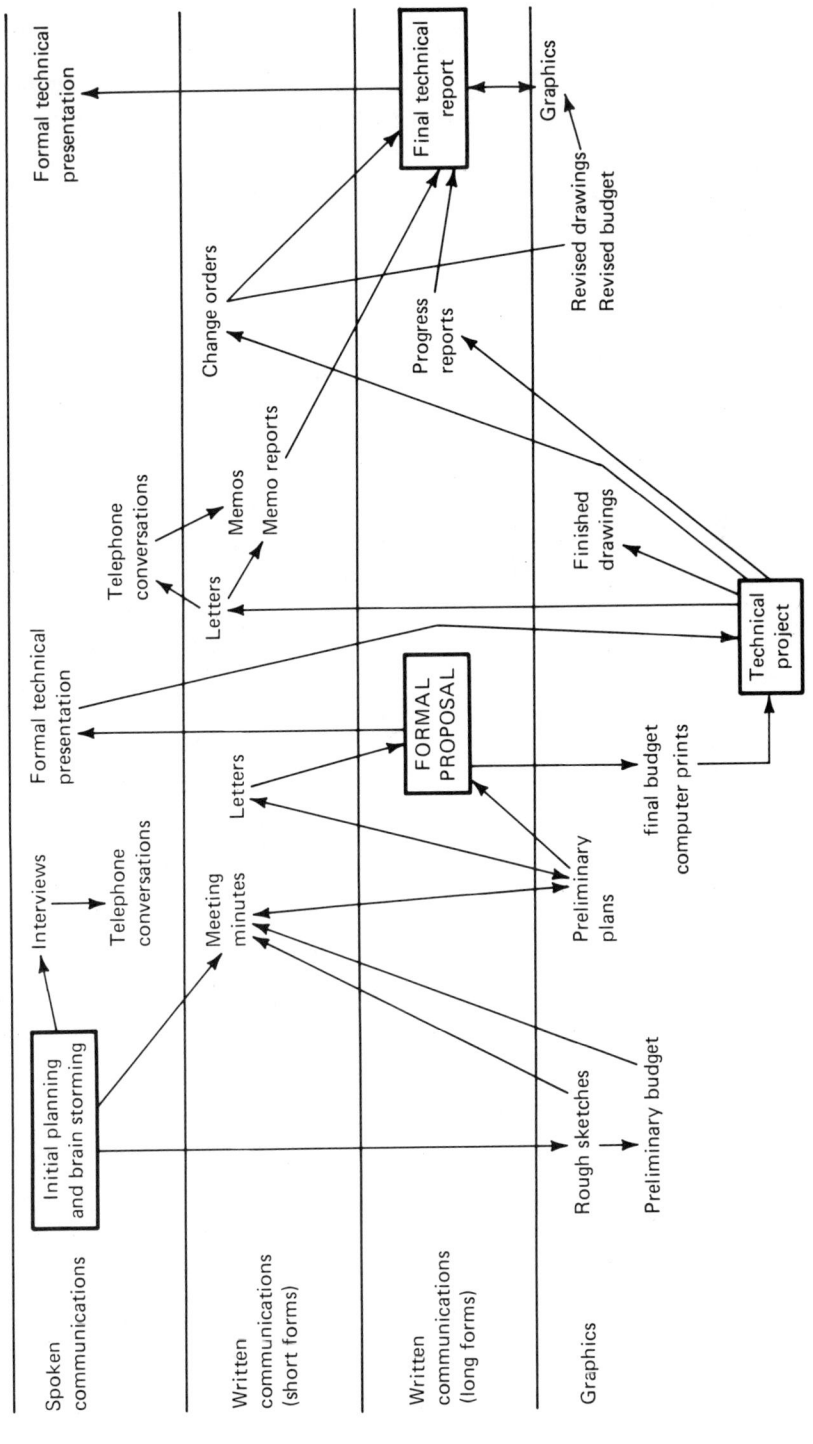

chapter 1

Technical Writing Functions

Technical writing is a part of almost any job an engineer or technologist (E/T) does.

Technical writing is also a hidden but indispensable factor in our modern civilization. To illustrate that statement, I'll suppose that you are an E/T student living in an apartment building on the outskirts of a city. Let's begin on an average day as your clock-radio awakens you so that you can get up, bathe and dress, eat breakfast, and drive to school. You are aware that technology has influenced every step in that daily process; this chapter is intended to make you more aware of how technical writing functions as a hidden factor in the technology which shapes our lives.

To talk about the transistor as a technological development is easy and obvious if we begin with your clock-radio. The transistor, which caused the shift away from the vacuum tube, is one of those major technological changes which alter the directions of industries, the course of careers in technology, and the way we live our lives day by day. It is easy to imagine how the transistor set off a chain reaction of memorandums, letters, short reports, long recommendations reports, scientific articles, and technical presentations—and even a Nobel Prize Acceptance Speech from those who developed it.

To examine fully the statement that technical writing is a hidden factor in technology, let's look more closely at your clock-radio. The housing is molded plastic, and a part of the housing forms a pedestal which tilts the face of the clock slightly upward to give you a better view. That little convenience is the concrete realization of some E/T's idea. Sitting at a work bench or a drawing board, the E/T may have said to himself, "Hey, I think I have an idea for a better way to design this housing." Next the E/T called over a supervisor or a senior designer to say, "Look at this sketch. This housing can be cast in one piece which includes this little pedestal; then it's easier to see the clock." The senior designer says, "I think you may have a workable idea there. Write a report to go with your drawing and I'll present it to our department management the next time we have a conference."

The most important thing to note here is that the E/T's process of invention goes only two steps before the idea must be given written form. How well one is able to translate an idea from a creative concept to a drawing, and then to written form, may be the key to whether the idea ever goes into production—and the E/T goes on to a promotion and a better-paying job.

Your clock-radio is playing, you're still struggling to wake up, and already we are into a series of technical-writing tasks. For the design process does not end in the transaction between the E/T and the supervisor; someone must present this idea to a group of others who may be, not E/Ts, but people with administrative responsibilities: someone who authorizes procurement and allocation; someone with a degree in finance who must consider budget questions; someone in management who assigns personnel to projects; and someone in marketing who predicts whether the design will sell. By this time, the memorandum report may have grown to a bulky package of memorandums, letters, short reports, and finally a lengthy recommendations report. The president of the company, who may have two dozen other projects to deal with, may read only the Executive Summary, or perhaps the Summary, Conclusions, and Recommendations of the long report before he decides the ultimate fate of the young E/T's innovative idea.

The vehicle for that entire process is not the idea, however brilliant and innovative; it is not the drawings and specifications, however clear and accurate. It is the word: the written and spoken communication which links the innovator to the supervisor to the design team to the financial, managerial, and advertising personnel and finally to the top-level management of the company.

Such thoughts of the technical communications tasks associated with such an everyday item as a clock-radio are enough to trouble your sleep, so you may as well get up to begin the day.

You go into the bathroom, switch on the light, and turn on the tap in order to splash cold water on your face; again you've encountered interesting examples of technical writing as a hidden factor in our lives. Both the design of the light fixture and the design of the faucet were subjected to much the same process we considered in the case of the clock-radio. Each of these designs began with an E/T saying, "Hey, I think I've got a better way to do this."

How the electricity and water were actually installed in your bathroom provides other examples of the function of technical writing. E/Ts designed the electrical systems in your building—how the power is led onto the property, into the building itself, and from the central electrical system to each switch in the building. The diameter of the conduit, the direction of the conduit upward to each floor and along the walls to your bathroom, the size and composition of the wire, the dimensions of the switch box, and the type of switch plate—each step formed a part of an E/T's task. The written documents produced at every step of this process included a long report describing the entire electrical system, the codes, the specifications, and the costs. Perhaps the senior electrical engineer offered the clients who own the building several options for the design, at different prices. Perhaps that report included not only the immediate cost of the installation but also the long-term considerations of the cost and availability of electricity. But the important point to remember here is this: before the building ever took shape, the electrical system existed as a design and within a written report. They are the communications tasks which accompany technical work of any significance.

Long before anyone, even you, knew that you would get up today and switch on the light in your bathroom, that simple action was made possible by a long series of technical documents. A letter ordering parts and materials from a supplier, a second letter if those were delayed, a letter to the client if the materials were delayed so long that the project fell behind, still more letters if the cost of the copper wire suddenly rose—any or all of these documents may have been written before the first conduit held the first inch of wire to the switch in your bathroom.

During construction, someone may have decided that the switch should be 40 inches above the baseboard, not 42. This decision set off a chain of change-order letters to architects, engineers, technicians, and contractors. In the final stages of construction, interior designers may have gone through an elaborate process of selecting the light fixtures. Members of the design team insisted that a certain fixture had great beauty and utility for its price, but the overall project manager or perhaps the clients said, "No, that's too expensive. We cannot spend more than this amount for the fixtures." So, the interior designers had to go back to their catalogues of bathroom lighting fixtures: another example of technical writing. As with your clock-radio, a lengthy process took place before you could switch on the light in your bathroom.

Before you splash more water on your face, consider the designs of the plumbing fixture, the basin, and in fact the entire plumbing system; they all took the same path as the design of the electrical system. Before ground was broken for your apartment building, geotechnical engineers did an investigation of the site. First they bid for the job in a proposal describing their services and their fees; this document combined a highly technical description of the job to be done with a sales appeal to convince the clients, or project engineers and architects, that their company could do this better than any other (or as well, and more cheaply). When their proposal won the contract, the geotechnical engineers then took core samples and applied their knowledge of the soil, rock, and water formations in the area. They determined how much weight this site could bear; how the weight should be distributed on bed rock, pilings, concrete slabs, or other supports; where to connect the plumbing in relation to soil and rock formations; and, depending upon the severity of the winters, how deep to lay the pipes.

Once the geotechnical E/Ts took off their hard hats, wiped their sweaty foreheads, and eased their feet out of their muddy boots, they wrote a report to describe the site, to give the results of their technical investigations, and to make recommendations to the client. If the random pattern of their core samples missed an underground spring or an old land fill, let's hope they explained, in writing, that a random sample cannot reveal every possible feature of what lies beneath the surface. If they determined that two or three different sizes and types of piling could each serve to support the proposed structure, they may have described each option's advantages and disadvantages. They then committed themselves to a professional and technical judgment about the best foundation pilings for your apartment building. Whatever the geotechnical E/Ts decided, their decision went into a written report for the engineers and architects (a technical audience), and for the legal record. If the building settled 3 feet on the northwest corner as the third-floor structure was being added, the report of the sub-surface investigation would be one of the first documents the project manager referred to.

If you have been standing too long in one position, musing upon technical writing as a hidden factor in our daily lives, a pain in your trick knee may remind you to take the drug prescribed by your orthopedist. That drug is the end of a long route mapped by technical writing : a large body of literature on the knee, on injuries to the knee, and on various drugs, therapies, surgical procedures, and prostheses used in treatment.

If you injured that knee while jogging or playing soccer, you went to an internist or general practitioner to say "Doc, I've got this pain in my knee." The physician translated that non-technical description into medical terminology and referred you to an orthopedist, who made an even more technical diagnosis. A telephone call and a follow-up letter, case notes, x-rays, lab reports,—all of these forms of communication followed you as you limped with a cane from one office to another.

A physical therapist may have received your case and, with the professional work, the task of communicating in a pivotal position: in highly technical terms to the attending orthopedist who supervised your treatment, and in clear and persuasive terms to you as a non-technical audience: how, when, and how much to exercise. "And be *sure* to eat something before you take this drug." The printed flyer in the package with the prescription describes the chemical composition of the drug, records the treatment's success under controlled conditions, and lists possible side effects. At every step in the process there was technical writing.

If you filed an insurance claim about your injury, your ability to write a letter or talk accurately and persuasively to a claims adjuster may mean the difference between compensation to pay for your medical treatment and having to limp along as best you can with your bills.

At this point you may want to run the tub brim-full of hot water and soak your tired bones, or perhaps your reflections about the hidden factor of technical writing have taken so much time that you'll be late for school. You may have to dress hurriedly and even skip breakfast, but don't forget our hypothetical story of the design process which produced your clock-radio: every appliance in your kitchen underwent the same process. Even the foods you eat were subjected to a process which included technical writing as a large component: soil analysis, climate analysis, real estate investment, irrigation systems, use of pesticides and fertilizers, use of harvesting machinery and labor, processing and preservation, U.S. Department of Agriculture or Food and Drug Administration regulations, advertising writing—a long series of communications guided your breakfast to the table.

Perhaps you are, by this time, eager to skip breakfast so you can dash out to your car and away from this network of technical documents which underlies everything you lay hands on, but the reminders continue as you step off the elevator and leave the building. You see a traffic helicopter in the distance over the highway, and recall a humorous news report about technical writing: when the Army recently called for proposals to develop a new helicopter, one company presented a multi-volume design document heavier than the aircraft proposed. If you see a large airplane flying overhead, you may recall a bit of technical-writing trivia: the Lockheed C-5A can transport M1 tanks, but it cannot carry as a cargo the letters, memos, reports, manuals, and other related technical documents generated by its production and operation. If it is so early in the morning that the moon is still visible, you may think of another bit of technical-writing trivia: the documents generated by the Apollo moon shot would reach to the moon and back.

As you drive to school, your way is paved with the written word, communicating the technological knowledge of your world; so hurry—you can hardly afford to be late for writing class.

chapter 2

Technical Writing Systems

Technical writing is formal rather than imaginative, and the documents an E/T produces are written in a formal setting, rather than by a writer working alone. To explain and illustrate that statement in the context of a major project: a writer who has to write two dozen letters a week would waste an enormous amount of effort if each letter were treated as an item individually hand-crafted to fit into a unique situation. If the geotechnical engineer who did the sub-surface investigation on the site of your apartment building took off his muddy boots, leaned back in his chair, and contemplated his toes until he felt inspired to write a technical report of his findings, the progress of the entire project would be slowed.

The image of the creative writer who writes in solitude does not apply to the technical writer. The writing required in a technical project cannot be produced in a random way. Individuals write as the job requires, and most technical writing takes the form of a specific type of document designed and produced for a specific purpose within the project. So many writers and such a volume of documents must function in some systematic way.

To illustrate how writers work in a communicationns system, let's continue the story of your apartment building. For this phase, you are working on the project as an E/T for the engineering firm of Stuart and Lee.

Let's say that the idea for developing a second phase of that property originated in a meeting of investors. In tracing that idea from the point of origin

Technical Writing Systems 11

to the day the owners take possession of the finished project, we see communications taking many forms: oral, written, technical (calculations, drawings, and specifications), and visual. Over a period of perhaps a year, that project will generate many different kinds of communications as the concept of the apartment building takes its final form in reality. Let's track that system of communications at least part of the way through the project.

- The investors begin with a conference.
- A memorandum records the minutes of that meeting for each person present, and for the files.
- A request for proposals (RFP) is published in newspapers or in trade papers.
- Someone at Stuart and Lee reads the RFP during the daily review of publicationns which contain leads on possible projects.
- After a phone call, the marketing director of Stuart and Lee then mails a brochure and a letter to the investors.
- The investors and the representatives of Stuart and Lee have a conference to discuss the possibility of their doing the electrical, plumbing, and mechanical systems for the building.
- A memorandum records the minutes of the meeting and copies are distributed to all present, with copies for the files of each group.
- Stuart and Lee electrical engineers begin work on a design proposal.
- Ms. Boyd, your immediate supervisor, asks you to locate information about specifications and prices for the light switch to be used in the individual apartments.
- You locate the information and report it to her in a memorandum.
- Ms. Boyd calls Mr. Polk, who is working downstairs on the HVAC design, and discusses with him what the peak load on the electrical system is likely to be.
- He notes the substance of their conversation in his telephone log and confirms their agreements to her in a memo; he retains a copy for his files and a copy for the files on the entire project.
- As Ms. Boyd, Mr. Polk, and their staffs of engineers and technicians are at work on the technical design, Mr. Stuart and Mr. Lee hold a series of conferences with the architects Hill, Hood, and Johnston. The minutes of those meetings are written as memorandums; copies are distributed to those who were present and later filed by each firm.
- (After dozens and perhaps hundreds of such transactions, the prosposal is written.)
- The proposal is technical in its content, but it is also produced on heavy, expensive paper; the marketing illustrations and technical illustrations are in color; and it is, in every feature, the work of an experienced team of competent professional architects and engineers.

- Carrying copies of the proposal, Mr. Stuart, Mr. Lee, and Messrs. Hill, Hood, and Johnston meet with the investors to present their design proposal.
- In the technical presentation, Mr. Stuart acts as chief spokesman. He has a display of drawings to show the overall design of the project. There are charts to show consumption of electricity and water over a projected ten-year period. There are color slides to show exploded drawings of the parts of the HVAC system, there is a cut-away drawing to show the construction of the building from the brick wall inward, through layers of insulation and interior walls, to the steel girders which form the structural skeleton of the building.
- During a question-and-answer period, the investors raise questions about initial cost; Mr. Hood as an architect argues persuasively for a high initial investment, justifying it by the profits the investors can expect from a beautiful project.
- After presentations by many firms, the investors discuss the proposals and decide to hire Stuart and Lee as the engineers and Hill, Hood, and Johnston as the architects. A letter announces their decision.
- Contracts are drawn up by attorneys for both groups, they are signed, and work begins.
- Stuart and Lee publishes a request for bids and proposals on the sub-surface investigation for this project.
- Hill, Hood, and Johnston engages the services of Jackson, Inc., a large interior design firm, to plan the plumbing fixtures for the kitchens and baths and the light fixtures, carpets, window dressings, colors, and wall coverings throughout the building. The two companies together will determine what hardware to use on all doors in the building.
- Hill, Hood, and Johnston and Jackson, Inc. coordinate their work with Ms. Boyd's electrical design in a steady flow of letters and phone calls.
- Southern Illinois Geotechnical Engineers has a staff member who reads the trade papers to select projects for which the firm may write proposals.
- Hidden Hills Apartments Phase II seems like a promising project, so Stuart and Lee receives a brochure with a cover letter describing SIGE's capabilities.
- SIGE is hired to do the sub-surface investigation and a general survey of the site of the project.
- A team of geotechnical engineers visits the site, sinks a random series of holes to get core samples, and assesses the site.
- The team leader prepares a report of findings with information about rock, soil, and water formations in that area and on that specific site, and renders professional opinions: the building can be set on any one, or a combination, of three types of pilings; on the type of soil in the area, paving for the parking lots should be of a specific weight; grading of slopes in that type

of soil should take into consideration certain specified factors. The team also offers SIGE's services in supervising the process of sinking the piles, so that the work as proposed in her report will be done according to strict specifications.

We are already several weeks or months into this project, and the only signs of it on the site are the holes left in taking core samples and the prints of the geotechnical engineers' muddy boots. But by this time there are dozens of technical documents. Before the project is complete there will be hundreds more.

A major sub-system of the entire communications process will be Stuart and Lee's final design report on the electrical, plumbing, and mechanical systems.

Such a report of perhaps 400–600 pages will be the work of a team of dozens of people. Each person does not lock herself up in an attic room to commune with herself about the flow of hot air in a building; instead, the members of a team know that they must work systematically to produce a long report. They know that the report must include an executive summary, a table of contents, an introduction, a glossary of terms (if they decide that their audience needs one), chapters on each of the systems, graphics to illustrate parts of chapters, tables of data to illustrate certain parts of chapters, a summary, conclusions, recommendations, and an appendix (or a series of appendices).

As a young E/T it may fall your lot to write three pages on the electrical switches in each apartment, but your senior electrical engineer will have general responsibility for the report's chapter on the electrical system. The senior engineer who works as project manager may write the conclusions and recommendations and the executive summary before the entire report is proofed and sent off to a printer who will turn out an expensive illustrated volume to present to the client.

The finished report *looks* like a book, but a single reader will not settle down to read it before a crackling fire with a good cigar and an after-dinner brandy nearby. Instead, the report will be a working document randomly accessed by different readers with different needs. The investors may rely primarily upon the Executive Summary, or on the Summary, Conclusions, and Recommendations. The accountants for the investors will concentrate on the chapters related to costs. The technical chapters will become working documents to coordinate the work of plumbing contractors, electricians, and HVAC mechanics. Carrying the report with you to refer to your three pages about the light switches, you as an entry-level E/T may have to wear a hard hat into the construction site to talk with the foreman who supervises the electrical workers.

After months of work, punctuated daily by communications, a construction review may be one of the final documents. As a memorandum report, it will include page after page of specific notations: a switch plate is missing in the foyer of apartment 603; a pipe fitting is loose and leaking in the extreme southeast corner of the basement, and on and on until every detail of the building construction is reviewed.

Meanwhile, the interior designers are concerned that everything be completed in time for the grand opening. The insurance companies write to ask for a specific date when one type of insurance will lapse and another become binding. The real estate broker needs to know how to prepare the rental leases and the ad campaign for tenants. There are city and county codes to be met and licenses to be applied for. The elevators must be tested and inspected so that the inspector's letter may be displayed.

Within the system, communication may be oral and informal (as in a brief conference over a drawing board or as in a telephone conversation); oral and formal (as in a conference, or a technical presentation followed by questions); entirely technical (as in drawings and specifications, engineering calculations of the flow of hot air, tables of projected costs, tables of projected consumption of KWH); artistic (as in the colors, layout, and paper quality of a proposal); legal (as in contracts, and letters which enter the legal records of the project); or graphic (as in the illustrations which accompanied the proposal and the technical presentation).

Whatever the form, the formal channel for this flow of communications is the written word. Any idea, any oral agreement, any technical solution, any legal factor—none of these was operative in our hypothetical project until it took written form.

chapter 3

How Technical Professionals Write

The function of technical communication as it underlies technical projects should be clearer now that you have read this far in the Introduction. You should also have some sense of how technical communications are produced within a larger system as a technical project takes shape. Our next step in understanding more about technical communication is to define technical writing in contrast to writing in general. It is especially important for you to understand that technical professionals who communicate learn specifically how to write for special purposes.

Describing how to write for special purposes can best be approached by dividing the writing task into its component parts: invention, arrangement, style, grammar, and mechanics. Traditionally, these components are involved in any writing. Different types of writing, however, call for different emphases upon the component parts. Understanding how to do technical writing depends first upon knowing how each component relates to the writing an E/T does.

INVENTION

In writing terms, invention can be defined as "thinking up something to say." In scientific and technical terms, invention is closely related to the creative

15

idea of the "Ah, ha!" response at the moment of discovery. As technical writing fulfills a special purpose, invention does not enter into the writing process for the E/T as it does for the creative writer. Before an innovative E/T sits down to report the results of research into fiber optics, the creative (inventive) technical work is complete; the writing itself is not creative, but communicative.

Misunderstanding the writing process, many E/Ts say, "I can't write; I've just never been very creative." E/Ts are supposed to be creative in their technical work, but the writing they do need not be creative in the same way that we think of poems and stories as being so. Technical writing almost by-passes the invention step altogether; technical writers write because they already have something to say about their technical work.

ARRANGEMENT

The essential emphasis in technical writing is on arrangement. Other terms which make the meaning of the term clearer are *rhetoric*, which means "the arrangement of material," or *logical order*, or even *outlining*. An E/T who has had a creative idea which solves a problem then has the task of writing about that idea in some orderly, logical way. Rhetoric is the subject which teaches writers how to do that. The thought patterns which rhetoric describes are very much like the systematic methods E/Ts use to state and solve a math problem or to set up a lab apparatus to perform an experiment.

If an engineer asks, "What is this?" rhetoric answers with specific instructions about how to write a definition and what kind of definition to write for a given situation.

If a technologist asks, "How does this work?" rhetoric answers with specific instructions about how to describe a process: historical, mechanical, physical, logical.

The sections in this text concerning the memorandum, the short report, the proposal, the paragraph, and the outline will give a more complete discussion of rhetoric as a primary tool for the technical writer. For introductory purposes, apply the following questions to any subject of your choice:

what is this?

how is it similar to something else?

how is it different from something else?

is it part of a larger whole?

what are its component parts?

how does it work?

what made it happen?

what will happen as a result?

Any E/T must ask and answer such questions in the process of doing any technical task. Rhetoric demonstrates how to write the answers to those questions about technical subjects. Each question in the list above represents a traditional rhetorical pattern: definition, comparison, contrast, classification, division, process, cause, effect. Educated people have learned these techniques of rhetoric since the time of Aristotle.

Even highly educated E/Ts sometimes say, "I know what I want to say, but I don't know how to say it." That statement can be more accurately stated, "I know what to say, but I don't know how to arrange it in a logical, systematic way." Neither creativity nor eloquence as a writer offers an adequate answer to the person struggling to get thoughts into words. Knowing the principles of rhetoric will almost always work for any writer—and for technical writers best of all.

In writing about any subject, you can identify these rhetorical tasks:

Definition

Communication begins with—and cannot happen without—a mutual agreement on what a given word means. Your rhetorical tasks always include determining what words you must define for your reader and defining them to ensure understanding between you.

Description

Whatever object or process you deal with exists in space. Place your project into its proper spatial relationship; that is, your rhetorical task is to establish a space where the problem is solved. (The renovation of a building offers a good example of how your project exists within a space.) Describing furnishes a fixed reference point.

Comparison and Contrast

If you offer the client a series of alternatives, the rhetorical task then becomes matching up the features in an orderly way. Some of you may recall seeing a newspaper picture of two boxers with little insets that show their respective height, weight, neck size, length of reach, chest measurements (expanded and unexpanded), and other physical characteristics. Your rhetorical task in offering alternative recommendations to a client follows much the same logical pattern—and a great deal of money rides on that decision too.

Classification

Do you need to relate this one problem to a larger problem within the entire project, to the current state of the art, to a general technical concern? (Any project having to do with energy offers a good example of how classification works.) Your task is to provide perspective.

Division

Divide the project into its logical parts. Presumably the parts of a process or a system in an engineering context exist in some logical, effective relationship to each other. Your rhetorical task is to state and describe the relationship of the parts. (An electrical, plumbing, or HVAC system provides a good example of how division works.)

Process

If the technical task is to figure out how something works, that translates into the rhetorical task of process analysis; that is, into explaining and illustrating that process in a report. How does the process work? Why does it work this way? (Any mechanical or electrical system fits this.)

Cause and Effect

Often the E/T is hired to find out what caused a certain situation, and even to predict what the long-term consequences of its continuation would be, in physical, mechanical, or economic terms. The rhetorical task is to apply the conventions—and avoid the pitfalls—of cause-and-effect writing.

To sum up, then, there are standard methods for analyzing a rhetorical task. As with technical tasks, some outlines follow routine, cut-and-dried methods. Others require innovation. Only to the degree that you understand and can use the routine methods can you innovate when necessary. There is no mystery about rhetorical thinking and planning of this sort. It is the same as the E/T's technical labor over planning and design down to the last minor detail. Rhetoric is the tool which translates technical problem-solving into technical prose.

STYLE

Explaining style is rather like explaining electricity; it is easier to describe its effects than how they are produced. Style is a composite of invention, arrangement, grammar, and mechanics which carries a writer's individual stamp. A poet, essayist, or dramatist strives for a style which becomes his hallmark. The technical writer need not, indeed should not, strive for a highly creative and individualistic style. Instead, technical style should be clear and unobtrusive. As an analogy, consider eye glasses; their function is to enable a reader to see the print on the page. The glasses do their job best when the reader is entirely unaware of the lenses, their color, or a bifocal correction. Technical style works in exactly the same way; the technical writer tries to communicate information in clear terms, uncolored by personal style.

The E/T who says, "What I write seems too *boring*," shows a misplaced concern with style. Technical subjects are not communicated to surprise and delight the reader, nor is technical communication a stylistic self-indulgence for the writer. Technical style is almost entirely functional and almost never ornamental. The task of the technical writer is to inform the reader, not to entertain.

GRAMMAR

Grammar is the basic internal structure of how words in any given language are arranged in sensible order. The word *grammar* itself refers to the most basic definition and application of terms and information in any subject. To know the symbols, terms, and definitions in algebra or geometry might be called grammar in that sense of the term. And certainly knowing the fundamental order of words in the English sentence is the beginning point for writing.

Although many students no longer learn traditional English grammar, technical writing is quite conservative in the demand for precise, grammatical statement. In recent years, the number of non-native readers of English has increased the demand upon writers of English to produce more nearly grammatical prose. Computer applications also allow less latitude for non-traditional patterns in writing. There are now some computer programs which do sentence diagramming, a traditional tool of grammatical analysis now being revived to make the grammatical structure of English clearer.

However a technical writer learns the structure of English—through traditional instruction in grammar, from a computer program, or by extensive reading and listening—the requirement for grammatical accuracy underlies all technical writing. To meet that basic need for a secure foundation in language skills, this text includes an appended handbook on words, sentences, paragraphs, and outlines, as an aid to students who need to strengthen and improve basic writing skills.

MECHANICS

Matters of punctuation, spelling, and other conventions of English comprise the mechanical aspects of writing. Standards for technical writing are also quite conservative about spelling, precise use of punctuation, and all matters relating to the production of clear and accurate prose in final written form. Technical readers who are accustomed to giving attention to decimals and milimeters can be far more influenced by variations in mechanics than readers of imaginative literature, editorials, and essays.

This catalogue of the components of expository writing makes it clear that invention, arrangement, style, grammar, and mechanics are part of all writing. Technical writing, as writing for a special purpose, makes different demands upon writers than other types of writing; the relative degree of emphasis upon each component of the writing process differs as well.

chapter 4

How Technical Writers Use Form

While all the basic writing skills are essential to technical writing, form is the greatest determinant in what E/Ts write and how. Arrangement is basic to any writing, but the formal arrangement of the parts in a memo or a long report is more strictly prescribed in technical writing. Style is a feature of all prose, but formal style in a letter is very strictly prescribed by convention and technique. Although individual judgment enters into any decision about writing, technical writers are always guided by certain constants in the process.

THE ELEMENTS OF FORM

Form

"What to say?" translates for the technical writer into "what form to use for this situation?" An E/T should know that a memo communicates in-house and a letter communicates to external audiences. An experienced E/T knows that one situation requires a proposal as a defined form, but another situation requires an informal memo report (with or without recommendations). Because of prescribed form, an E/T almost never has to fall back upon basic writing skill as the only aid to technical communications.

Design

Each form in technical writing has a specific design. A memorandum *looks* a certain way on the page, and the writer's first task is to design that memorandum to those specifications. This is not a writing skill at all but a combination of two skills: a) designing according to specifications and b) knowing how to read and follow directions.

Conventions

Apart from its content, a letter operates under conventions governing spacing, date, inside address, salutation, complimentary close, signature, and typed name and title. In addition, there are established methods of beginning a letter and established ways of ending a letter. The skill which a writer is asked to bring to such a task is not great talent as a writer but the ability to learn and use established methods. Any student in a chemistry lab learns certain procedures; any student in a surveying class learns standard methods; any entry-level E/T who must begin to write on a job should know the conventions of the formats in which technical writing is done.

The memorandum and the letter are short forms which offer examples of how a form operates within a system. Elsewhere in this text, these forms and others are discussed in detail. Their purpose here is to show briefly what you might do in a situation which required you to produce a letter or a memorandum.

EXAMPLES

Memorandums

Let us say that you are a technician in the electrical engineering department of the firm which is designing Phase II of your apartment house. Your assignment is to find the suppliers for the switch boxes to be used on all the bathroom switches in the building. Once you have found that information, you write a memorandum reporting it to the higher-level engineer.

This text contains a page which shows the design and specifications for a memorandum which will be typed on an 8½-by 11-inch page. Most companies will have a similar set of specifications; some have a printed form for memorandums so that you have only to fill in the date, recipient, writer, and subject. Your first step in writing, then, should not be difficult; instead, it is mainly a matter of following a simple set of directions.

Before you say, "But a typist will do all that," remember that a typist will type from the draft you produce. The finished product which you initial will be your work and responsibility as if you had written, typed, proofed, and performed every task associated with producing the memorandum.

At this point, you know the formal design and some of the conditions of production, but you do not know what to write as the first sentence. This is determined, not only by your skill at expressing yourself, but also by 1) the audience who will read your memorandum, 2) how the reader will use what you have to report, 3) the characteristics of the subject matter in your memorandum, and 4) your purpose in fulfilling this assignment. These four factors, more than your talent as a writer, determine how you will write the memorandum.

1. Your audience is a higher-level engineer; this means that you can write to her in technical terms. She is in your organization; this means that you need none of the conventional courtesies of a letter.
2. Your reader will use the information in this memorandum in a larger design process. She needs to be able to see the data clearly displayed on the page, refer to it, perhaps walk to the next drawing board and discuss it with a colleague, perhaps send a copy to someone else with whom she discusses it in a telephone conversation. She will also need to file a copy for the records. This memorandum is not a work of art, then, but a working document.
3. The information in your memorandum about switch boxes includes manufacturers' names, code numbers, and costs per unit. A paragraph in which you list all of those in three consecutive sentences may *look* attractive as a paragraph on the page, but the engineer will have to search through the sentences to find the information she needs. To tailor the form of your memorandum to the function, arrange the information as a numbered list of items down the page.
4. Finally, your purpose was to locate and report information. Most memorandums are for the purpose of conveying information in-house. You were not asked to give an opinion; you were not asked to write a page of deathless prose. You were assigned a specific task, and technical writing provides the training to enable you to do the writing component of that task in an informed, systematic way.

Your finished memorandum should look like the following example:

Memorandum: Example 4.1

STUART AND LEE, CONSULTING ENGINEERS

February 22, 1990

MEMORANDUM

TO: Ms. Belle Boyd

FM: John Benet

RE: Switch Boxes for Bath Rooms
    Hidden Hills Apartments
    Job Number: 803

In response to your February 21 request for a list of available switch boxes, I located the following items in the General Electricians' Catalogue:

1. United Electric: A3265-22; $2.00 @; $1.72 @ per hnd.
2. Great Northern: 798-B5; $2.05 @; $1.68 @ per hnd.
3. Pacific Power: 8390; $1.98 @; $1.71 @ per hnd.

A check of the electrical codes for Taunton County shows that each of these will meet the code. In addition, each of the switch boxes carries the UL stamp.

There are many variations of the basic memo form: construction reviews, meeting minutes, short reports, and even some proposals. This illustration and discussion on the memorandum serves, however, to illustrate the statement that technical writing is formal writing. Documents are designed in a prescribed way and produced to specifications when and as a job requires.

Letters

The letter is the most varied form in technical writing. The content ranges from the most technical and routine matters of requesting a change in specifications for the switch boxes, to the task of placating a client, to the most subtle arts of persuasion in sales and in legal matters. An entry-level E/T will usually begin by writing a routine letter.

Again using Phase II of your apartment building as an example, suppose that Ms. Boyd uses your memorandum on switch boxes to make a decision. She will include in her electrical design for the bathrooms the switch box manufactured by Pacific Power. In a meeting of the entire team of electrical engineers, she has learned further that the building will use the same switch box for all light switches, not only in the baths, but in the kitchens, foyers, bedrooms, and throughout the building.

Your assignment then becomes to follow Ms. Boyd's instructions to write to Pacific Power for the following information: whether they have a price per 1000 on the switch boxes; whether that price includes shipping and delivery from their suppliers; and whether they have a supplier in the immediate St. Louis area.

This assignment is slightly more complex than your initial assignment to write a memorandum, but you are still dealing in a task which can be defined as formal. Knowing the audience, the purpose, and the characteristics of the subject matter help to guide you, but there are further guides about what to do to address a specific situation in an effective way.

For the letter to Pacific Power, you know these things before you begin to write:

1. the design specifications for how the letter should appear on the page.
2. the appropriate conventions for the external apparatus (everything above and below the body) of the letter.
3. the standard techniques for the first and last paragraphs.
4. that your audience is a large manufacturer, and that you have three specific items of information to request from it.
5. that your three requests may be referred to more than one person within that organization and to a supplier in your geographic area. It is most likely that this letter will not be read and laid aside; like your memorandum to Ms. Boyd, this is also a working document which will serve as a list to be referred to and worked with.
6. that the clearer your letter is, the more likely it is that Pacific Power will send a prompt response.

Your letter should look like the following example:

Letter: Example 4.2

February 22, 1990

Construction Parts Division
Pacific Power
1902 Sunset Boulevard
Venice, California 90111

Gentlemen:

Our firm is designing the electrical systems for a large apartment development in St. Louis. We are seeking a supplier for the switch boxes to be used throughout the complex.

Your switch box #8390 seems to meet our design specifications. It is listed in the General Electricians' Catalogue at $1.98 each or $1.71 per hundred. To make our final decision we need the following information:

1. For large orders, can you quote a price per 1000 on this item?
2. Do your prices include shipping?
3. Do you have a supplier in the St. Louis area? If so, we would appreciate having their name and address.

Please send this information to:

Attn: Ms. Belle Boyd, Senior Electrical Engineer
Stuart and Lee, Consulting Engineers
909 Battery Avenue
Lexington, KY 50213

We will appreciate hearing from you soon so that we can maintain our schedule on this project.

Sincerely,

John Benet, Technician

Writing a letter is not a totally cut-and-dried, cookbook process; however, a large percentage of the effort involved can be eliminated if the process is learned and practiced as a routine skill.

A high level of skill and sophistication in letter-writing would be required if problem situations arose from this apparently simple situation of gathering information about one small part of a larger electrical-systems design. What if all the agreements had been made to have Pacific Power supply several thousand switch boxes, and the shipments were late? What if a strike closed down the supplier of metal for Pacific Power? What if a revolution in a small country closed off the supply of ore to the metal supplier? What if there were a significant increase in cost between the order for the first 1000 switch boxes and the order for the second 1000? What if the switch plates ordered by the interior design team would not fit over the switch box when it was recessed in the walls of the apartment building, or what if the plates would fit in the baths but not in the foyers? What if the clients began to inquire insistently about the delays in construction?

Any one instance on that list of things that could possibly go wrong would call for a letter or a series of letters. Higher-level managers write such letters, not entry-level technicians—and that is a large part of the reason why they get more money and more gray hair. But the manager who must write letters through a crisis on a project also has guides to follow. Whatever the subject, purpose, and audience for the letter, these factors remain constant: the design on the page, the external apparatus, and the conventions and techniques of the first and last paragraphs. Beyond that, a manager who is well trained in the strategies of technical writing would know the methods to apply in given situations.

Understanding form, knowing how each part of a form operates within the whole, and knowing how to produce each part to specifications—these are the skills which produce effective technical communications. A perfect command of English grammar will not enable you to write a report; a gift of literary eloquence may even operate against the effectiveness of a technical letter. Not how one writes in a general sense, but how to plan and execute the use of a particular form for its particular function—that is the skill of the technical writer.

Manuscript specifications for all forms appear in Appendix A, pages 463–487.

chapter 5

How Technical Professionals Write for an Audience

The writer's relationship to the reader is perhaps *the* distinguishing factor which makes technical and business writing different from all other types of writing. Consider the following statements to support that general statement:

1. A letter or memorandum is most often addressed to a specific person within a specific situation. While the document may be *about* a subject, it is addressed specifically *to* someone who is immediately involved with the subject. Therefore, this kind of writing differs from a book or an article. Technical writing exists precisely because it is directed to one specific reader or group of readers. As such it is always specifically tailored to the needs of that known audience.

2. A technical report may be a detailed and expert discussion of a subject, but most technical reports are written to satisfy the needs of a specific and well-defined reading audience, and therefore are confined to what that particular audience needs to know.

3. The documents E/Ts write are expressions of individual work, but they are also part of a direct transaction with the reader. If the reader does not understand the manuscript, the necessary transaction does not take place. Therefore, an E/T must always write with an audience in mind. In addition, considerable imagination and sensitivity are often required to analyze an audience and to direct a letter or a report to that audience's need for a solution to the problem.

How Technical Professionals Write for an Audience

Deep thoughts

$\Sigma = MC^2$

Furrowed brow

Closed eyes—
quiet, genius
at work

???

The E/T writer in communion with himself about what he knows

Reading audience

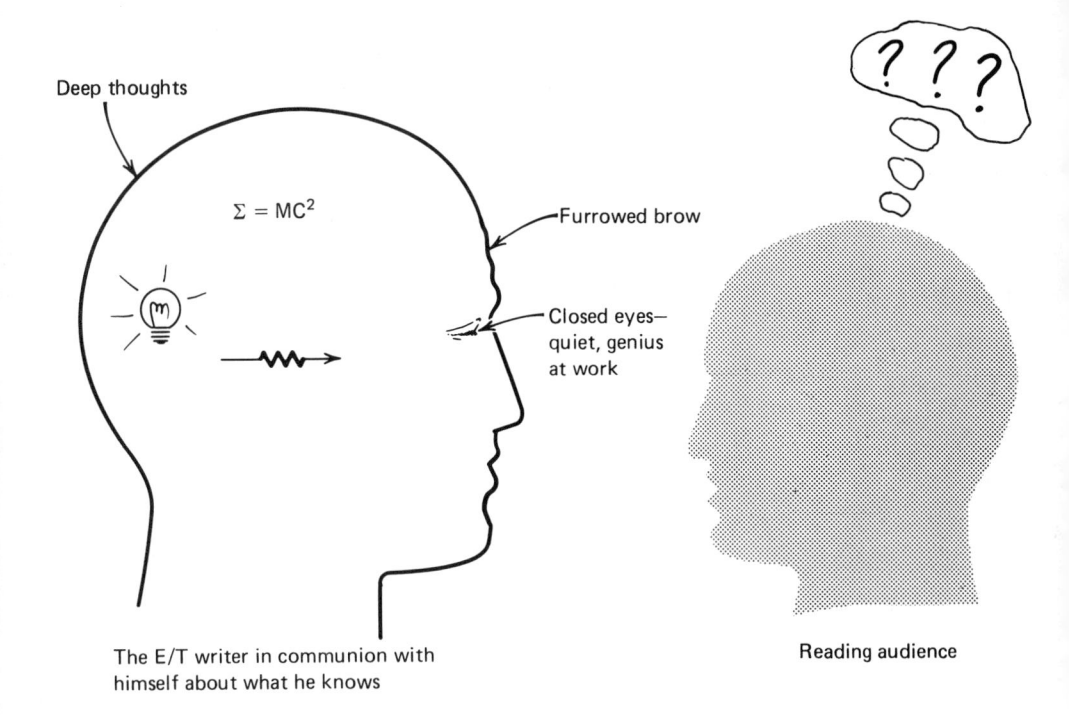

An E/T can ill afford to be bound up in his own head with his ideas, and lack an overriding concern for the reading audience.

A widespread misconception about technical and business writing is that any doubt or disagreement about what to do must always be settled by applying the "correct rule." Many rules of grammar and mechanics do apply to writing, but, many times, looking outward to a reading audience's needs will supply the most effective answer. "What's the rule about underlining subheadings or indented subheadings?" does not have as direct a relationship to the needs of readers as "What can I do to design a page which my intended reader can quickly *see* and refer to?" Many writers regress to an academic notion that an applicable rule must be found among the writer's own resources, when in truth it is best to look outward and perceive the audience's need.

To look outward, to analyze and perceive a reader's needs to know and to communicate in terms he can understand, to look upon a technical or business document as a working document directly used by a reader—that view of the writing task in terms of its intended audience will answer the great majority of writing questions E/Ts have.

To determine subject, purpose, occasion, and audience is one of the oldest and most traditional ways of planning to write anything. The E/T writer, however, should break down those larger considerations into a specific inventory

How Technical Professionals Write for an Audience

E/T writer absorbed in remembrance of rules past — Page number where? At all? Table of contents? RULE 2 spaces? 3 spaces? RULE Number? Letter?

Audience — I can't *see* this page!

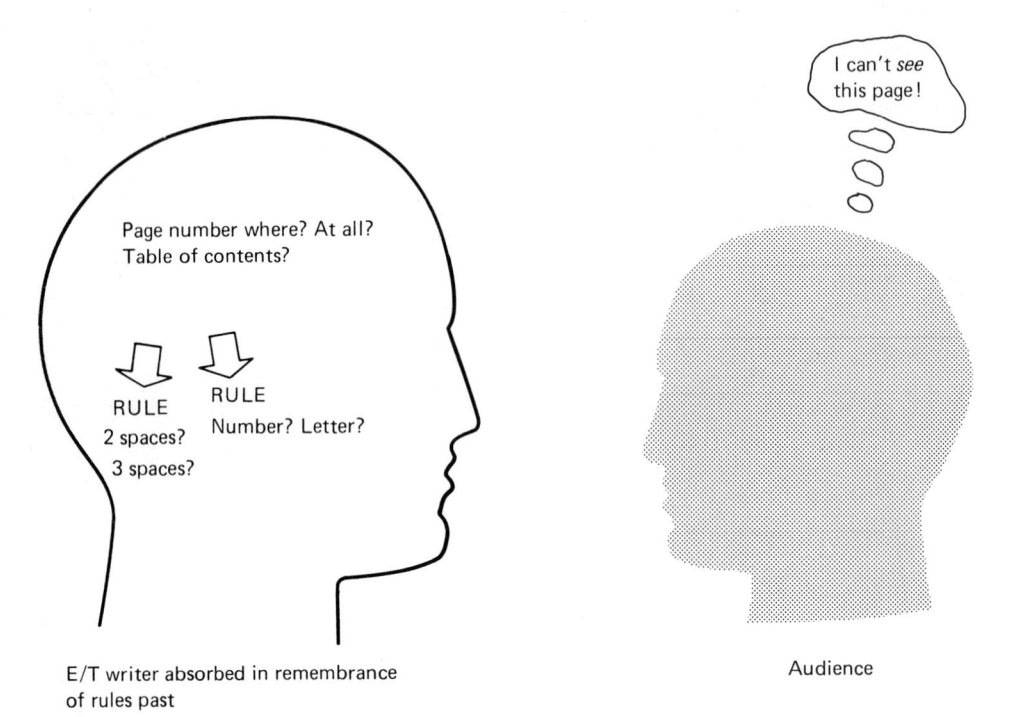

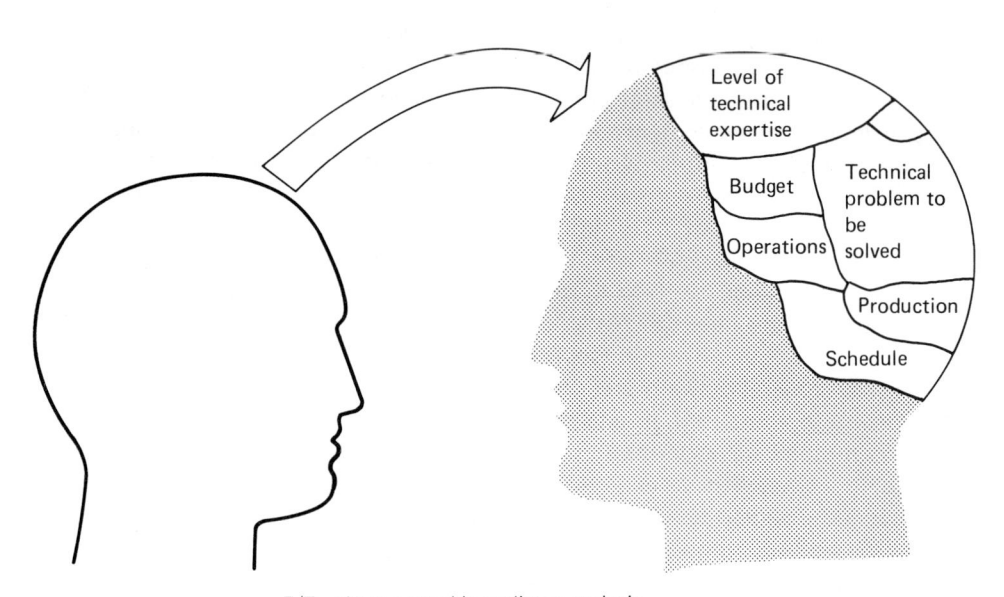

E/T writer engaged in audience analysis — Level of technical expertise, Budget, Operations, Technical problem to be solved, Production, Schedule for each memorandum, letter, or technical report. For the experienced writer, such an analysis will become almost automatic in writing one of the short forms. Even a highly experienced writer will make a thorough and careful audience analysis before writing even one word of a long technical report.

AUDIENCE ANALYSIS

Memorandums

Especially if you are a relatively inexperienced writer, ask these questions before writing a memorandum:
1. What was I specifically asked to write in this memo?
2. If there was not a specific assignment, what is the situation which this memo addresses? Within that larger picture, what detailed part does my job here require me to cover?
3. What general type of memo have I been asked to write? (administrative, directive, meeting minutes, etc. Refer to the general types of memos in the chapter on memorandums.)
4. What is the best logical division of the subject to make it useful to the reader? (Again, refer to the section in the text which describes various forms of organization in a memo; also look at the section in the writer's handbook on the logical arrangement of data.)
5. What is the most useful and effective design of the memo on the page to reflect the logical order of the writing?

Even when a memorandum is a routine assignment for an in-house document, the beginning writer, especially, should sell himself and his work by analyzing the reader's needs, and then planning, designing, and writing the memo accordingly. The reader may not fall back in his swivel chair, hold the memo at arm's length, and exclaim, "FAR *OUT!* What a super memo!" But such a careful audience analysis as is listed above can have a very positive, if unconscious, effect upon an audience, even in the most routine situations.

Letters

Audience varies with each letter and each situation. The chapter on letters divides the parts of the letter-writing task into constants (design, convention, and technique) and variables (situations in which letters are written). The reading audience adds a human variable to which a writer must always be alert. No form in business or technical writing requires more careful and sensitive audience analysis. Especially if you are a beginning letter-writer, ask these questions before writing a letter:

1. Who will read this letter—only the recipient, or is it likely to be duplicated and distributed? Will this letter be an important part of a project file as a part of a legal record? Overall, in writing any letter, keep in mind the multiple purposes served by a single letter to a single reader.
2. What does the person need to know? And what does she already know that she should be reminded of, or have redefined so that reader and writer occupy the same ground? How much does the reader need to know to understand fully what this letter discusses, to be fully informed on the technical background, and to have enough information without being inundated with irrelevant detail?
3. What special techniques apply to this letter: in the opening paragraph, in the format of the letter (lists, underlining, short paragraphs), in specific situational techniques (breaking bad news, giving an explanation, motivating to action by a specific date), or in a sales-oriented last paragraph?

To put yourself in the reader's place, and to anticipate and address the reader's need, is an imaginative and essential way to analyze an audience. In discussing audience analysis with a group of young women in a design firm, I frequently use the analogy of writing a letter and inviting someone to dinner. A careful hostess or host is alive to the needs of a guest: favorite foods, an ash tray conveniently placed beside a chair, a favorite flower, a special soap or color in the powder room. People in social, and especially in romantic, situations are basically performing thorough and effective audience analysis. Thus, to the degree that a writer is alive to the needs and preferences of a reading audience, to that same degree communications (or *sharing with*) will be effective.

Proposal

A proposal, which describes the scope of services of a firm or which bids for a project, presents a particular need for good audience analysis. A prospective client with a technical problem to be solved is like a patient who goes to a physician with a complaint phrased in layman's terms ("I have a stomach ache," or "I have a cough and am running a fever.") The physician examines the patient (gathers data about the technical problem), processes that data in her mind (in high-level technical terms which are a kind of shorthand understood only by other physicians), and then diagnoses and prescribes in a mixture of layman's terms and technical terms. A good physician who makes you feel good about the situation informs you about specific medical terms, reassures you by discussing the larger context of specific symptoms and carefully explains the effect of the prescribed course of treatment.

Audience Analysis

Any competent professional who makes a proposal does appreciably the same thing: looks at the problem which the client has explained in non-technical terms, translates that problem into a technically expert understanding of the problem, and then re-processes the data for the client to explain in technical, but understandable, terms what the problem is and how his company can solve that problem.

Before writing a proposal, ask and answer these questions:

1. What is the problem *as the client describes it*?
2. What is the problem as I understand it technically? (By problem here we mean a "subject for study" or "a project to undertake.") What work is to be done?
3. How can I best explain this to the client? Who will read this proposal? What is the level of technical expertise? Should I explain the unknown by the known with the use of analogies or comparisons with other projects of a similar kind? Who has final decision power in accepting this proposal? What are the special concerns of the decisive readership of this proposal?
4. How should the proposal be designed and produced, not only so as to be attractive to the reading audience, but so as to be logically organized and arranged on the page in such a way that the proposal is easy to *see* and to refer to?

Technical Report

Since technical reports are likely to be long documents with many readers, no group should begin planning any report without first doing a thorough audience analysis.

1. Who will make up the audience for the report (managers, legislative officials, financial officers, production personnel, engineering personnel, a lay person, a lay committee)? The answers to those questions will, in turn, answer the questions about how to state and narrow the topic of the report, how to include and discuss technical information, and where to include it (in the body, in appendices, in graphics, in discussion).
2. What specific technical problem is to be addressed? What specific work is to be accomplished? What have the technical professionals been asked specifically to do? If you have no specifically assigned task, how does your technical understanding of the problem help to define and specify what subject(s) the report should address and thoroughly discuss?
3. Have you been asked to make recommendations based on the report of the technical problem?
4. Who has final decision-making power to authorize the work recommended

in the report? While a technical report may contain great masses of data (tables, charts, graphs, computer print-outs), the decision-maker is the ultimate audience for a report. Data alone will not carry the expository or persuasive weight of the report unless the writers also communicate (i.e., share data and understanding) with the principal readers who must make decisions based on the report.

5. What is the most effective logical organization for the mass of data in the report? (For this, note especially the section on the outline in the writer's handbook.)
6. Are there special concerns about page design and graphics design within the text, and supporting data in the appendices? (Refer to the discussions of graphics and of when to append data.)
7. Are there special concerns about sales effectiveness and public relations? Even when a technical report represents a routine job, any technical project is a major sales campaign as well. As the text discusses from time to time, especially in the section on proposals, the E/T lacks the sales techniques so widely used by other segments of our society. However, in the planning, design, selection of detail, and actual writing, a technical report is a major sales document. The reading audience as consumer is therefore all-important.

Concern about audience analysis is the great division between writing in an academic situation and writing on a job. As a student technical writer you are almost entirely concerned with the technical content and accuracy of what you write. Your accountability to an audience is limited almost entirely to a professor or a grader. This makes student writers usually apt to think that, once the technical thoroughness and accuracy of a manuscript have been assured, the work has ended.

Audience analysis introduces a human variable into any technical or business document you write. When you write on a job, your concern for the reading audience equals or even exceeds your concern with technical content. You must always be alive to the rightful needs and demands of the reading audience in any technical or business writing you do.

How writing skill, knowledge of technical documents, and audience analysis combine can be illustrated in a slightly fictionalized version of a conversation I had with an engineering technologist. His writing assignment was part of a rather routine technical assignment.

A company was using several World War II-vintage generators to supply power to one unit of a manufacturing facility. The generators usually performed well, and had long since paid for themselves because they had been in use for so many years since the initial investment for their purchase. From time to time, however, a part on one of the generators would fail, the generator would shut down, and power would be cut off to a part of the manufacturing plant. Replacement parts for generators of the World War II era were very rare, though parts

Audience Analysis

could be specially tooled to repair the generator and return it to service. Individually tooled parts are expensive and, in addition to that cost, the company lost money when a unit of its facility had to be shut down while the replacement part was being prepared.

The management of the company decided that they needed to make a decision about the World War II generators: should the company go on tooling replacement parts and shutting down a part of their facility while a generator was out of service, or should they make a major capital expenditure and replace the old generators with new ones? The E/T had already performed the invention step. He knew the general subject and he had thoroughly investigated the client's problem. He said, "I could write them a *book* on generators." He also knew the design of a short report. What he could not decide was how to write the report. Acting as a communications consultant, I suggested that he give more attention to audience analysis:

CC: What did the audience ask you to do?
E/T: Make a study of the generators.
CC: What is your report supposed to tell them?
E/T: About the generators.
CC: What are they supposed to *do* about the generators?
E/T: Either keep repairing the old ones or buy new ones.
CC: Did they hire you to make the decision for them?
E/T: I know what *I* would do if they were *my* generators, but they did not hire me to make the decision.
CC: Did they hire you to make a series of alternative recommendations about what to decide?
E/T: No, all I'm supposed to do is to make a study of the generators.
CC: OK, let's imagine that the management of the company has your report; they are sitting around a conference table reading it to each other. When they finish, what will they know that they asked you to find out? What will they then be able to do that is not your job to do?
E/T: They will know how long the World War II generators are likely to remain operative in their current condition, the frequency of breakdown, the down time, and the costs to the company for parts and down time. I've done a projection for two years. They will also know how much it will cost to replace the old generators with new ones and how many years it will take for the cost savings to return their initial investment for new generators.
CC: What would you do?
E/T: Oh, I'd replace them. With old equipment like that, even if they don't build machinery that good anymore, the parts situation can only get worse.
CC: But did the management *ask* you to give that opinion?

E/T: No.
CC: Then your purpose is *only* to inform, not to persuade.

(That conversation illustrates that the E/T was a well-informed expert in his field with a specific task to do. He could not write, however, until he had first defined in very specific terms his purpose in writing, the form in which to write, and the needs of his audience. Even if he knew what the company, in his expert judgment, should decide, he was not hired to make recommendations.) Our dialogue continued:

CC: Your rhetorical task is to compare and contrast.
E/T: OK, old generators compared and contrasted to new generators.
CC: Right. Now set up a basis for comparison: initial purchase price, cost of operation, frequency of down time, projected conditions during the next two years. Go down that list with your data about the old generators; then go down that *same* list with your new generators.
E/T: OK.
CC: With that overall design, your major rhetorical technique will be process writing—how this system operates, and according to what mechanical and physical principles. And your major process will be cause and effect: if A occurs, B will follow. But be aware of all that you cannot predict with complete accuracy in cause and effect writing: if B follows, C is likely to occur, and under certain conditions, allow for the possibility of D and E, and qualify your statements accordingly.

The E/T still had the words to put on paper—along with drawings, tables of data, and budget figures. The report had to go to the typist, back to the E/T for proof reading, and then to another unit of the firm to be bound before it was hand-delivered to the client. The E/T may even have had to meet with the management of the client company to field questions or to present the report orally. The path toward the completion of that communications sequence was blocked, however, until he had analyzed the specific needs of the audience, and then determined the overall design of his report according to design specifications and traditional rhetorical patterns. Neither his knowledge of the formal design of a report nor his basic skills in style, grammar, and mechanics could come into play until he had fully exercised the thinking skills which the study of rhetoric gives us. Hence the importance of arrangement as a crucial step in technical writing. Hence also the analysis of audience as a primary concern.

As a happy ending to the story of the struggling E/T, I would like to be able to say that an English teacher as consultant saved the day and helped the E/T make the world a better place through clear thinking and clean, clear, communicative English prose. No such moral to the story exists, however. What the story does have is this: a bottom line.

Cost savings is the ultimate end to the dialogue between E/T and consultant, for while the brief problem-solving session saved anguish over outlining and writing, it more importantly saved time and money. If the E/T had contracted with the client for three days of work at $45.00 per hour, that is a flat fee for the job of $1,080.00, including the time required to write the report. If the E/T's indecision over what to say required two hours, that reduced the profit on the job by $90.00. If the E/T spent 20 minutes talking with the consultant who was paid $60.00 per hour, the $20.00 that cost the company actually saved money. The less time required for the E/T to write a clear, effective report, the greater the profit for the company.

After we discuss the function of technical writing, the systems, the forms, and the requisite skills, cost considerations always draw the bottom line on technical writing. Every word that you write on the job carries a price tag. So learn to write each word efficiently and well.

chapter 6

Technical Writing and Professional Ethics

WHY A WORK ETHIC IS IMPORTANT TO A TECHNICAL WRITER

Up to this point in the introductory section we have been discussing the functions of technical writing in the world we live in day to day. We have seen that technical writers most frequently operate in a community of writers in a large organization or on a large project. *Communication* means, in fact, "that which is shared." Therefore, at this point in your reading you should have a vivid sense of how technical and engineering professionals, members of other professions such as medicine and allied health, agriculture, foods, and nutrition, and workers in a multitude of government agencies all participate in the work generally called technical writing.

We have discussed the work of the technical writer in terms of writing to specifications of design and content. In the discussion of the skills required in technical writing we have seen that the most classical discipline of rhetoric relates closely to how engineers and technologists think and, in turn, write technical documents.

Audience analysis then identifies technical writing as a kind of writing intimately bound up with the concerns of the writer for the reader. The technical writer, then, is associated on many levels with other people in community. Unlike the poet or novelist who may have only an indirect connection with his society, a technical writer acts as a worker in community with others.

Therefore, our discussion of accuracy and quality as indispensable components of technical work and technical writing leads logically to a discussion of the ethics of technical communication. As a writer with a direct effect upon others' lives, and as a worker with direct responsibility for how that work affects society, the technical writer in engineering, technology, architecture, agriculture, allied health, and all other technical disciplines should develop some concept of the ethical issues of his work.

A DEFINITION OF THE ETHICS OF TECHNICAL WRITING

Ethics is the study of right and wrong. That standard definition leads us into immediate difficulty with definition, for it is difficult to say what is absolutely right or, equally, wrong. The study of situational ethics—the theory that what may be wrong in one situation would not be wrong in another—does not always offer sufficient answers either. Speculating upon the philosophical dimensions of ethics may still leave the technical or business writer in search of a workable definition of ethics.

Two techniques of rhetoric offer us a way through the difficulties of defining ethics for the writer. The first is to make a distinction between one kind of ethics and another, or to stipulate the sense in which we will use a given term in a given situation. If we define ethics for a writer as a part of the work ethic, then we have reduced the concept of ethics to a more manageable scale: ethical questions which arise from writing situations relate to how a writer works. Whether he loves his neighbor, or how he treats his silver-haired grandmother, are not the points at issue in this discussion of a work ethic, however pertinent they may be in a total view of capital-E Ethics.

The second rhetorical technique which helps us with this question of ethics is classification and division. To *classify*, ethical questions for a writer are *a part of* the larger question of the work ethic. Our question becomes "Is the writer a good worker?", not "Is he a good person?" The subject becomes more manageable if we then *divide* the concept into its *logical parts*: how a writer works depends upon 1) the value he sets upon himself as a worker, 2) the value he sets upon his work, 3) the value he sets upon being one of a community of workers.

A student operating under a deadline to produce a report by next Friday—or tomorrow—may consider the grade on the report very important indeed. He may, however, take a much less serious view of himself as a worker. After all, it's only an English course. Or it's only a class exercise. Besides, there are all those reports in the fraternity file, or all those books in the library, sitting there simply waiting to be copied. Once I've graduated, when I have a real job—if indeed E/Ts really do have to write as much or as well as people say they do—there will be time to do it as expected, there will be time to do it right. For now, the deadline looms. Achieve the desired end; never mind the means.

Such an approach to assigned work raises the question of ethics. How any writer approaches his work depends upon how mature that writer is in his view of himself as a worker. Does he *act*, out of his inner sense of ethics or does he *re*act to authority which has made an assignment and which will assign a low grade if he does not produce? Does he *re*act to the possibility that he will be punished in some way for plagiarism? Does he allow the responsibility for how well he works—or whether he does his work at all—to lie outside his own control and to rest entirely with an authority which controls all the rewards and punishments? These questions about a mature, responsible approach to an assigned task underlie the ethical question of whether it is ethical to plagiarize.

A student who assigns an adequate value to himself as a worker will consider, first of all, that it matters to *him* how well he does a writing assignment. Perhaps what he writes will not be deathless prose, and perhaps his report will repose in the campus incinerator at quarter's end and not on a library shelf. But the writer will have derived value from the exercise: increased writing skill, increased management skill, a strengthened sense of responsibility. The student as worker will have been *active* on his own behalf rather than *re*active in a way that says, "I and my work have little value."

THE FUNCTION OF THE WORK ETHIC IN TECHNICAL WRITING

For the long term, this view of the student as worker is especially important. To place a low value upon writing assignments may make little or no difference, at the time, to a student who is able to survive by plagiarizing or by doing sub-standard work. Ultimately, the price for that approach to work comes later: when the student enters a job with an immature attitude about his relationship to authority, or when he operates at a disadvantage because he has an immature attitude about taking responsibility for himself as a worker. To work poorly as a student is only to delay the time when one is able to work well as an adult wage-earner in the real world.

To set an adequate value upon oneself as a worker cannot be separated from setting an adequate value upon the product of your work. Engineering students labor over the finest details of an engineering drawing, architecture students spend days on their designs, and every math student knows the consequences of a misplaced decimal. Those same students may not give the same attention to accuracy and quality in a technical report which explains and sells an engineering project or an architectural design. Drawings and specifications are the medium of communication between technical experts; an accurate, well-written report is the medium of communication with most paying clients. That view of the verbal dimension of any engineer's work makes it important for a student writer to assign the same value to written work that he would assign to an apparently more technical, and therefore apparently more important, project.

Mastery of techniques, and knowledge of standards and specifications, are as important to a writer as they are to a designer. Especially in a professional situation, a writer also has the added dimension of copyright law to consider, just as a professional must work within certain prescribed codes. In the same way that an engineer knows specifications and codes governing technical work, a writer avoids problems of ethics; he knows what plagiarism is and what copyright laws are, and, in particular, he knows how to avoid problems by using established techniques.

Using references in a report is, first of all, a management problem: deciding which resources are needed and handling them in such a way that material cited can always be traced directly to the source. If the material is managed in such a way that the writer always knows how to trace a reference to its source, writing and making citations is mainly a matter of technique. A writer who manages his material well will not find himself with a quotation on a photocopy— and with no clue to indicate what book he copied from. A writer who can summarize a paragraph for the key idea will not fall into the trap of copying verbatim long passages which make his report an edited patchwork of others' writing and not his own composition. Many a hapless student has been penalized for violation of ethics in report writing when in fact his sins were the failure to master technique, failure to give adequate attention to detail, and in a word, failure to set an adequate value upon doing work of high quality. Knowing how to write to specifications is the first defense against problems of ethics in writing.

We read of lawsuits over copyright and trademark violations. A newspaper story which was awarded a Pulitzer Prize turned out to be fiction, and the community of journalists was profoundly shocked. Less well known is that the great body of scholarly writing is not protected under copyright; instead, masses of data and ideas are protected only by a concept of ethics within a community of professionals. The penalty for violating that code is not a poor grade, or even loss of a job, but loss of trust among one's colleagues, loss of the credibility which makes a professional's judgments ones which other professionals accept and respect, loss of the respect which makes an engineer a person with whom others want to do business. Such issues may seem quite remote to the undergraduate under pressure to do the task at hand. Ethics, however, is related to "the *habit* of right conduct." The earlier a student can conceive of himself as preparing to be a professional within a tradition and within a community of professionals, the better will he be prepared to make the ethical choices which all professional work requires. To function in the working world, a student must learn a practical working definition of ethics: "what you do when nobody's looking."

As a study of right and wrong, ethics relates to a whole range of personal, social, and professional activities. For our purposes here, we have stipulated that ethics for the writer relates most closely to the work ethic. Dividing ethics for the writer, we have discussed how the writer values himself as a worker, how he values the work, and how he conceives of himself in relation to other workers. Applying the techniques of rhetoric has produced, if not an ultimate answer to all questions of ethics, a workable definition for the writer.

Another technique of rhetoric, analysis for the root principle, provides a conclusion for this discussion of ethics. From the three divisions of the work ethic we may derive the root principle of ethical behavior for the writer: mature, responsible, high-quality work.

Ultimately, your success as a technical writer will depend, not upon your talent as a writer, but upon how skillfully and how conscientiously you work.

THE STANDARD CODE OF ETHICS IN TECHNICAL WRITING

THE SOCIETY FOR TECHNICAL COMMUNICATION CODE FOR COMMUNICATORS

As a technical communicator, I am the bridge between those who create ideas and those who use them. Because I recognize that the quality of my services directly affects how well ideas are understood, I am committed to excellence in performance and the highest standards of ethical behavior.

I value the worth of the ideas I am transmitting and the cost of developing and communicating those ideas. I also value the time and effort spent by those who read or see or hear my communication.

I therefore recognize my responsibility to communicate technical information truthfully, clearly, and economically.

My commitment to professional excellence and ethical behavior means that I will

- Use language and visuals with precision.
- Prefer simple, direct expression of ideas.

- Satisfy the audience's need for information, not my own need for self-expression.
- Hold myself responsible for how well my audience understands my message.
- Respect the work of colleagues, knowing that seldom is only one communications solution right and all others wrong.
- Strive continually to improve my professional competence.
- Promote a climate that encourages the exercise of professional judgment and that attracts talented individuals to careers in technical communications.

Reprinted with permission from the Society for Technical Communication.

part 2

Short-Term Writing Projects

7. Memorandums
8. Letters
9. The Informal Report
10. Correspondence Systems

The short forms in writing like the memorandum and the letter are usually associated more with business writing than with technical communication. Working as a technical professional, however, requires all the skills of a business writer who works for an insurance company or for a company whose business is retail sales. Therefore, this major section of this text is devoted to the short-term or situational writing all technical professionals must do.

The memorandum is rather routine as a form and in its writing conventions, but it is a very versatile document, used to communicate the most routine information or to announce significant company policies. The letter is the chief means of communicating with external audiences in any business transaction. Although engineers usually think in terms of lengthy reports, the letter records the daily and weekly transactions arising from any major project. It appears in combination with the memorandum to create a short, informal report.

To illustrate the movement of communications within a project, this section concludes with a discussion of how correspondence functions as a coherent system within a large project.

Manuscript specifications for the letter and the memorandum can be found in Appendix A.

chapter 7

Memorandums

The memorandum (or "memo") is a document sent within an organization, not to external readers. In business and technical writing a routine memo is thus the first writing a beginning engineer or technologist is likely to be assigned, although some memos are used for high-level executive communication. The skills involved in technical tasks—visual aptitude, technical expertise, and the ability to devote attention to precise details—can and should be applied to writing a memorandum. By the time an E/T has a job which requires writing a memo, a technical education has already provided many of the requisite skills, even if it did not include instruction in memo-writing.

As with most tasks, an E/T can best write a memo by breaking the task into its component parts and approaching them systematically.

COMPONENTS OF THE MEMORANDUM

Function

The key word for the memo is *information,* for the chief function of the memo is to inform. Within an organization, the impact of a memo will be determined almost entirely by how effectively and how efficiently the memo communicates information.

Design

The memo is a written document, but it is also a design within an 8½- by 11-inch space. As the summary example shows, the memorandum design has two parts: the heading and the body. The heading functions to communicate and to record: date, reader, writer, and subject. The body design functions to make the content visually easy to see, physically easy to handle, and logically easy to read. Writing memos requires writing skill, but just as important are these related skills: design skill, in using lists and underlining; and thinking skill, in creating a logical order.

Convention

The memo contains none of the conventional usages which characterize the letter: salutation, tone of first paragraph, tone of the last paragraph, and complimentary close. Adherence to prescribed form, a plain style, and factual content are the conventions of the memo, along with attention to titles within the corporate structure. In the memo, as much as or more than in any other written document, your work speaks for itself in the formal, accurate way you report it.

Technique

Since the memo does not call for a formal introduction and conclusion as the letter does, techniques for producing a memo are primarily 1) adherence to form, 2) logical thinking, 3) a page design which reflects the logical order, and 4) precise and objective communication of information.

Organization

Usually a memorandum does not require a writer to do extensive outlining as would be required for a work of greater scope. Instead, a technical writer should look for the inherent order of the technical work which the memorandum reports and discusses. As the sample memorandums show, the order is likely to reflect the order in the external situation which the written document describes. Thinking in such a systematic way is a characteristic skill of students in technical disciplines. Your major task in planning a memorandum is to translate those orderly thought patterns into written form on a well-designed page.

Basic Writing Skill

Beyond the skills in design and organization which we have noted, the additional skill which a memorandum requires is the ability to write clear, effective, well-constructed sentences. All other formal aspects of a memorandum are so routine that writing sentences is likely to present the greatest challenge to a beginning writer.

Style

Most technical writing calls for the "effaced writer." That is, the writer's expertise and evidence of the writer's work, not the writer's style and personality, are central to the communication. This principle of technical writing is nowhere more applicable than in the memorandum.

EVALUATING MEMOS

The memorandum is a simple and basic form, but its simplicity is deceptive. Perhaps a good introduction to just how exacting a simple form can be is a demonstration showing how a number of freshmen wrote their first memos. This is how the demonstration works:

1. The assignment explains what the students will be doing both in class and out of class, so the reader knows the background of the memorandum.
2. The memorandum assignment is itself a memorandum. The student writer is asked only to reproduce the design on the page.
3. The assignment has three advantages of audience and purpose:
   a. the professor as audience is familiar.
   b. the audience has a specific need to know and the student has information which meets that need.
   c. the student has the added incentive to sell his competence and the quality of his work to the audience.

4. The assignment has these advantages of subject and organization:
   a. the subject is well defined.
   b. the parts of the subject are provided in the assignment.

How can anyone miss on such an assignment? A close reading of the examples of memorandums will show how various writers respond to exactly the same set of directions with varying degrees of writing skill and quality workmanship.

See Appendix B on evaluation checklists. It contains a note on using them, and a specific one for memorandums.

Original Assignment

Assignment due November 17 in your completed folder

Subject: For the next 4 class meetings, I will be lecturing on rhetoric and individual rhetorical patterns: definition, example, comparison and contrast, classification and division, process, and cause and effect.

Your Job In Class: To listen, take notes, ask questions.

Reading: To read in your rhetoric text and handbook the sections designated for each of those rhetorical principles. This will give you sufficient background reading for the lectures.

More Reading: Assemble a series of examples (at least three for each of the patterns listed above) from your general reading and place them in your folder. Do this in this way: 1) clip an example from a newspaper or a magazine, tape it to an 8½- × 11-inch page, give the source (title of publication, author, page, date, volume, number) and label it (definition, comparison, contrast, etc.) 2) Xerox a page from a text book or some other book, clip the sample, and repeat the identification specifications listed above. Compile your own reference of rhetorical patterns from your general reading.

Writing: Using the topic of your diagnostic essay for this class, write an example of each of the rhetorical patterns. (2 pp. max, per pattern)

This assignment is not designed to be a burden added to everything else you have to do. Instead, it is designed to be a systematic way for you to organize your study, to use the lectures and class readings, and thus to build your skills in increments. What to do and how to do it will become clearer as we go along. How you manage your time is ultimately up to you; I would suggest that you spend an hour per day on this assignment. Human nature being what it is, you may wait until the last minute to begin. However, my job is not to see how you do it or that you do it, but to evaluate the skills you display in other writing assignments in class. My job is to give the assignment and evaluate the results; you are responsible for covering the distance between those two points.

Progress Report Assignment

November 5, 1990

MEMORANDUM

TO: English 1001 Class

FM: Maxine T. Turner, Ph.D.

RE: Progress Report on Assignment Due November 17

Write a memorandum reporting your progress on the assignment due November 17. The heading should conform to the form you see on this page. The progress report should include the following:

1. The process you are using to collect material.
2. Your schedule.
3. Your sources.
4. A discussion of the kinds of rhetorical samples you are locating in the sources and their usefulness to you.
5. A discussion of the text resources related to the assignment.
6. Any problems you are having in your work on this assignment.

This assignment is to be handed in Tuesday, November 10.

A Checklist for Evaluating Memorandums appears in Appendix B on page 489. Manuscript specifications appear in Appendix A on pages 464–467.

Student Memo: Exercise 7.1

November 10, 1990

MEMORANDUM

TO: Maxine T. Turner, Ph.D.

FM: Georgette P. Burdell

RE: Progress Assignment Due on November 17, 1990

1. PROGRESS

   I am collecting material by reading articles from various sources. I read the article first and then I classify it according to the rhetorical pattern it follows.

2. SCHEDULE:

   I allow myself at least two hours every other day for reading to find examples of the rhetorical patterns. I am now in the process of writing an example of each of the rhetorical patterns, using the topic of my diagnostic theme.

3. SOURCES:

   The sources listed are those I have used thus far. I plan to use other sources to vary my reading content and to get an overall view of different writing styyles.

   (1) CHEM ONE by Jurg Waser, Kenneth N. Trueblood and Charles M. Knobler.

   This text was instrumental in helping me to find patterns such as process and definition.

   (2) THE TECHNIQUE newspaper

   I found a few definition and example type rhetorical patterns in this publication.

   (3) SCIENTIFIC AMERICAN magazine

   I was able to find at least one example of almost every pattern in this publication. I found examples of definition, process, cause and effect, and classification and division.

4. TEXT RESOURCES:

The reading assignments were very instrumental in helping me to distinguish between the patterns. There was an excellent discussion of the patterns in Hairston, Chapter 3. The appendix in McCuen gave excellent examples of the patterns in the example student themes.

5. PROBLEMS:

The only problems I am having are writing my own examples for each of the rhetorical patterns. Other than that the assignment presents no problem.

Student Memo: Exercise 7.2

November 10, 1990

MEMORANDUM

TO: Maxine T. Turner, Ph.D.
FM: G. P. Burdell
RE: Progress Report on Assignment Due November 17

I feel that my progress on the assignment due November 17 has been very good. I have worked hard and at a steady pace. At this time, about half of the assignment is finished. In order to give a better explanation of how much I have done and also how I have done it, the following paragraphs will discuss six basic areas of progress.

The first area is the process I use to collect material. I feel that the best way to learn to write is to read; therefore, I read samples of good writing. I then review the material, looking for rhetorical patterns. I select good samples of each pattern that will be included in this project. Finally, when I have collected enough material, I will read the samples for a third time and then pick three of the best for each pattern. The criteria which I will use in choosing these will be their quality as rhetorical samples, a diversity of topics, and a diversity of sources.

My schedule, the second area, has been very flexible. I spend at least one hour, five days a week working on English. The hour may be used for reading background material, looking for rhetorical samples, writing my own samples or finishing class warm-up exercises. I will be finished writing and collecting samples by Wednesday, November 11. This will leave sufficient time to revise writings, re-write material and organize the material into a usable booklet.

The third area includes my sources. Because much of our work is and will be related to technical writing, I have tried to use sources which include that type of writing. I am using Scientific America and text books right now. I plan to also use other scientific books and magazines before I am finished.

The fourth area is about how the samples of rhetorical patterns have been useful to me. To date I have found at least one sample of each of the patterns. They have been useful to me in two ways: I have become both more familiar with the patterns within the paragraph and with the use of the pattern within an essay. It is interesting to read through an article and then divide it up into simple rhetorical patterns.

Fifth, I have found the text resources to be extremely useful. Each of our three texts—plus the lectures—defines the patterns in a different way. This difference is very helpful because it gives us at least four ways to view a rhetorical pattern. The more ways of looking at something, the better the chances are for one to understand it.

The sixth area has to do with problems. Right now I have not had any major problems. I have worked very efficiently and I have learned a great deal from what I have done.

Student Memo: Exercise 7.3

MEMORANDUM

TO: Maxine T. Turner, Ph.D.

FM: Georgia P. Burdell

RE: Progress Report on Assignment Due November 17

---

1) Process: The process I am using to complete this assignment includes the use of my free time between classes on certain days to write and cut out examples of the rhetorical patterns.

2) Schedule: My schedule consists of free time between 2:30 p.m. and 4:00 p.m. on Tuesdays and Thursday and also between 2:00 p.m. and 4:00 p.m. on Wednesdays.

3) Sources: My sources include Scientific American, Technique, and textbooks.

4) Many examples of definition, comparison and contrast, and process can be found in these sources, but few examples of cause and effect and classification and division can be found.

5) Many rhetorical patterns can be found in textbooks, especially Biology and Calculus.

6) Problems include trying to complete this assignment along with other homework and studying and trying to find enough time to do everything.

Student Memo: Exercise 7.4

MEMORANDUM

In order to finish my assignment I have been relying heavily upon the articles in Scientific American magazine. Thus far, I have taken all my examples from the magazine because of the wide variety of articles in it.

When looking for examples, I scan an article and try to locate something that fits the description of one of the rhetorical patterns. I then read it over several times to make sure it fits and then I clip it out.

Because of the heavy workload of my other classes, I sit down only once or twice a week for several hours and work on the assignment. I seem to be managing my time alright because I'm over half way done with the assignment and it hasn't become a burden to me.

I have located examples of the following rhetorical patterns: definition, example, classification and division, and cause and effect. They have helped me to see how other writers develop their ideas through the various patterns. I have tried to use this knowledge to develop my own examples of the rhetorical patterns.

The only texts I have used are my English texts. I have used them to help clarify the patterns in my mind so I know what to look for in the various articles.

The main problem I am having is confusing definition, example, and classification & division with one another. I am also unsure of what process is. However, I'll find out in class Tuesday and I'll hopefully be able to finish the assignment by November 17.

Student Memo: Exercise 7.5

Progress Report on Assignment Due November 17.

I have been collecting material such as newspapers and my text books. I have set some time aside each or every other day and sat down to find the various examples that I need. I'm finding it quite hard to find examples of classification and division in the newspapers especially. My definitions are coming out of my text books. After looking through enough material, I find it easy to pick out the certain examples.

I have not yet begun to write examples; most of this week will be set aside for that. I would say that I have found half of my examples so far.

Summary Evaluation

Now that you have read several examples of memos in response to a single assignment, if you were a manager and these writing samples appeared as a part of an employee evaluation, how would you as a prospective employer judge:

1. ability to following directions?
2. attention to detail?
3. thoroughness and specificity in giving information?
4. logical arrangement, including visual design?
5. clarity of style?
6. mechanical accuracy?

If you were a manager who needed the information in the memo as one set of data to use with others in making a decision or in taking an action, which memo would be most helpful?

Which memo writers would you carry on your payroll to follow your directions, provide you with needed information, and represent your company with written documents?

To return to the initial question about memos: if the memo form is so simple, then why worry? What can go wrong? As these memorandums show, many things can go wrong if a technical writer does not devote to writing the same attention to accuracy of design and detail she must give to any technical work.

Prepared by your critique of several memos, you should be able to see the distinctions among the four major types of memos, which follow the Summary Example, illustrating the general format of the memorandum, in the next section.

MAJOR TYPES OF MEMOS

Summary Example 7.0

Month day, year

MEMORANDUM

TO: (Name of recipient, with title if appropriate)

FM: (Name of sender, with title, if appropriate, with initials penned)

RE: (The reference line has the same function as a title of a report; it is an aid to the reader, for reference, and for filing.)

1. <u>Form</u>: The memorandum should appear on the page <u>exactly</u> as these instructions appear on the page. That is, a memorandum should be (a) <u>visually</u> easy to see and refer to (b) <u>physically</u> easy to handle and refer to (c) <u>logically</u> easy to follow. The qualities of good writing are required (i.e., clear sentences and coherent paragraphs), but a part of the communications burden is distributed to "external" or "mechanical" typographical features such as you see here.

2. <u>Audience</u>: Memorandums are sent in-house to subordinates, superiors, and colleagues. Letters are sent to external audiences.

3. <u>Purpose</u>: The purpose of a memo is to communicate and record factual information, to state policies, and to state professional opinions and judgments.

4. <u>Situation</u>: Depending upon purpose and audience, memos generally address the following situations:

   A. <u>Informative</u>: As the term suggests, the informative memo contains factual information and explanations. Such a memo is classified as "expository" in that the primary purpose is to explain.

   B. <u>Directive</u>: Again as the term implies, the directive memo is a series of instructions to a subordinate or to anyone in the firm. It is important to divide such writing into steps and to use <u>so that</u> explanations to interpret why a given step or the task as a whole should be done.

   C. <u>Meeting Minutes</u>: The primary purpose of the meeting memo is to record 1) who was present (insert "Attending," and a list after the "RE": line in the heading), 2) when and where the meeting took place, and 3) what was discussed and decided at the meeting.

   D. <u>Administrative</u>: Such memos have two general purposes: 1) to state a policy or 2) to render a professional opinion or judgment. Either the Informative or the Directive memo can contain elements of policy statements or opinions to give context and perspective for the otherwise expository matter; Administrative memos, however, more nearly fall into the classification of "persuasive prose."

5. <u>Style</u>: In keeping with the purpose of a memo, the style is objective and unadorned. A memo requires none of the conventional courtesies which characterize a letter. In a memo especially, avoid masses of unprocessed data. Use form as a way to deliver content in some logical, systematic way.

Informative Memorandum: Example 7.A.1

April 26, 1990

<u>MEMORANDUM</u>

TO: Edward R. Lee

FM: Jon Jackson

RE: 1990 Modifications H & H Water Systems

I visited the Davis Building today to observe the work done by Smith Electric in relocating the panels and transformers from the basement to the second floor.

Please review the following notations:

1. Smith Electric should provide cover for conduit at the transformers.

2. Jones Contractors should patch the concrete where the conduits pass through the floors.

We should check these items again before final inspection to assure that all work is up to code.

Informative Memorandum: Example 7.A.2

April 27, 1990

<u>MEMORANDUM</u>

TO: John Scott, Superintendent of Spinning

FM: Lance Hilton, Quality Control Technician

RE: Twist factor considerations for styles:
    P39-148, P39-149, D49-153, D49-155

This memorandum reports the results of fatigue tests at 250 cycles/minute on the comptiss apparatus; the following twist factors (linear density ½ × turns/cm) achieved maximum strengths and elongations:

Style	Linear Density (Gm/1000m)	Spinning Systems	Maximum Strength Twist Factors	Maximum Elongation Twist Factors
P39-148	35.7	Conventional	55.3	42.7
		Open-end	58.5	47.4
P39-149	25.0	Conventional	56.2	40.4
		Open-end	63.2	47.4
D49-153	20.0	Conventional	56.2	40.4
		Open-end	64.8	53.7
D49-155	20.0	Conventional	53.7	39.5
		Open-end	61.6	50.6

Directive Memorandum: Example 7.B.1

May 5, 1990

MEMORANDUM

TO: Edward Phillips

FM: Jon Jackson

RE: Unit D
 Hidden Hills, Phase II

Below are some items for which we need answers in order to complete one design. Please send us this information as soon as possible so that we can meet the schedule.

1. We need a site plan showing all paving and landscaping schemes.

2. What is to be done to the picnic area on the West end of the project?

3. We are still awaiting final architectural drawings for the entrances and and foyers.

Please let us know what should be done regarding the required fire doors, particularly in Building D.

Please respond to these items as soon as possible because our work depends on your reply.

Directive Memorandum: Example 7.B.2

April 18, 1990

MEMORANDUM

TO: Fitting Staff

FM: Sally Lund, Chief Designer

RE: Standard Measuring Method

It is imperative that we give our customers a perfect fit in all garments. This is not always an easy task since most figures are not perfect. However, measurement precision enables us to achieve a perfect fit.

The key element for a perfect fit is a sloper which has the identical measurements as the customer. Although there are several methods for drafting a sloper, we are adopting the "Head" method for uniform use throughout the company. According to this method, the ideal figure is 9 heads tall. This method defines the vertical distance from the hair line to the bottom of the chin as 1 head. The body is then measured relative to this head measurement. Theoretically, the customer's ideal figure should correspond to the following chart:

hairline to chin	1 head basic measurement
chin to shoulder line	⅓ head
chin to bustline	1 head

shoulder line to bustline	⅔ head
bustline to waistline	1 head
waistline to hipline	⅔ head
hipline to pelvic line	⅓ head
waistline to pelvic line	1 head
pelvic line to mid thigh	1 head
mid thigh to knee	1 head
knee to smallest part calf	1 head
calf to floor line	1 head
	10 heads

Example of an ideal figure:
hairline to chin measurement = 6 inches
    1 head = 6 inches
Theoretically: chin to bustline = 6 inches

Fitting Staff
April 18, 1990
Page 2

    Since a given client rarely is an ideal figure, we must compare the customer's ideal figure with the customer's actual figure. The actual figure is obtained by measuring these distances on the customer's body. Use the method described here so that the forms can be constructed uniformly and especially so that the cutting room staff will operate on a uniform standard.

Example of an actual figure:
hairline to chin measurement = 6 inches
1 head = 6 inches
Actual: bustline to waistline = 9 inches

    The comparison of these data will indicate whether the customer has a short waist or a long waist, a long neck or short neck, short legs or long legs. The actual figure example indicates that this person has a long waist. It is our job to design for these figure irregularities; that is, we must maximize the good points and minimize the bad points.

    Adoption and uniform applications of this standard may cause some of us to change more familiar ways of doing our work. However, as we grow as an organization we must adopt more efficient methods. The head method should result in cost savings while we also are able to give the same top quality designs to our clients.

Meeting Memorandum: Example 7.C.1

May 1, 1990

MEMORANDUM

TO: Richard M. Stuart, Chief Electrical Engineer

FM: Jon Jackson

RE: Stuart and Lee—Hill and Hood meeting to discuss Hidden Hills, Phase II

Attending: Edward P. Lee, Stuart and Lee
 Edward Phillips, Stuart and Lee
 Jon Jackson, Stuart and Lee
 Howard Hill, Hill and Hood
 Mark Edwards, Hill and Hood

At a meeting of Stuart and Lee representatives and architects Hill and Hood, those attending discussed and agreed upon the following:

Safety:

1. Sprinklers and standpipes will not be required. The architects will specify fire extinguishers: specifications, placement, service schedule.//
2. The first floor level is at 12, or equal to the flood plane. The engineers will design basement space and any crawl space for underwater service.

Security: Engineers will design a security lock system for all outside entrances with a control console at the reception area and in the resident manager's office. This will include a clock system.

Communications:

1. Architectural specs call for a paging, intercom, and background music system.
2. A central dictation system for the management offices will not be required. A change order will be issued with a credit to Stuart and Lee of $9,500.

The meeting was adjourned at 5:30 p.m. An additional meeting is scheduled for 3:00 p.m., Tuesday, May 15, at the offices of Hill and Hood. Mark Edwards will be responsible for distributing the agenda for that meeting.

Meeting Memorandum: Example 7.C.2

April 16, 1990

<u>MEMORANDUM</u>

TO: Members of Sigma Sigma Sigma Fraternity

FM: Paul Scott

RE: Minutes of Fall Rush Committee Meeting

Attending: Joe Calhoun
              Peter Marks
              Mack Thompson
              Stephen Davis

The Fall Rush Committee met at 8:00 p.m. on April 15 in the Sigma Sigma Sigma House Library. The following agenda items were decided:

<u>Dates</u>: The 1990 Fall Rush will begin on Tuesday, September 16 and end on Saturday, September 27, 1990.

<u>Budget</u>: We have a budget of $3,230 to fund rush parties and outdoor rush activities.

<u>Proposed Party Schedule</u>:

Open House	Tuesday, 16
IFC Band Party	Thursday, 18
Swampwater	Friday, 19
Four Winds	Saturday, 20
Wine and Cheese	Sunday, 21
Drown Your Sorrows	Wednesday, 24
Daiquiri Party	Friday, 26
Beachcomber	Saturday, 27

This party schedule will be voted on by the brothers at the next general chapter meeting. Once this schedule is confirmed by a vote, the committee will work out details for each party, including budgeted items for each.

Administrative Memorandum: Example 7.D.1

May 17, 1990

MEMORANDUM

TO: Robert S. Jones, Director of Management Engineering Dept.

FM: James C. Smith, Administrator

RE: Improved Utilization of Nursing Personnel

Problem Statement: Continued inflation necessitates finding new means of cutting expenses. Personnel costs currently constitute 45% of this hospital's operating budget. Of this 45%, 90% are nurses' salaries.

Discussion: Because nurses' salaries make up such a large part of the budget, it is important to avoid inefficiency in this area. It is my opinion that improvement of nurse utilization could lead to significant savings in personnel costs.

Proposed Solution: Please have your department initiate a study to discover where underutilization is a problem. I would like the following information:

1. which units have a problem
2. the extent of the problem
3. the savings to be generated by solving the problem
4. the approximate cost of solving the problem.

Do this on a unit basis.

Conclusion: Once I have reviewed your findings I will meet with you to discuss the units for which it is economically feasible to develop improved staffing patterns. For this phase of the project your department will work in cooperation with each nursing unit since this should facilitate the acceptance of the changes. I would like your findings within the next six weeks.

Administrative Memorandum: Example 7.D.2

May 17, 1990

MEMORANDUM

TO: Marvin Muscle

FM: Polly Pectoral

RE: Loss of monopolar needle electrodes

Problem: We have recently lost needle electrodes which were sent to Central Supply for sterilization.

Discussion: Following the first incident of losing needles, we began a log to sign the needles in and out of the Department. Recently, however, 7 monopolar needles were sent to Central Supply and only 3 were returned. This happened even though the Physical Therapy Aide had packaged the needles, marked them as belonging to the PT Department, and placed them on the cart according to procedure.

The loss of these needles represents a monetary loss of $46.60. In addition to the cost there is the result of interruption of patient care and inconvenience to the physician.

Action: To avoid further problems with the loss of needle electrodes, I recommend the following action:

1. Keeping a log to record the location of needles at any given time.

2. Packaging and marking of needles which are sent to Central Supply for sterilization.

3. A log at Central Supply to facilitate handling and distribution of equipment and supplies from hospital departments.

4. Distribution and posting of directions for logging, packaging, marking, and distributing of the needles so that all staff will learn and follow a uniform procedure.

Administrative Memorandum: Example 7.D.3

February 14, 1990

MEMORANDUM

TO: All Faculty

FM: Edward D. Deason, President

RE: Report of Special Committee on Faculty Rights

The Special Committee on Faculty Rights has developed the following statement on dismissal of faculty members. Please read this statement carefully, both to inform yourself and also to bring to the Committee's attention any needed discussions or revisions.

A tenured faculty member, or a non-tenured faculty member before the end of his contract term, may be dismissed for any of the following reasons provided that the institution has complied with procedural due-process requirements:

1. Conviction or admission of guilt of a felony or of a crime involving moral turpitude during the period of employment—or prior thereto if the conviction or admission of guilt was willfully concealed;

2. Professional incompetency, neglect of duty, or default of academic integrity in teaching, in research, or in scholarship;

3. Sale or distribution of illegal drugs; teaching under the influence of alcohol or illegal drugs; any other use of alcohol or illegal drugs which interferes with faculty member's performance of duty or his responsibilities to the institution or to his profession;

4. Physical or mental incompetency as determined by law or by a medical board of three or more licensed physicians and reviewed by a committee of the faculty;

5. False swearing with respect to official documents filed with the institution;

6. Disruption of a teaching, research, administrative, disciplinary, public service, or other authorized activity;

7. Such other grounds for dismissal as may be specified in the Statutes of the institution.

chapter 8

Letters

Writing letters is a task which can be broken down into these components: function, design, convention, technique, basic writing skill, and situation. Some of these components require the ability to verbalize rapidly, accurately, and effectively. Other components on this list are not writing skills at all, but design, thinking, and management skills. Writing effective letters depends upon understanding and mastering the component parts of the letter-writing process.

COMPONENTS OF THE LETTER

Function

A letter is a written document which communicates to a reader. A letter also has a variety of other functions: 1) it becomes a part of a legal record of a transaction, 2) it is a record of data which can be retrieved, 3) it is a sales and public relations tool, 4) it is a routine record of an oral transaction, 5) it is a shopping list to place an order, or 6) it is a set of directions for the reader to follow. Whether a letter is postmarked at a given day and hour may have a bearing upon its legality; how it is delivered may also influence its effectiveness. The variety of functions makes a letter more than a written document which communicates to a single reader; it is one of the most varied and useful tools in business and industry.

Design

For all the variety of its uses, a letter produced by an individual in a company has a standardized design based on a set of specifications. Writer and typist work together to design and produce a visual configuration within an 8½ by 11-inch space. The draft is like a rough sketch; the finished copy is like a finished drawing; and typist and writer together check every detail of the finished letter just as they would a blueprint. This process requires not writing skill, but design skill and the ability to attend to the smallest detail in quality control. An important part of learning to write letters is learning to transfer to a writing task many skills which technological curriculums teach.

Convention

One who designs a letter operates within strict limits which define standard design. A convention also describes right and wrong, but one convention does not prescribe a solution for every case. Instead, a writer learns a range of conventional approaches to letters, and also learns how to make the appropriate decision about using a particular convention. Later on, this chapter gives a detailed discussion of what a convention is and how conventions operate in letter writing. At this stage, it may be more useful to draw an analogy between social conventions and conventions in letters.

In the United States people usually greet each other with, "How are you?" and answer, "Fine, how are you?" That exchange is so conventional as to be almost meaningless, but everyday courtesy prescribes some such exchange. In the same way, *Dear* at the beginning of a letter and *Sincerely* at the end are conventions. That is not something you stop to ponder over; it is done that way, so do it—almost without thinking.

Not all social or letter-writing conventions are automatic, however. Dealing with different people in different situations calls for knowing what conventions to use when. Presumably you greet your best friend's father with a conventional handshake and a "How are you, Mr. Jones?"—not with some of the more creative and unconventional handshakes and greetings you know. Americans usually shake hands with the arm almost fully extended; Germans bend the arm almost at a right angle to place those who greet each other at a closer physical distance.

In other countries, men embrace and kiss. In the news we see our public officials adapting our conventions or adopting the conventions of other countries as a matter of courtesy and as expressions of good will.

Conventions in letter writing operate in much the same way. Some are prescribed and almost automatic. Others a writer learns in a textbook like this one, and still others an E/T learns through experience in writing letters on a job.

The appropriate convention appears in writing as the final form, but skills gained through observation and experience determine what a person writes. This discussion of convention reinforces the point that all problems with writing a letter are not necessarily writing problems.

An essay which I wrote, for the June 1980 issue of Delta's in-flight magazine *Sky*, explains how and why conventions operate in letter writing.

"From date line to signature line, a letter writer can pause at almost any point to spend time trying to decide the right form to use. Asking the advice of someone else in the office may only lead to disagreements: one writer cites one authority; a typist points out just the opposite in a handbook; and the boss invokes the half-remembered rule of an English teacher at old Lee Junior High.

"The greatest problem with such questions about business letters is that they seldom have to do with right or wrong. Instead, doubts and disagreements usually have to do with matters of convention.

"Like rules, conventions may be fixed and arbitrary. Unlike rules, conventions are continually changing, and they may vary between one good authority and another. Sometimes conventions are based on practical reasons for using a particular form. Some are set for legal or record-keeping purposes, while others reflect matters of appropriate taste for a given situation. Still other conventions exist for no good reason except that most writers simply do it that way.

"The important thing to remember about conventions is that they are used not to win arguments, but to settle them, so that the important—and expensive—work of letter-writing can proceed unhindered.

"This review of the parts of a routine business letter shows how conventions operate to address questions which can arise in the letter-writing process:

"*Letter or Memo:*   By convention, letters go to external audiences; memos are sent to communicate or to record facts in-house—and the more concise and unadorned the better.

"*Date Line:*   The date line raises the question of how to space the letter: date and signature line to the right, centered, or flush with the left margin?

"Writer and typist may while away the better part of an afternoon discussing this, but neither could successfully prove that one spacing is inherently better than another or that one way is right and another is, by definition, wrong.

"Such a question, however offers an ideal example of how a convention operates; the answer lies in what works most conveniently for most people, and the electric typewriter provides the answer: hitting the return key makes the full-block (flush-left) style the simplest, most efficient spacing for a letter.

"The fact that that works for most people does not mandate it for all, but conventions do tend to develop out of workable solutions to the writing process.

"*Inside Address:*	The record-keeping function of business letters mandates the inside address to show the full name and company affiliation of the recipient, even if you write several letters to the same person each week. That's how a convention can operate in a fixed, very arbitrary way for reasons having little to do with the immediate writing and reading of the letter.

"Another fixed convention is to assign a courtesy title to the recipient: 'Mr.', 'Miss', 'Mrs.', 'Ms.', or whatever professional or military title applies. An official title can form a part of a necessary record; for instance, if you write to both physicians and academics, shift from titles to degrees to make the distinction between John Meade, M.D., and Margaret Mead, Ph.D.

"The question of when or whether to use 'Ms.' provides one of the best recent examples of how new conventions are formed.

"Why this happens is that changes in social conventions are always reflected in language. Usage can then enter a period of flux when nobody is exactly sure what to do, there being no fixed convention as yet. What to do in the absence of such a convention is to establish one which works best for you or your company in most situations most of the time.

"If you are in doubt about whether to use 'Ms.', you may want to seek a practical answer in this way: 1) If you are replying to a letter from a woman, check to see whether she has indicated the title she prefers. Many women decide what title they prefer and indicate that by placing the title in parentheses before the typed name beneath the written signature; for example, (Mrs.) Belle Boyd. 2) If you do not know the preferred form, or even whether you are writing to a woman, my own arbitrary conventions are a) to use 'Ms.' when in doubt, since it can serve either for 'Miss' or 'Mrs.'; b) to use 'Gentlemen' as a form of address for a company, assuming that a woman recipient serves as the agent or representative for a larger group properly addressed as 'Gentlemen'; and c) for androgynous names like Tracy, or initials, to use 'Mr.' and hope for the best.

"The reply to your letter should answer any questions about form, but this range of initial choices makes it all the more important to set conventions to govern such decisions and get on with the business of writing. Such conventions should not be chiseled in stone; rather, they should be reviewed periodically to determine whether the convention 1) remains the current usage you normally read and hear; 2) is appropriate for your business; and 3) is a workable solution most of the time.

"Instead of meaning that you have found the correct answer, adopting a convention more nearly means that you have made a decision. Judgments governing such decisions should be based on building good will for your business, not upon political conviction. If, as in the case of Ms., conventions are not yet established, it is better to reflect conservative trends rather than try to set new ones. Once you've done the best you can, don't spend any more time fretting over the one 'right' answer when there isn't one.

"*Salutation:*	*Dear* is one of those conventions so common that people use it without thinking. Even if you are writing to threaten legal action, begin with *Dear So-and-So.*

"*Names Within a Letter:*	Referring to a person in a letter is a part of the record-keeping function of a letter. 'Jon' may make perfect sense to you and the reader this week, but a year from now you may need to have 'Jon Jackson' on file for your records. So use the full name at first reference and either first or last name thereafter.

"**Complimentary Close:**   Sincerely is the form now almost universally used in the U.S. Since in the passage of time conventions tend to become simpler, Yours truly and Yours very truly are not used as much as they once were.

"What matters here is not so much what you decide, but that you do decide upon one form to use consistently. In this, as with all conventions of the letter, a company projects the same image by using the same conventions. More importantly, it is a waste of time for typists to keep track of Mrs. Hood's preference for one form and Mr. Johnston's for another.

"**Signature:**   Authority and courtesy influence the signing of a letter. Both are involved if you have someone sign your name and add initials, for your message may be, 'I have no time for you.' More significantly, when you yourself do not sign, the reader may question the actual source and authority of the message.

"Conventional courtesy also applies if you address the recipient as 'Dear Joe'; then sign only your first name: 'Sincerely, Mark'. This may not always work for superior-subordinate or subordinate-superior communications. Decide such questions about tone and the appropriate level of formality in this way: if you write the first letter, you set the tone and form of address; if you are answering, adopt the tone and form of address of the sender's letter.

"Further advice for less-experienced writers includes these matters of taste: 1) Given the legality of signatures, avoid nicknames. For business purposes, 'Buster' just won't do. 2) Sign your name legibly. Until you've made it to the top, hold off on using a squiggle and a dot. 3) Sign your name parallel to the typed name below. Wait until you sign autographs to use a slanted flourish at an angle up the page. Do not ascribe a title to yourself. Except for professionals like physicians who pen letters after their signatures, letters and official titles for other writers should appear after the typed name.

"**Typed Name:**   Although a title can at times speak more powerfully than the entire letter, the distinction I draw here—mainly as a matter of taste in dealing with my own academic letters—is this: a title should lend power more to what is said than to the person who says it.

"That is, if the reader needs to know your official identity, not so much as an individual but as an officer of your company, a title is both necessary and appropriate. If a letter contains technical data or professional judgments, letters to indicate professional registration of academic degree give the needed authority and validity to the substance of the letter.

"Like all conventions, however, such conventions shift with time and circumstance. Perhaps the only unchanging certainty about letters is the expense of writing and typing them. That means that the time spent debating with yourself and others about what's best to do ultimately adds to production costs—too often without improving quality control.

"Beyond determining cost and quality, however, an unchanging human factor I observed is this: whatever the doubts and disagreements about letters, most writers share a very real and earnest desire to know and to do what is right—or at least a desire not to be tripped up and embarrassed by some obscure rule that the reader may know and he may not.

"Understanding more about how conventions operate is thus both helpful and reassuring if a writer knows 1) that questions about business letters frequently have no correct answers; 2) that answers can be discovered through observing what others do; and 3) that an organization can adopt its own conventions. Establishing conventions saves time and money, as well as grief, for there is also much to be said for feeling more confident about letter writing."

My essay from *Sky* does not deal with one of the most important conventions which govern language: effective use of standard English, and especially the use of idiomatic English and current conventions in business English. To improve your command of conventional usage, pay particular attention to the discussion and the exercises in the Writing Lab on Words in Section Four of this text. Knowing the conventions which govern language has far-reaching technical, business, and social importance.

Techniques

As the discussion of design and convention has shown, the letter is a somewhat formal document governed by constants of design and convention. The work of writing letters in a job situation is also a repetitive daily routine for many writers. Types of letters vary from situation to situation, as further discussion in this chapter will show, but writers are best served when they learn a series of standard techniques for writing some parts of letters.

Writers frequently delay beginning a letter, not knowing exactly how to begin or feeling that each individual letter calls for a new and original approach. Writers frequently neglect the ending of a letter by cutting off the letter with the paragraph which completes the discussion of the subject or by adding a cliche like "If you have any questions, please do not hesitate to call." More effective techniques can make a letter function more effectively.

Learning a series of techniques for such formal parts of the letter as the beginning and ending can save time and effort while increasing the effectiveness of letters. Learning and applying a technique here is not a limit upon creativity but a way to operate more freely within the formal limits of a letter.

Just as important, learning techniques requires one of the most important qualities you can bring to any work you do: a willingness and an ability to follow directions and perform according to standards and specifications.

Function, not style, is the overriding consideration in beginning or ending a letter. The beginning paragraph in a letter states the subject. If you devise an effective way of doing that, use that technique again and again; vary that technique for different situations, but vary only slightly. An analogy for that advice can be drawn from sports: once a pitcher perfects his motion or once a golfer perfects her swing, the best advice is to maintain that motion and rhythm. In the same way, a letter writer who must write several letters a day wastes time needlessly if she tries to devise a new and original beginning with each new letter.

Components of the Letter   75

The same principles applies to the last paragraph of the letter. The purpose is to summarize, to re-state the most important point of the letter, or to motivate. The purpose is *always* to sell and build good will. A narrow range of techniques can be learned to fulfill each of these purposes. Master those techniques which work and make them work for you again and again.

The following sections summarize the basic techniques for beginning and ending letters:

How to Begin a Letter: Routine Requests, Reports, Responses

Purpose: The first paragraph should state the subject for a busy reader who needs information immediately. The first paragraph is you-centered; that is, it exists for the reader.

Problems: Under the pressure of time, these problems are most likely to arise:

the trite beginning: "per your request"; "enclosed please find"; "enclosed herein"; "presented herewith".

the passive beginning: "The drawings completed by us are enclosed herein." "Your letter has been received by us."

the runaway beginning: "We are in receipt of your letter describing the pumps on the above referenced project which have malfunctioned due to there being a difference between the information on the plates and the actual size of the generator."

Solving the writing problems: Address the writing problem by better defining your task:

response: If the letter is a response to the writer, say so in a positive, sales-effective way:

In response to your letter
In reply to your letter
To confirm our conversation

reminder: Jog the reader's memory and make the letter a useful record:

of February 6
about the referenced DOR project

<u>report:</u> State the subject of the letter: clearly, simply, directly

this letter addresses the questions you raised
the purpose of this (answer/letter) is to clarify

<u>request:</u> Make a request clear and persuasive. Use <u>so that</u> motivation

I am writing to (ask/request) that you _____ by _____ (date) so that (I/we/the contractors) can _____

<u>Solving the time problems:</u>   Remember that the writer is probably as much in a hurry as you are.

<u>peg opening</u>: who, what, where, when, why—all or a combination of these will most often tell the reader what he needs to know in paragraph 1. The first paragraph of a news article usually reflects a peg opening; concentrating on how the first sentence or two work in a news story is a good way to learn how this is done.

<u>list:</u> if all else fails, simply list the topics in the letter.

How to End a Letter

<u>Purpose:</u>   The last paragraph, more than any other part of a letter, <u>belongs to the writer</u>: to restate and strengthen what has been said in the body of the letter, to build personal good will, to increase sales effectiveness.

<u>Problems:</u>   Not only the pressure of time, but failure to appreciate the importance of the last paragraph create these problems:

<u>the throw-away line</u>: "If I can be of further assistance, please do not hesitate to call."

<u>no last paragraph</u>: the letter simply stops without the usual stab at an effective ending.

Components of the Letter

<u>Solving the writing problem</u>: Rather than labor over original endings, use one of the following endings to suit your needs:

<u>Sell your work</u>:

"This letter should answer the questions you raised; however, if you have any further questions, please call."

"This letter should provide sufficient information to solve the problem you describe; if I can help you further, however, please let me know."

"This letter ought to provide all the information you need at this stage of the project. If you need more information, let me know."

(These endings reinforce what you have done in the letter: answered questions, given information, solved problems. They are a way of saying to a reader that you have done your job.)

<u>Motivate the reader to act</u>:

"We would appreciate your sending the _____ so that the _____ can proceed with their work."

(Use <u>so that</u> motivation to let the reader know what the consequences of his action will be.)

<u>Motivate the reader to act on time</u>:

"We would appreciate hearing from you by _____ so that the next stage of the project can begin."

(Use a specific date whenever possible because that will register more strongly than "by the end of the month.")

<u>Sell yourself and the company</u>:

"We appreciate the opportunity to work with you on the _____ project and will look forward to hearing from you again."

(All writing involves risk, but to be courteous, to buy good will, and to sell yourself and your company is an effort you should always make.)

Points on how *not* to begin and end a letter appear on pages 359–360.

Basic Writing Skills

I've said it several times already in this text, and I'll be saying it several more: you must be able to construct sentences in clear, concise, idiomatic standard English. No one will care whether your writing lacks poetic imagination, or even the mildest spice; all that matters is that your writing convey what you want to communicate, clearly, efficiently, and naturally. Just as you might consider it imaginative and amusing—once—to see a drawing of a flower growing out of the ground in an electrical schematic, but would find it silly and boring when you had to refer to that schematic again later, just so will any "creative" adornments in your prose style detract from the usefulness of your letters as reference documents.

Situation

The major variable in letter writing is the situation which calls for a letter. While design, convention, and technique in letters are relatively constant, and standard English is certainly constant, any business or technical situation can vary in entirely unpredictable ways. Deciding what to write to a given reader in a given situation thus introduces a demand for a wide range of communications skills: the ability to express oneself accurately and effectively, management skills which produce sound judgments, experience which enables a writer to assess the risks in a situation and the possible impact of a letter on a reader.

A text can describe the design of a letter down to the placement of a byte, or single character of type, on the page; a text can establish the acceptable range of conventions and techniques. No text can reasonably prescribe exactly what to do in every letter-writing situation. Situations which require letters can, however, be reduced to three broad categories: information, action, and money. Letters related to those situations can be divided into a list of basic situations:

A. letters which request information
B. letters which give information
C. letters which request action
D. letters which report action
E. letters which direct action
F. letters which sell

This list of situations represents routine situations: one letter which brings one reply. No technical document exists in isolation, however. As illustrated by the large construction project which the Introduction describes, communications are exchanged among a large group of writers and readers within a total system of communications. This section on situation establishes basic approaches to routine situations. The chapter on correspondence systems which follows later in the text expands the discussion of situation to include a much broader range of variables a writer is likely to confront when actually writing on a job.

Note: How to use the examples of letters in this section:

The section first presents three summary examples of letters; a summary chart on conventions follows, along with a summary chart of how situations determine form in letters.

The examples of letters in this section of the text are divided into three parts; each one discusses a different letter-writing situation:

1. *Design:* The model shows how your finished letter should look on the page and how the complete letter should deal with a given situation.
2. *Construction:* The instructions specify how to write each part of the model to specifications.
3. *Situation:* The discussion of the context for the letter in a working situation describes the major variables in a given situation.

SUMMARY EXAMPLES OF TYPES OF LETTERS

Routine Business Letter: Summary Example 8.1

Month, Day, Year

Full Name of Addressee, Title if appropriate
Company Name
Street Address
City, State Zip Code

Dear Name:

The first paragraph of a letter is the "subject paragraph." In it you let a busy reader know that you will do one of these things: report, request, reply, or remind.

This first "body paragraph" explains the subject stated in paragraph 1. Begin here to give the necessary details to explain and expand the subject. Strike a balance between giving a full explanation and selecting the details your reader needs to know. In that way avoid the "total mind dump."

Ideally, a business letter is no longer than a page. For a longer letter, break the page with lists, underlining, and other typographical features which will make the letter easier to read and refer to.

Paragraphs can and ought to be of uneven length and shorter than paragraphs in traditional expository writing.

The penultimate paragraph usually deals with the consequences of the data explained to this point: cost, time, action required of the reader, date for an action or a response, or problems which remain to be solved. Not all letters, of course, contain this section.

The final paragraph should be sales oriented. Re-state your request, let the reader know how to get in touch with you, offer to give further information, reinforce what you have said in the body of the letter. Leave a good impression because every letter is a sales letter.

Sincerely,

Typed Name, Title if appropriate

INI/tials

cc: copies to Name(s)

Enclosures:

Routine Technical Letter: Summary Example 8.2

Month, day, year

Full Name of Addressee, Title if appropriate
Company Name
Street Address
City, State Zip Code

RE: The subject as a title with key words useful for reading and filing

Dear Name:

State the subject of the letter in a sentence or two. Write this for a busy reader who needs to know immediately why you are writing.

In the second and succeeding paragraphs give the necessary information to explain the subject. Does the reader need to know the background of the subject? Does he need technical information? Are you asking him to do something? If so, explain in this paragraph. If there are several parts of this explanation section, use typographical emphases:

1. List items you request.
2. List actions you request.
3. List objects you are describing.
4. List steps in directions you are giving.

You gain by using such design features because a list saves you much of the difficulty of writing paragraphs. The writer gains in that the letter is easier to read. This technique also recognizes this fact: technical letters are <u>working</u> documents which must be easy to read and refer to.

Maximize the effect of your last paragraph by writing, "This letter should give you the basic information you need for writing and designing a technical letter; if you do have questions, however, please let me know." Such a tone in the last paragraph is preferable to the usual cliches.

Sincerely,

COMPANY NAME IN ALL CAPS

Typed Name, Title if appropriate

INI/tials

cc: copies

Formal Letter Re Technical Project: Summary Example 8.3

Month, day, year

Company Name
Street Address
City, State Zip Code

Attention: Full Name of Recipient

RE: The subject as a title with key words useful for reading and filing

Gentlemen:

State the subject of the letter in a sentence or two. Write this for a busy reader who needs to know immediately why you are writing.

In the second and succeeding paragraphs give the necessary information to explain the subject. Remember that a letter headed "Attention" and with the salutation "Gentlemen" is written from you as an agent of your firm to the agent of another firm. The content is technical, official, and contractual, and your relationship to the recipient is thus entirely professional.

In discussing the technical content of the letter, make use of lists, underlining, and other typographical features to:

1. Give technical data such as details of plans and specifications.
2. List items to be changed, with necessary reference numbers.
3. List parts with necessary reference numbers.

You gain by using such design features because a list saves you much of the difficulty of writing paragraphs. The reader gains in that the letter is easier to read. Very often the substance of a letter like this one contains technical details which do not lend themselves to treatment in traditional paragraphs.

Instead of using a throw-away line like, "If I can be of further assistance, [...]", write, "This letter should give you the basic information you need; if you do have questions, however, please let me know." Even the most routine and technical letter is always a sales letter as well.

Sincerely,

COMPANY NAME IN ALL CAPS

Typed Name, Title if appropriate

Manuscript specifications for letters appear in Appendix A, pages 468–473, 486. An Evaluation Checklist appears in Appendix B, page 490.

Summary Examples of Types of Letters 83

The following charts summarize for quick reference the items discussed in the CONVENTIONS and SITUATIONS sections of the letters chapter.

Letters

Conventions:	Function	Form	Convention	Tone
DATE	record keeping	spell out month	civilian: month day year	
INSIDE ADDRESS ...	record keeping	as address appears on envelope	socially or formally, spell out *Street, Avenue,* state	
TITLES	record keeping courtesy routing	military, religious or professional as appropriate	*Mr.* as a courtesy *Mrs., Miss, Ms.* if the preference is known; *Ms.* at first use if not.	
SALUTATION .. NAME		*Dear*	*title and surname #first name as appropriate (see note at Signature below for appropriate use)	Writer of the first letter sets the level of address and tone
PARAGRAPH I	to state subject record keeping	1½-3½ lines 4-5 line MAXIMUM	avoid cliches; READER-CENTERED; be concise	incisive appreciative
BODY PARAGRAPHS	to explain/ clarify to describe to direct	paragraphs of varying length indent, list, underline	avoid a conversational tone and a leisurely and discursive style. DICTATION IS *NOT* CONVERSATION	logical factual expository VISUALLY clear
LAST PARAGRAPH	to summarize to reinforce to build good will	1½-3½ lines 4-5 lines MAXIMUM	avoid cliche, throwaway lines. reinforce; re-state dates, actions WRITER-CENTERED	cordial helpful
COMPLIMENTARY CLOSE		*Sincerely,*	*Cordially* in some instances for close business associates	
SIGNATURE ..	legality courtesy	legible parallel to typed line	*full name #first name *ONLY*, if you address the reader by first name. avoid nicknames	Another's signing your name and adding initials can be impolite.
TYPED NAME	record keeping	full legal name	never ascribe a title *before* the name.	

Conventions:	Function	Form	Convention	Tone
TITLE/ LETTERS	assignment of role and responsibility		to lend weight to the content of the letter, not to the writer	professional, not political
INITIALS	record keeping	CAPS/lc	signer's initials/ writer's initials/ typist's initials in large organizations	
COPIES	record keeping courtesy	cc:	sometimes xc:	
ENCLOSURES	record keeping sales reinforcement	*Enclosure:*		

Letters

Situation:	to facilitate understanding	to facilitate a response	to motivate a response
To Request Information	be clear and specific about what you request	list items requested with full information to identify each item include name and address within the letter	date and *so that* if needed
To Give Information	be thorough, accurate, and clear; select details according to your analysis of what the reader needs	use lists, underlined subheadings, and a clear logical division of information	
To Request Action	divide the action(s) into clear and specific steps	use lists and other format devices to make the letter a *working* document	interpret the request with *so that*, explanations; use specific dates for completion
To Report Action	refer to the original request and date process and interpret the report of action; avoid the mind dump		
To Direct Action	divide the action(s) into clear, specific steps interpreted with *so that* explanations	use lists and other format devices to to make the letter a useful set of instructions	use specific dates for completion supported by *so that* motivation
To Sell	rely upon specific data of past experience allow facts to argue	relate the specific data to the prospect's needs express a technical understanding of the prospect's needs	express willingness to explain further be clear about when and how you may be contacted

Letter Which Requests Information: Design

December 6, 1990

American Consulting Engineers Council
1455 15th Street, N.W.
Washington, D.C. 21002

Gentlemen:

This is a request for one copy of each of the following documents:

    No. 737/SIG (FR Nov. 20, pp. 56050-56058)

    No. 720.08/SRA

Please mail them to:

    Edward R. Lee
    10 North Frederick
    Rockville, MD 11863

A check for $2.00 is enclosed. If there are further charges for mailing and handling, please let me know.

Thank you for making this material available.

Sincerely,

Edward R. Lee

A. LETTER WHICH REQUESTS INFORMATION

Construction

P.1. Just as the first paragraph of every letter states the subject, in this paragraph state your request for information (the subject) simply and clearly.

P.2. Assume that the letter will be treated as a purchase order. List clearly on the page the item(s) you request with all necessary reference numbers, dates, pages, etc.

P.3. List your name and mailing address as it should appear on the return envelope so that the reader is spared the trouble of assembling those bits of data from the letter head and the signature line.

P.4. Indicate that payment is enclosed, for your letter also serves as an record of the transaction. If there has been no information about costs, at least offer to pay any charges associated with your request. And, of course, always say "thank you."

Situation

The letter which requests information is one of the simplest, most routine letters a worker writes. The details of construction, however, show how you can enhance your company's image and your own image as a worker by using a pleasant tone and design elements which make your letter easy to see and handle as a working document.

Letter Which Gives Information (Technical Design)

JACQUELINE HOUSTON, M.D.
GENERAL ORTHOPEDIC PRACTICE

November 4, 1990

Charles M. Claus, M.D.
Internal Medicine, Inc.
4602 West Broad Street
Denver, Colorado, 66608

Dear Dr. Claus:

In response to your request for information about your patient John Pelham, this letter reports a history of treatment to date:

Previous Condition: Mr. Pelham has been under my care since June 1989. He reported then that he had had neck problems since an automobile accident in 1985. X-rays taken in June 1989 revealed advanced degenerative changes at the C3-4 and C4-5.

Previous Treatment: Treatment in 1985 was conservative but with good results. Physical Therapy Associates in Denver can supply a full report on their work with Mr. Pelham.

Current Condition: Symptoms were significantly exacerbated and aggravated by an automobile accident on September 1, 1987. Since that time, Mr. Pelham has had neck pain, stiffness, headaches, and right upper extremity pain and paresthesia. There was evidence of reflex changes in the right upper extremity and radicular paresthesia of the right ring and little fingers.

Results of Tests: A cervical myelogram performed at Mountain Hospital on October 15 showed abnormal filling in the area of the C4 root to the right.

Current Treatment: Upon consultation with Dr. John Alan, Neurologist, he and I agreed that further conservative management is indicated at this time. PTA is again in charge of treatment under my supervision.

I trust that this report will enable you to complete and update your case record for the patient under your care. If I may help you further, please let me know.

Sincerely,

Jacqueline Houston, M.D.

B. LETTER WHICH GIVES INFORMATION: TECHNICAL

Construction

P.1. Give sufficient information (e.g. date and full name) to tie this letter to the writer's file and the reader's primary need to relate this reply to the original request. Getting the information he needs, not merely a cordial first paragraph, will make him happy and thus create the desired good will.

P. 2-6. Divide the technical matter into logical steps. Make this readable and usable as a working document by using headings and the proper spacing. Breaking the page into parts helps the reader to find what he needs. As an internist and the patient's physician, he may already know the history of previous treatment and may need to know only one item on the list to do his job (e.g. "Current Treatment"). The rest of the letter is a record which may or may not be needed for the future, but it is complete.

LAST P. The last paragraph is formulaic, but it nevertheless expresses the writer's understanding of the reader's need to have a record and states that that need has been satisfied. The letter ends with an entirely conventional—though entirely valid—offer to help further if need be.

Convention

Use M.D. after the name to indicate medical doctor. Dr. could refer to any of a number of academic degrees.

Signature Line

The writer's title lends appropriate authority to the content of the letter.

Situation

The relationship of writer to reader is secondary to the communication of technical data. The writer may not be asked to render an opinion but simply to report information. Fulfilling that function and subscribing to formal, conventional courtesies which characterize such letters are what this situation requires. E/Ts and other professionals write a large number of such letters. One can hardly carry that burden of correspondence *and* create an arrestingly original way of expressing technical data with every letter. To attempt that is to misunderstand what the job calls for and also to be indifferent to the time and costs involved.

Letter Which Gives Information: Professional (Design)

December 12, 1989

Maxine Turner, Ph.D.
English Department
Georgia Tech.
225 North Avenue, N.W.
Atlanta, GA 30332

RE: 1990 APTA Annual Conference
    Atlanta Hilton Hotel
    Atlanta, Georgia
    June 10–15, 1990

Dear Dr. Turner:

Thank you for consenting to participate in the 55th Annual Conference of the American Physical Therapy Association. Correspondence with Harold Smith of our Conference Committee indicates that your presentation is scheduled for Wednesday, June 13, 1990—1:30 to 3:00 PM and 3:30 to 5:30 PM.

Would you please complete and return the enclosed biographical data sheet to me by January 19, 1990. This information will be used both for publicity in our Journal, Physical Therapy, and in our Conference Program of scheduled events.

To accompany the biographical data, I would also appreciate having a black and white glossy photo of you by the same date, January 19, 1990. Good Polaroid shots are acceptable.

Concurrent sessions are scheduled throughout the course of our Annual Conference. It would be appreciated, therefore, if you would for publication purposes provide for the attendees a short abstract of no more than 100 words of your anticipated presentation on "Writing Skills for the Clinician."

Maxine Turner, Ph.D.
December 12, 1984
Page 2

The form for your abstract is enclosed; if possible, we would greatly appreciate it if you could forward this also <u>January 19, 1990</u>.

The Association presents honorariums to its speaker participants. By action of the Board of Directors there is a "fixed minimum" in certain "Participation Categories" which must be extended to the speakers.

Requested honorariums above the fixed minimums are negotiable within the limits of the total budget of the APTA's Conference Committee.

If you have an established "minimum honorarium," the Conference Committee would appreciate knowing about it in advance in order to prevent possible embarrassment at a later date on both our parts.

In addition, APTA honors travel (air coach), and provides hotel accommodation for two nights and food per diem at $24/day for two days, if needed, or unless other arrangements have been agreed upon.

APTA will support the cost of <u>handout materials</u> essential to complement a speaker's program. The committee encourages this as long as the cost of the materials is "within reason."

If you develop any question about handouts and/or have any other questions regarding your part of the program, please write or call me collect: (202) 466-2070.

Thank you again for consenting to share your expertise with our membership. I look forward to working with you these next six months.

Sincerely,

Frank L. Allender, Director
Program Development

FLA/mm

C. LETTER WHICH GIVES INFORMATION: PROFESSIONAL

Construction

P.1. This paragraph states the subject directly for the reader and for the record.

P.2–9. Several body paragraphs give specific detail which the recipient needs to know and, in particular, specify what material the recipient is to send in return. Note how the short paragraphs and underlining highlight the information. Also note the use of repetition of the due date.

The series of short paragraphs related to payment to the speaker, expenses, and the cost of materials is especially effective. Some professional situations require restraint and good taste in the discussion of money. Here the writer relies upon quoted phrases from his organization's regulations and sets up reasonable limits within which to negotiate money matters.

LAST P. This part of the letter is divided into two paragraphs, one offering to answer questions and highlighting the phone number, and the other expressing good will and anticipation.

Situation

I include this letter from my own correspondence files because it is unusually thorough. Such attention to detail expresses the thoughtfulness required of any person who serves on a committee to conduct a conference, or invites an individual speaker. Such a model should be immediately useful to those working in student organizations.

Letter Which Gives Information (Sales—Public Relations) Design

---

JONES BROTHERS PHARMACEUTICALS

October 16, 1990

Sister Naomi, OSH
Convent of St. Hilda
19 Paraclete Drive
San Jose, California 94004

Dear Sister Naomi:

Thank you for your letter of September 29 asking about Jones Brothers Revivify Protein Capsule.

In answer to your question, a single 500 mg. capsule contains 10 calories.

We appreciate your interest in our product and wish you the best of luck in persevering with your diet.

Sincerely,

L.A. Ridley, Manager
Customer Relations Division

LAR/jbj

Enclosure: Jones Brothers Products Brochure

D. LETTER WHICH GIVES INFORMATION: SALES–PUBLIC RELATIONS

Construction

P.1.   For the record, give the date and form of communication. For the record, name the product, service, action, or person. This sentence is pleasant but it also contains the very businesslike features of the form of communication this letter addresses, the date, and the inventory item in question.

Letter Which Gives Information: Sales—Public Relations 93

P.2. Acknowledge and answer the question the writer's letter asked—and by indirection, the fact that the writer is answering. The facts are not only specific and technical but also understandable and useful to the non-technical reader.

LAST P. Build good will by expressing appreciation and adding a personal note. The first part of the sentence is conventional and the last part is a half-humorous personal reference to the reader's letter. The first half of that sentence can be learned by rote; the second half requires developing experience and sensitivity to the individual to whom the letter is addressed.

Convention

Correct form of address and title for someone who does not have a conventional social or professional title. Small courtesies based upon a knowledge of "doing the right thing" are necessary in many business writing situations.

Enclosure

The reader evidently uses the product which is the subject of the letter. To enclose further information about the product line is appropriate.

Situation

The communication of information is secondary to the relationship of writer to reader in sales letters. To supply the information requested is the primary reason for writing, but the writer may carry his effort further to make an opportunity to build good will with the customer or client. Tone and manner are important components in writing to a lay audience which is a customer or client. The writer should not violate convention by being too personal, but learning to find the appropriate range to achieve a cordial tone is important. Some techniques of courtesy ("Thank you for your letter ...") can bridge the gap between a mechanical letter and an appropriate personal reply. So long as the specific information is provided, content is less important than the sales and PR function of the letter. To a market researcher, one letter may represent, by a given radio, perhaps several hundred buyers. Therefore, many companies are willing to spend the time and incur the costs to answer individual inquiries—and thus buy good will.

E/Ts deal with many situations in technical careers which require the tone reflected in this letter.

Letter Which Requests Action: Design

March 4, 1990

Mr. B. N. Forrest
32 East Bay Street
St. Louis, MO 42428

Dear Nate:

A contractor has reported a problem with a fire hydrant behind Unit B of the Hidden Hills Apartments, Phase II.

Please send someone from your plumbing unit to check this. You will need to contact Joe Polk of Acme Paving on site because the present location of the hydrant blocks their efforts to finish the drainage ditches and the sodding of the adjacent areas.

We need to clear this up by March 8 if at all possible so that we can coordinate the plumbing work with the paving and landscaping of the property.

This should supply you with sufficient information to take the needed action. I'll look forward to hearing from you soon.

Sincerely,

STUART AND LEE, ENGINEERS

Edward R. Lee

E. LETTER WHICH REQUESTS ACTION

Construction

P.1. State the subject/problem, giving all available information. The first paragraph of almost all letters states the subject. Most thinking about technical solutions begins with a statement of the problem. Your first paragraph should fulfill both of those specifications.

P.2. A plan of action to solve the problem is usually the next step in technical thinking. This paragraph should give the reader enough specific information to take the necessary action to initiate the problem-solving process.

Letter Which Requests Action 95

P.3. People tend to work better when they relate their action to other actions rather than to operate in a vacuum. People also tend to work better if they have some idea of the results or consequences of their actions. This third part of the letter casts the specific request for action into a more understandable context for the reader. Show how the requested action relates to other actions, especially with the use of *so that* motivation.

P.4. According to the techniques of writing the last paragraph, reinforce this point: your letter has given the person the necessary information to act upon. In addition, motivate by expressing your anticipation for a reply—by a specific date, not in general terms such as "by the end of the week" or "a.s.a.p."

Situation

Your relationship to the audience (superior to subordinate, subordinate to superior, technical to technical, technical to lay) determines how you will approach making your request for action.

Letter Which Reports Action: Design

March 6, 1990

Mr. Edward R. Lee
Stuart and Lee, Engineers
10 North Frederick
Rockville, MD 11863

Dear Ed:

In response to your March 4 letter about the location of a fire hydrant behind Unit B of the Hidden Hills Apartments, Phase II, I am reporting the following investigation and action:

1) The problem was that the hydrant was located down the embankment and into the rip-rap lining the ditch.

2) Joe Polk of Acme Paving met Sarah Cochran of our plumbing unit at the site and together they solved the problem.

3) The hydrant will be moved back approximately 36 feet west of its present location and set on the south side of the main on top of the embankment.

> Cochran altered her drawings to reflect this change and her crew will have completed the work by March 8. No additional cost will be incurred by the client.
>
> This should solve the problem, but if you have further questions, please let me know.
>
> Sincerely,
>
> FORREST & COMPANY
>
> B. N. Forrest

F. LETTER WHICH REPORTS ACTION

Construction

P.1. Without repeating everything that the reader said in his call or letter, briefly re-state the problem as the reader expressed it.

To save time, reduce this to a formula: "In response to (this is positive) your (*date*) call/letter about (*state problem*), I am reporting the following investigation and action:"

P.2. Reduce the body paragraph to a list which includes a suitable combination of *who, where, when, why, how*.

Save yourself and the reader time and make this easy to read by using a list: a) statement of the problem, b) discussion, c) corrective action.

The important thing here is to translate the account from a narrative of something that happened to a logical presentation of a problem, the discussion, and the action. This is not a short story, so don't say, "You wrote and then I called Joe and then we went over there and stood around in the mud and then we decided..."

LAST P. A routine technique for the last sentence will suffice.

Situation

To report the activities of your work is an important part of your work. To a technical reader, clear technical exposition is the strongest selling point. A prompt and accurate response answers the reader's request; it fulfills the purpose of the reader's initial letter, and your ready response should make his job easier. Such routine, even understated, exchanges between E/Ts literally make large projects *work*. The image you want to project here is of someone making a prompt, well-organized response to the reader's request. The professional credibility you gain with a letter of this type rests primarily upon your ability to state a problem, and provide a report of a solution, in logical form.

Letter Which Directs Action: Design

November 30, 1990

Hill, Hood, and Johnston
1200 Union Street
Atlanta, Georgia 30303

RE: Electrical Specifications
    Hidden Hills Apartments, Phase II

Attention: Ms. Jane Pendleton

Gentlemen:

For the referenced project, please issue a change order on the following items:

1. Re. Specifications, Section 12208, Paragraph 2.10-A, 9: Add Ill. Electric, I.T.E., or Pacific Power as equal manufacturers.

2. Re. Drawings and Specifications: Delete all references to padmounted transformers and secondary service.

We estimate that this change order should result in a $12,000 credit. We will appreciate your taking care of these changes. If you have any questions, please let me know.

Sincerely,

STUART AND LEE

J. B. Stuart

G. LETTER WHICH DIRECTS ACTION

Construction

P.1.　　Make this first sentence an automatic formula: "For the referenced project, please issue a change order on the following items:" so that you do not use unnecessary time in beginning the letter.

P.2.　　Include complete and accurate information. List items for ease of reading; this is a working document. Remember that the reader is likely to use the information *as a list* which is likely to be distributed to other readers. Complete and accurate information will ensure against delay or inaccuracy in making the changes ordered.

P.3.　　Both for the reader's information and for your records, include a statement of the consequences of the change ordered. State factors such as cost, time, and personnel which this change order will affect.

LAST P.　　Express your appreciation and your willingness to give further help or information. This is also an appropriate place to use *so that* motivation. If having these changes made in a certain way by a certain date is important *so that* another part of the job will be facilitated, motivate the reader at this point.

Convention

Ms. Pendleton is acting legally as the agent of a corporation; legal convention overrides social usage here, although the trend for some companies points toward "Ladies and Gentlemen."

Reference Line

Note the record-keeping and filing function of this line.

Situation

A narrowly and precisely defined understanding of the relationship between writer and letter is required here. The weight of this letter rests, not in the communication between writer and reader, but in the writer's communication of a directive for actions which have legal or contractual significance. The recipient does as the letter directs to fulfill a technical requirement of the referenced project. Although such a letter can and should motivate the reader to act, the reader is bound by the contractual nature of the letter to act. Remember that even so routine a letter as this can have sales/public relations value. Giving clear and accurate information makes you easier to work with. Explaining why or to what purpose makes your letter thorough. Adding a courteous word, however routine, makes you more pleasant to work with.

Letter Which Sells: Design

April 26, 1990

Mr. Leon K. Polk
1864 Bull Street
Savannah, Georgia 31863

Dear Mr. Polk:

It was a pleasure discussing our firm with you last Wednesday and learning of Polk and Company's plans and interests for the future.

As I mentioned to you, we have recently designed a project for Crater Company of Petersburg, Virginia, which is very similar to the one we discussed with you. It features a 42,000 square foot controlled-temperature facility, which contains 36,000 square feet of beverage warehouse, a 1,200 square foot keg cooler, and 6,000 square feet of office space. This is an example of the kind of design we can produce to meet your specific needs.

The enclosed brochure further illustrates Stuart and Lee's range of capabilities in design. The material on HVAC design should be of particular interest to you. I would also like to point out the training and experience of our personnel in this field.

I would appreciate your giving John McPherson a copy of the brochure. At any time convenient to the two of you I would appreciate the opportunity to discuss your firm's needs in greater detail. I will look forward to hearing from you.

Sincerely,

STUART AND LEE

Edward R. Lee

H. LETTER WHICH SELLS

Construction

P. 1.   One sentence will do to express your pleasure in meeting the prospect and to cite the time and the subject of the meeting. The sentence should recall your meeting to the prospect's mind in an easy and gracious way—like a smile and a handshake of greeting. It is direct, simple, and pleasant.

P. 2.   Reinforce what you said at the meeting by giving the concrete details of a design you cited. Recall to the reader and reinforce in his mind what you talked about to place on the record your major selling point: a design very similar to the facility he is considering. Relate your previous experience to the prospect's stated plans or needs. Point out, however, that the example is only one sample of your work, and that, in any case, your design for him will be specifically directed toward meeting his individual needs.

P. 3.   Enclose a brochure *and* direct his attention to pertinent parts of it. It is all too easy for a reader simply to lay aside an enclosure, so for that reason it is important to let him know that the information in it is useful to him. By its nature, a brochure is general; specify which part of the brochure is of particular interest to the reader.

Last P.   The last paragraph must strike a delicate balance between begging for the job and assuming that you'll get it. It is appropriate to do two things:

1. *Be receptive:* State your willingness to give further information or to spend time discussing this in more detail.

2. *Be available:* State your willingness to meet again. Because your range of action here is somewhat limited, your statement has to be carefully calculated. Your task, then, is to make it easy for the prospect to act upon your proposal.

Situation

The selling of technical services or products is far more restrained in tone than other selling techniques so familiar to our society. As stressed in the chapter on proposals, a presentation of facts and clear explanations carry the major persuasive force of sales.

Every letter is a sales letter in that the quality of the communication and the accuracy of content, in themselves, carry out an effective sales strategy.

CONCLUSION

The ability to translate your knowledge and opinions into words on a page is but one of several skills required in letter writing. Designing a letter is closely akin to the visual and mechanical aptitudes E/Ts exercise in their technological and scientific curricula. Conventions and techniques are verbal in their final form, but these factors in a letter require learning skills and good work habits more than they require an aptitude for writing. In these parts of the letter-writing process, the willingness to learn and the ability to work can compensate for the absence of high verbal aptitudes and well-developed writing skills. The situation which demands a letter cannot be subjected to exact definition, but a letter writer can develop management, decision-making, and planning skills. The skill of thinking clearly can carry an untalented writer a long way toward an effective letter. Above all, any written communication must meet the specifications of basic writing skills: words, sentences, paragraphs.

To learn to write good letters: 1) analyze the task and divide it into its component parts as we have done in this introduction; 2) master each task as a separate component and in relationship to the other components; 3) learn the relative difficulty for performing each task; 4) learn the relative importance of each component in a particular kind of letter for a specific situation. Remember that the ability to write letters does not rest solely with your ability to write. Letter writing is a series of interrelated skills.

Master the process. Manage it well. In a word, divide and conquer.

Exercise

The letter to give information is one of the most important in any technological career. To reinforce the discussion and examples on this type of letter, evaluate the student letters which follow. The assignment itself is the first letter in the exercise. It requests information and gives instructions on how to reply. Use the letter evaluation form to assess how well the student writers responded to the assignment letter. The evaluation checklist can be found in Appendix B.

Assignment Letter

January 10, 1990

Burdell and Associates
1863 State Street
Mechanicsville, PA 19109

Attention: Ms. Georgia P. Burdell

RE: Report due March 4.

Gentlemen:

In response to your offer to answer questions about the project underway, this letter lists our remaining unanswered questions.

We would very much appreciate your discussion of the following items:

1. Select a term from your report in progress and define it fully. Please demonstrate your ability not only to define, but to punctuate the words and phrases that characterize such writing.

2. Knowing that you are pressed for time on this project, we ask that you answer A or B below:

    A. We are in some doubt as to just what property of ours your report concerns. Describe the space or site.

    B. Select a part of your report and describe how the process works and why. This can be a procedure, a mechanical function, a chemical reaction, or any process involved in the report.

In your reply, please follow the form of this letter so that our records will appear in a uniform format. This letter should give you all the information you need to write a reply; if you do have questions, however, let us know.

Please let us hear from you by 11 a.m., February 14, 1990, so that we can continue with our work on this project.

Sincerely,

STUART AND LEE

Edward R. Lee

Conclusion

Student Letter

February 13, 1990

Stuart and Lee, Inc.
2001 Stuart Avenue
Leesburg, PA 19110

Attention: Mr. Edward R. Lee

Re: Request for information on project 80–101.

Gentlemen:

Thank you for your letter dated February 10, 1990, requesting the classification of certain terminology used in describing our project. I regret that Ms. Burdell is no longer working in our research division; however, as Project Engineer I will be happy to answer any questions you may have.

In your letter you asked for further explanation of the nature and function of the lift tube assembly.

1. The lift tube is the device in which the actual pumping of the water takes place. It consists of a long cylindrical tube with an inside diameter of 20 centimeters. The lift tube is positioned vertically extending from near the bottom of the water jacket to a height of 70 cm. above the water surface. At the submerged inlet of the lift tube are 2 vapor nozzles through which air or, in the case of our project, pentane vapor, is bubbled. The outlet of the lift tube terminates in an inverted, bell-shaped catch basin to which is attached the pipes that carry away the pumped hot water.

2. In operation, vapor bubbles escaping from the vapor nozzles rise directly into the inlet of the lift tube. These bubbles, which are about the same diameter as the inside diameter of the lift tube, are less dense than the surrounding water, and consequently exert an upward buoyant force on the water above. As the bubbles rise in the tube, the column of water in the tube above the bubble is forced upward. When the water column reaches the free end of the lift tube, the water spills over into the bell-shaped catch basin and in turn runs into the attached pipes. Meanwhile, the pentane vapor continues to rise into the condensor assembly above the lift tube. The rate of bubble formation is fixed at a low rate so that the period of time between the release of bubbles

Mr. Edward R. Lee
February 13, 1990
Page 2

from the vapor nozzles is sufficient to allow the lift tube to refill with water.

I hope that this has clarified the system's mode of operation; however, if you have further questions regarding this or any other phase of the project feel free to contact us. If you are in the Summerville area, drop by the lab and take a look at the working prototype.

Sincerely,

INTEGRATED ENERGY SYSTEMS

Alan C. Edgar

Student Letter

May 29, 1990

Stuart and Lee
1864 State Street
Mechanicsville, PA 19109

Attention: Edward R. Lee

Gentlemen:

RE: Response to your letter of February 14, 1990.

In response to your request for answers to your, as yet, unanswered questions, I am happy to provide you with all the desired information you desire.

The term timing charts refers to a chart which is set up to show the sequence of events that are to happen within the circuit. From this chart many things can be learned. We can determine what IC chips we will need in order to carry out the desired operation. It also tells us what will be the output of the individual chips as well as the output of the circuit, as a whole, at any given instant of time. By comparing a predicted timing chart with the actual timing chart of the circuit, we can debug it.

The sequencer which you have asked about is a 74161 synchronous clock. The operation of this clock is not unlike that of a simple signal generator. The 74161 generates 4 clock pulses which are labelled A, B, C and D. The pulses that are used are square waves.

The clock pulse A is the basic pulse off of which the others operate. The A pulse is used to trigger the B pulse. The C pulse is triggered by the B pulse and the D pulse is triggered by the C pulse. These pulses are negative edge triggered. When the A pulse goes from a high to a low state the B pulse is triggered "on." When A goes from high to low on the next wave the B pulse is triggered "off". This drop from a high state to a low state triggers C. This process continues and eventually all pulses are activated. The A, B, C and D pulses operate at 2 MHz, 1 MHz, .5 MHz and .25 MHz, respectively.

I hope these explanations answer the questions in the way you wanted. If any questions still remain please do not hesitate to get in touch with me.

Sincerely,

BURDELL AND ASSOCIATES

Wayne Jones

Student Letter

February 28, 1990

Stuart and Lee
Atlanta, GA 30332

Attention: Edward P. Lee

RE: Terms in Report on the Electronic Guitar Tuner

This letter answers questions 1 and 2B of your letter. These questions were: 1) define a term used in our report, and 2B) describe a process used in the report.

1) A notch filter is an electronic device that eliminates a certain frequency. For example, the notch filter described in our report will turn a light off when the exact frequency of the guitar string is reached. Notch filters can be tuned to eliminate any frequency, and this principle is used in the electronic guitar tuner by setting the notch frequency to that of a correctly tuned string.

Mr. Edward P. Lee
February 23, 1990
Page 2

2B) To continue in the vein established in number 1 above, the actual process of how the light turns on and off will be described.

The output of a tuned notch filter is fed to the gate of a trail, a form of electronic switch. Another electronic component, a light emitting diode (LED) is the device that is turned on and off by the trail.

When the guitar string to be tuned is struck, it sends a certain frequency to the input of the notch filter. If the string is not in tune, the notch filter will pass the frequency. This frequency will excite the gate of the trail which will turn the LED on.

If the string is in tune, the notch filter will not pass the frequency. Therefore the gate of the trail will not be excited and the LED will not turn on.

The terms above will also be described in the final report, where they will be elaborated on through the use of diagrams.

Thank you for your continued interest in this project, and feel free to contact us with any further questions that you may have.

Sincerely,

MSM ELECTRONICS

Mark Cunningham

Student Letter

May 29, 1990

Stuart and Lee
1980 Church Street
Atlanta, Georgia 30134

Attention: Mr. Edward R. Lee

Re: Electronic Scoreboard Design Project

Gentlemen:

We apologize for not answering your questions promptly, as your letter of February 14, 1990 requested. This letter should answer any remaining questions you have on the Electronic Scoreboard Design project.

The term microprocessor, as indicated in the proposal, refers to a small processor which correlates the input and output of a system. Perhaps we need to define not one term, but three: processor, input, and output.

The inputs of a system are the data or information the system needs to do its task. In this project, the inputs are the buttons on the scorekeeper's console. These buttons feed in what the scorekeeper wishes to take place, and act as an input to the microprocessor.

Similarly, the outputs of a system are the events that occur as a result of the inputs. In this project, the outputs of the total system are the lights on the scoreboard which tell the audience what the scorekeeper says the score, remaining time, inning or quarter number, or other information are for any given time. The outputs of the microprocessor are the signals to be fed into the amplifier for output to the scoreboard.

The term processor, however, is more complicated to define. Microprocessors are small computers that can accommodate only one set of data at any given instant of time. What is the difference between a microprocessor and a computer? The microprocessor is not as complicated or versatile as a computer, and is limited in the fact that it can only perform a specific group of operations that a large computer handles with ease. Both machines are processors, because they process information; but the microprocessor, a specialized computer, is a small processor relative to a computer.

Mr. Edward R. Lee
May 29, 1990
Page 2

Your letter also states that you would like to know how and why we use the Georgia Tech Cyber 74 computer (Cyber) to develop the microprocessor program.

Cyber, a large and very versatile machine, is equipped with a microprocessor simulation language developed by the Motorola Corporation called MPS. As in any computer language, the appropriate words in the appropriate places give the desired output. In this case, the proper program written in MPS simulates the actions of the microprocessor. As an example, the instruction ADD REG simulates the addition of the AC register with the variable REG and places the final value back into the AC (accumulator) register.

Why do we develop the program in Cyber instead of writing the program directly into the actual microprocessor? The answer, as proven in past experiences, is that Cyber is more efficient in debugging, or correcting mistakes in programs, than the microprocessor. To correct a mistake in a microprocessor, a new Read Only Memory (ROM) must be manually reloaded with the entire program, and the ROM must be resoldered to the microprocessor's circuit. This task is easier on Cyber. A correction can be made in a program in less than 30 seconds, and usually in no more than 2 minutes. Cyber also provides a method of following the program's process step-by-step, which also saves time, energy, and money.

We hope this letter clears up any questions you have. If not, feel free to call or write again.

Sincerely,

Eugene Frank

Student Letter

May 29, 1990

Stuart and Lee
2901 Hunter Street
Atlanta, Georgia 30334

Attention: Mr. Edward R. Lee

Re: Project for Stuart and Lee

Gentlemen:

Because of any confusion that you and your business associates may have about our project for your company, we will try to clear up any points that may seem ambiguous and confusing.

First of all, we would like to apologize for forcing you to refer back to us because of any problems caused solely by us. In order to clear up any confusion caused by us, we would like to go back and define one of the key terms that was often used in our report to you. We will also describe the site intended for our project.

1. The term <u>Frequency Band</u> is a key term in communications that describes what frequencies and audio or video signal can be produced. When using an amplifier to increase the power of the signal, the frequency band is important here because the signal may only be reproduced if it is operating between certain frequencies. If the frequency is too high or too low, the receiver will not pick up the signal. Therefore, no picture or sound is produced. The range of frequencies at which the signal is produced is the frequency band.

2. The site intended for this project is at your Atlanta division. We feel that our work should be done where you are so that you can make evaluations from what you have seen, instead of having to talk long distance and make further decisions on something that you cannot visualize. We have already sent 3 of our best project engineers to your location in Atlanta, and they would like to suggest that we install our communication system in the new wing of your complex and let you evaluate it yourself. Our crew from Rome would be able to start within the week of your decision of our suggestions.

I hope that this clears up any points not stated well in our report and helps you to speed up your decision on our continuing this project.

Sincerely,

BURDELL AND ASSOCIATES

Basil Jarvis

chapter 9

Informal Reports: Combining the Letter and Memorandum Forms

The memorandum and the letter have been described as very short documents generated both from large technical projects and from a whole range of other business and professional situations. Such situations can also require very long memos and very long letters, really too long to be effective memos or letters, and yet too short to warrant treatment as formal technical reports. A combination of the letter and the memo works to solve the problem of length and also to derive maximum sales and PR value from the communications situation.

LENGTH

Length is the first consideration about whether to write a memorandum report with a cover letter. If a letter or a memorandum runs 2½ to 3 or more pages, divide the forms into a memorandum report with a cover letter.

SITUATION

Situation is a further consideration in determining when to divide a long memo or letter into two forms. Following are three correspondence situations which are especially well suited to this method:

The Long Technical Letter

Most engineers write the following types of long technical letters. These are classified on the basis of the subject matter and the kind of thinking and planning each requires:

1. *Explanation of a Problem:* Rather than to spend time with a long letter or dense paragraphs, divide the thought process, use a form which reflects that process, and rely on memo format to demonstrate the explanation.
2. *Series of Actions:* Logical order is again the principle along with heavy reliance upon lists and other format devices.
3. *Construction Review:* Order lists of objects in some logical way, again using a clear page design to carry a large part of the burden of writing.

Remember that a letter is *to someone*; as such, it ought to contain sales and PR value. In contrast, a technical report is *about something*; as such, it ought to be factual in tone and content. Often the problem with a long technical letter is that it mixes the media of the letter and the report. Your task in such a situation is to divide and conquer.

The memorandum report of 2, 3, or more pages can be completely technical and objective in tone and content. It becomes a useful document that can be duplicated, distributed, and made a permanent part of the project record.

The cover letter compresses the more affective elements of the transaction into a paragraph or two on one page of letterhead. This is the means whereby you transmit and introduce (and also sell) the technical matter. A letter of transmittal allows you—as a report does not—the leeway to highlight a particular part of the report which especially concerns the reader. Such a letter can become a way of guiding and directing the reader through the report to assure that you get the kind of favorable reading you want.

The Technical Letter to a Personal Contact

Often a person's success in getting a job or solving a problem is enhanced by personal ties to someone in another firm. In such a case, one of the pleasures of business is to include a personal word at the beginning of a long letter or in the closing paragraph. To give an example of how that works, an engineer may address a lengthy technical letter to a former classmate who is now a government official, beginning, "Dear Sue, I so much enjoyed our recent lunch. It had been too long since we had a visit." Or one might end such a letter to a friend "Best regards to Marion. Charlotte and I hope to see you during the holidays."

Such an approach can be good business, for human beings write letters and they cannot always be all business all the time. When such a letter is copied, distributed, and filed for its entirely technical content, however, the personal and situational touches are not only quickly outdated, but the contrast in tone between personal and technical is inappropriate.

In this situation then, it is especially effective to confine the personal elements to the cover letter and the technical to the memorandum.

The Technical Problem or Business Disagreement

There are instances in which disagreements or misunderstandings must be settled with letters. Such a case may call for a straightforward, no-nonsense technical analysis or explanation; yet the conventions of a letter may prevent the writer from being so objective. The communications problem in such a situation arises from trying to perform two conflicting tasks in one document: 1) to be sufficiently courteous so as not to offend the reader and thereby lose the client and 2) to take a direct, entirely objective approach in establishing your own case technically. Here the issue is literally to divide and conquer by combining the hard-hitting memo report and the more diplomatic cover letter.

To consider a general example of how letter and memorandum work together, read the following letter from a designer to a client. Note the mixture of business and personal tone and how that approach obscures the facts to be established, the pressures of time, and the need for the reader's decision. Then pay particular attention to how the revision divided the letter into a memorandum report and cover letter.

Original Letter

February 9, 1990

Mr. Harvey B. Hill
61 Bull Street S.E.
Savannah, Georgia 31862

Re: Equipment Specifications

Dear Harvey:

We are hard at work on the construction drawings for your new offices. There are two areas that I will need your help on if they are to be included in the pricing package.

First is the library which has been forwarded to Jaque for her input. The importance in layout is due to structural requirements which I need to verify as soon as possible. I told Jaque to call me if I could be of any help to her and I will contact her again as soon as she returns to get her information.

The second area to be specified is the word processing equipment. I cannot accurately specify electrical or mechanical requirements until the equipment is chosen. If this decision has not been made by the end of February I can project some requirements that will allow the contractors to price an allowance as an option. The other option is to wait until your equipment decision is reached. Do you have a preference?

I also need your input on the telephone system. Ordinarily this would not affect the construction prices; however, should the system that you decide on require work performed by the general contractor we would not have it included in bid documents. This is not a very critical item but I want to bring it to your attention to insure that you are kept informed.

If I can clarify any of the above or aid in arriving at any of the decisions, please do not hesitate to call me.

Sincerely,

Edward R. Lee, A.S.I.D.

Revision: Cover Letter

February 9, 1990

Mr. Harvey B. Hill
61 Bull Street, S.E.
Savannah, Georgia 31862

Dear Harvey:

We are hard at work on the construction drawings for your new offices. The status of our work and our needs for your input are expressed in the enclosed memorandum report.

In the report I have tried to delineate how your time table for decisions affects our work schedule. We are better able to serve your needs—and to cut costs—if we achieve close coordination of our efforts.

Please call if I can help you with the decisions called for in the report. Meanwhile, I will keep in touch with Jaque Bureaugard.

Sincerely,

Edward R. Lee, A.S.I.D.

ERL/bb

cc: Mrs. Jaque Bureaugard

Enclosure: Memorandum Report: Equipment Specifications

Memorandum

February 9, 1990

MEMORANDUM REPORT

TO: Mr. Harvey B. Hill

FM: Edward R. Lee

RE: Equipment Specifications, Headquarters Office, Hill, Hood and Johnston

Harvey B. Hill
February 9, 1990
Page 2

Construction drawings are in progress for work on your new offices. However, continuing on schedule requires input from Hill, Hood, and Johnston. These decisions are especially needed if they are to be included in the pricing package.

Library: Structural requirements which we must include in the project depend upon your decision about placement of shelving. If you plan to use floor-to-ceiling shelves as divisions within the Library, this will necessitate supports on the floor below. If you plan to limit shelving to the walls, this will change the specifications both for lighting design and for insulation.

A decision from you on your needs for shelving will make possible our continued work on the drawings and specifications and a final, more complete estimation of costs.

Word Processing Center: Your final decision on the equipment to be installed determines, in turn, the electrical or mechanical requirements we include in our overall design. The options are to project what those needs will be or to wait until your equipment decision is made. We would prefer to have your decision, but we will make the projection if you prefer that course of action.

Telephone System: Your input on the telephone system would be helpful also. Ordinarily this would not affect the construction prices; however, should the system you decide on require work performed by the general contractor, we would not have it included in the bid documents.

Requested Actions: While allowing for the time you need to make the decisions discussed here, I am adding this overview of how these decisions relate to your scheduled priorities:

1. Library: Since this design includes structure, it involves the work of the architects and the contractors. Work cannot progress until we have a decision from Hill, Hood, and Johnston. We need a March 1 decision.

2. Word Processing: While less critical, being able to design with final accuracy is important here to avoid the possibility of later delays and additional costs. Let us know your preference by February 15.

3. Telephone System: As a part of the total design, a decision on this system will indirectly influence the design project and directly influence the allocation of contractor time and the details of the bidding process.

SUMMARY EXAMPLES

Following are Summary Examples of three types of long technical letters, each of which would be written as a memorandum report with a cover letter. As you study each example, remember in particular that the page design represents not so much a writing form as a thought pattern. Most letters which deal with explanations follow these forms and the process of logical thinking also follows these patterns.

Your major task in writing such a short report is to make the visible form of the report clearly demonstrate your thought process and the process you want the reader's understanding to follow.

An important distinction to make between writing, as we usually conceive of it, and engineering is this: writing is a logical but rather abstract process; engineering often deals in mathematical symbols which demonstrate the working out of a problem. A well-designed page of technical writing can also demonstrate the thinking out of a problem. Language is still the symbol-system you use, but you are not writing in a conventional sense; you are demonstrating engineering in the generally more accessible medium of words.

Cover Letter for Memorandum Report: Summary Example 9.0

Month Day, Year

Name of Addressee
Street
City, State Zip Code

RE: Reference Line

Dear Name:

Paragraph 1: Be positive (Avoid "Enclosed here," "per your request," and other tackies.)

First part of Sentence 1:
   "To answer your question about.../ To provide a further explanation about...." (—use this positive approach)

Second part of Sentence 1:
   "The enclosed memorandum report discusses the question about...... provides a record of ....."

Paragraph 2: In this paragraph direct the reader's attention to any one of these points:

1. an explanation of your effort and intention in writing this short report.
2. a particular feature that you think is strong and that you would like to point out. "You may be particularly interested to note that in our investigation we ....."

Last Paragraph: Again, be positive.

"Our report should provide you with the information that you need at this stage of the project; if we can help you further, however, please let us know."

Complimentary Close,

Signature

Name (Title or P.E. if appropriate)
Enclosure: Memorandum Report

Memorandum Report: Summary Example 9.1

Month Day, Year

MEMORANDUM REPORT: Explanation of a problem

TO: Name (and title if applicable)

FM: Name (and title if applicable)

RE: ("RE:" line functions as a title)

———————

Statement of your Problem: The assumption here is that somebody has asked you to explain something about a job. In this paragraph describe your understanding of the problem as the caller or writer has expressed it.. This is a communications base from which the rest of the report develops.

Summary: For a report of several pages, it is appropriate to include your concluding paragraph here. Write it last, then cut and paste it in this position.

Summary Examples 119

Method of Investigation: If you had to make an on-site investigation, describe here what you did. If you had to write, call, or collect data, describe what you did. (You can simply head this "Action:")

Solution: This details how your action solves the problem.

Explanation: Use this subhead instead of "solution" if you are merely giving a reason and not reporting a corrective action.

Memorandum Report:
Summary Example 9.2

Month Day, Year

MEMORANDUM REPORT: Construction Review

TO: Name (and title if applicable)

FM: Name (and title if applicable)

RE: ("RE:" line functions as a title)

Description of the Site: Describe the project under review, remembering that a given organization may have many sites. Also keep in mind the record function of a document of this type.

Authorization: What were you asked to do in making this review? Scope might, in fact, be a better subheading here, for that sets up the parameters: wiring, plumbing, what specific areas of construction?

Construction Review: Avoid long random lists: instead, set up some principle of organization as you make your notes:

    1. room by room

    2. system by system

    3. sequence of serial numbers

Write and organize this lengthy part of the report and then transfer that to the typed report.

Memorandum Report:
Summary Example 9.3

Month Day, Year

MEMORANDUM REPORT: Report of a Series of Actions

TO: Name (and title if applicable)

FM: Name (and title if applicable)

RE: ("RE:" line functions as a title)

_____

Statement of the Problem: Establish as a reference point your understanding of the question asked, the request made, the actions described by the writer or caller. To do so in this paragraph serves these purposes:

1. It clarifies your thinking as you begin to write.

2. It expresses your understanding of the other person's directions, and so equally distributes the burden of good communication between you and the reader.

Action: A narrative of a series of actions is easier to write if it is broken into steps—1, 2, 3, ...—rather than cast in the form of a paragraph,, with all the requisite logical transitions.

Avoid a total mind dump, however, when you are asked to report a series of actions. If step 1 is very important in the series, explain it with a sentence or two. If step 3 is a pivotal step, say so. Record the steps, but also interpret them, remembering that selection and emphasis is an important part of any art—and writing good technical reports is an art as well as a craft.

chapter 10

Correspondence Systems

Neither documents nor situations exist in isolation in a technical or business project. The memorandum which directs an entry-level engineer to make an on-site investigation of a problem originates with a telephone conversation between the engineering project manager and an architect; the investigation, in turn, leads back to the architect with a letter; the architect's change order letter to a contractor leads to changes in the project reported to the financial managers as budget changes. Each of these written documents is duplicated and distributed to others working on the same project, each one is filed, and all go to make up an entire correspondence system which is indispensable to the work's progress. The multitude of documents, and the complexity of the interactions which they represent, forms a vast communications system within a large project.

The next two sections serve to introduce the student writer to the many situations which a correspondence system handles. The first section deals with the "employment project": a small-scale sales project which a student plans and manages independently. As an introduction to a communications situation with important consequences for you, the employment project simulates on-the-job schedule factors. Requesting a recommendation on schedule simulates managing others upon whom your own work depends.. The careful design of letters of application illustrates the crucial necessity of giving even the most technical letter a sales appeal through good design, effective writing, and the accuracy and thoroughness of the data presented.

The second section is a narrative discussion of how correspondence works within a large technical project. Rather than reduce the situations to sample letters, this section by-passes design and construction notes to highlight major variables within letters in a comprehensive correspondence system.

THE EMPLOYMENT PROJECT

In the same way that we studied conventions and techniques for letters and memos, we can study in broad outline what techniques and conventions can be best applied to specific situations within a correspondence system.

Letter-writing is seldom a matter of one letter sent and one letter received in reply. Instead, a series of letters is much more likely to make up a system. An immediate example of how letters work in a system is the writing you do to apply for a job. You may have just begun to learn letter-writing; however, when you must produce and manage the flow of letters for a job application, you are suddenly a project manager and sales manager. Your project is to find a job; you are selling yourself. How well you manage this communications system will have a direct influence upon your success. That principle always holds true in letter-writing.

As with other writing tasks, your success with an employment project does not rest entirely upon your ability to write. Other important components are the ability to plan, design, and manage the employment project. Following is a list of your special concerns.

Components of Employment Correspondence

Audience

Knowing the audience of each letter will determine the tone of your letters, the details you include about yourself, and the amount of data you give.

Purpose

Your purpose in writing to someone for a recommendation will be different from writing to ask for a job application. You will be dealing with the same subject, yourself, but how you treat that subject in a letter will vary according to audience and purpose.

Design

How you design a letter for a page and how you place facts about yourself in a resume will have an influence upon an audience. You do not want a design flaw to cause a busy reader to pick up your letter, glance at it, and then lay it aside because it will be too expensive to extract the relevant information.

Convention

The job application and interview process can take on a somewhat social dimension. Therefore, in this situation as much as any other in business, you want to be aware of conventions and how to observe them in letter writing. This is no time to begin a letter "Per our recent conversation, please find enclosed..." Such cliches never make you appear more businesslike.

Timing

Writing a group of letters to different people makes timing important in the employment project. Make a schedule and keep it; do not request a recommendation from someone two days before it is due. Also make any application well within the set time limit. When you meet deadlines at the very last minute in the employment project, you do not create a good impression of yourself.

Filing

Keep careful records of what you have written, to whom, and when. A loose-leaf notebook can be an efficent project file for you. Divide it into sections where you file copies of the various types of letters you write. Keep a log so that you will know that you mailed your requests for recommendations on November 1 and your applications letter to six firms on October 28. Record the dates when the answers return. In this *systematic* way, you are aware at all times of the status of your correspondence project and how it operates to help you get a job.

Examples

In the following pages are examples of the types of letters you will want to write in an employment project and instructions for how to write them. Notice how each letter follows a general form, but notice in particular the variations in style, tone, and convention. Your eye and ear for such slight variations will help you gain flexibility and effectiveness within the limitations of the letter form.

Letter of Application: Example 10.A.1

May 6, 1990

Mr. John Green
3 Park Avenue
New York, New York 10016

Dear Mr. Green:

In response to your advertisement in the Cleveland Plain Dealer, I am sending you my resume for the position of Hospital Management Engineer at Highland Hospital in New York.

    My education includes study at Benedicta College and Western Institute of Technology. I received a two-year degree in Pre-Law at Benedicta; however, the education I received from Western-Tech has best prepared me to be a Hospital Management Engineer. I have taken the following courses which helped me obtain my degree in Health Systems Engineering: Hospital Management I, II, III, and IV; Data Processing; Operational Research; Health Planning; and Statistics. A Senior Externship has given me experience as a Hospital Management Engineer at Mercy Hospital.

My previous work experience was with Medved Paint Company, where I was a paint filler and equipment cleaner. I have been working for Wells National Company for four years. As a television installer for them, I work in hospitals installing televisions and renting them to patients. While working in these hospitals, I have become aware of the problems which exist. These two jobs have provided me with the funds to pay for my education and the experience of how industry and hospitals operate.

I would like to speak with you or one of your representatives regarding the opportunities at Highland Hospital. If you need futher information, I can be reached at the following phone number: 409-971-4113.

Sincerely,

Kenneth C. Clark
2267 Dayton Circle
Cincinnati, Ohio 40163

Letter of Application: Example 10.A.2

May 7, 1990

Mr. John Doe
Supervisor, College Recruiting
Major Industries, Inc.
2666 West Their Avenue
Their Town, USA 10000

Dear Mr. Doe:

May I ask you to consider my qualifications for a Sales and Marketing position with Major Industries in Atlanta? I have read many company brochures and publications about your new marketing program and feel that I can help contribute to its success.

By June 8, 1990, I will have completed requirements for a degree in Industrial Management, with elective concentration in marketing. As the enclosed resume shows, I have had four years of retail sales experience. I hope you will also note the positions of leadership in campus and community organizations which have helped me develop managerial skills useful in marketing.

Should my qualifications interest you, I would appreciate a meeting to discuss them further with you.

Sincerely,

George P. Burdell
Auburn University
Box 35767
Auburn, Alabama 36830
205-894-6743

Letter of Application: Example 10.A.3

May 5, 1990

Mr. Edward R. Lee
Stuart and Lee Engineering
2749 Darby Drive
Portland, Oregon 90067

Dear Mr. Lee:

In response to our recent phone conversation about summer employment at Lee Engineering, I have enclosed a copy of my resume as you requested.

The flier you had posted at school listed a number of prospective work areas that relate directly to my employment interests and technical background. Of particular interest are the areas dealing with stream analysis, groundwater hydrology, and hazardous-waste management. I completed a course in stream analysis and another in solid-waste management last quarter. Currently, I am taking groundwater hydrology and am working part time as an assistant in groundwater modeling studies.

The greater part of my work experience is not related to the kind of work I am interested in now, but it provided valuable exposure to the working world and matured me as an individual capable of working with others. I feel that this background, combined with the technical preparation that I have, would allow me to become quickly involved in the work your firm will be doing this summer.

I look forward to hearing from you; I may be reached at 892-2566 or at the address below.

Sincerely,

Vaughn S. Davison
251 10th Street, N.W. T6-148
Portland, Oregon 90348

A. LETTER OF APPLICATION

Construction

P.1. As with all letters, the first paragraph should state the subject, in this case the job you are applying for.

- If you respond to an ad begin, "In response to your advertisement for a _____ , ..."
- If you drop a name, "John Jones of your _____ Department has told me that there may be a job in the _____ Department at _____ ..." (Always use the company name rather than "your firm" or "your company"; as with advertisements, using the name itself is effective.)
- If a general inquiry, begin, "I am writing to ask whether you are accepting (applications for/have any vacancies for . . . .)"

P.2. The body paragraphs should discuss your education and experience, and how they will be helpful to you on the job.

P.3. Discuss first whichever is strongest, in your case now probably your education.

- If you include a resume, use these paragraphs to highlight points in the resume, to lead the reader into the resume, and to relate the resume specifically to the job. "The enclosed resume shows that I have had six courses in management related to the job as you describe it." "My resume gives a detailed list of my work experience. In particular, I have found working with _____ helpful; this should provide some background for the job at _____."
- If you do not include a resume or a completed application form, briefly summarize your work experience in a paragraph and your education in a paragraph. Do not discount *any* work experience. At this stage, no employer expects you to have been a company vice president somewhere. To have held a job, any job, shows that you can hold a job.
- You may also want to add a short paragraph about extracurricular activities, hobbies, and the like.
- While the letter should not be too long (1 page with a resume; without a resume, not more than 1¾ pages with plenty of room on page 2 for an uncrowded signature) it is at times helpful to note any special interest you have in a company. If it is located in a specific area and you have an interest in the area, say so. If it has many branches, express your willingness to relocate. Companies like loyal employees who are happy where they are. Don't overdo this, of course.

Last P. Letting it be known that you would like an interview but not asking outright for one requires tact and finesse. "I hope that this gives you a general idea of my education and qualifications for the job. If you would like more information, please let me know." Let the reader know how to get in touch with you: "Until December 10, I will be at Georgia Tech where my phone number is XXX-XXX-XXXX and my Post Office Box number is XXXXX; afterwards, I will be at home in Macon for the holidays (1048 Georgia Avenue, 31904)."

Resume: Summary Example

Name: full name, first name first; consider using ALL CAPS (skip 3 or 4 lines to set off your name at the top of the page)

Address: your permanent address or a mailing address where you can always be contacted

Phone: Be sure to include the area code in () and a - after the first three digits: (404) 894-2500

(In this first part of the resume, show who you are and how to get in touch with you. Reserve personal data for later in the resume.)

Education: (on a job application, place this after *Employment*)

Name of School   Location   Years Attended   Diploma/Degree

(Begin with the most recent and work backward chronologically. Be particularly specific about the name, city, and state of the school; Central High is insufficient information.)

(Provided you can keep your resume no longer than a page, a brief paragraph might be effective to describe your major academic concentration, your grade point average, or some special feature of the school's curriculum. Select details which will strengthen your application for a specific job or entry to a specific college.)

Employment: (for a school application, retain this order of presentation)

Name of Employer   Location   Dates of Employment (most recent first)

(After each job, you might include a brief job description: your duties, the title and name of the person to whom you reported and who supervised your work. Again, select details which strengthen your application for a specific job.)

Personal Data:

Date of Birth: (instead of age so that the resume will not be so quickly outdated)

Letter of Application

> Height: Weight: Health: (Write "excellent" unless you have a disability classification. If so, make clear that you are competent to perform the duties of the job.)
>
> *Hobbies and Special Interests:* (Select details here to reflect well upon your ability to perform well in the setting to which you apply)
>
> *Extracurricular Activities:* (For a school application these should appear after *Education*; for a job they should appear after *Employment* and *Education.*
>
> *Honors and Awards:*
>
> (DESIGN the page for ease of reading; SELECT DETAILS specifically related to the purpose of the resume; if at all possible, send an original of a resume on good quality paper, not a photocopy.)

Letter Requesting a Recommendation: Example 10.B.1

May 4, 1990

Mr. Max Powell
Southern Timber Company
Folkston, Georgia 30112

Dear Mr. Powell:

I am currently seeking a summer job with Eastern Electric in Savannah, Georgia. I feel that your recommendation would help me very much in my efforts to land this job.

When I worked for Southern Timber of Folkston during the summer of 1989, you were very kind to offer me a job for this summer when I left for school last September. However, as my graduate date draws near, I realize that I need work experience in the electrical engineering field.

Although my job with Southern Timber did not involve any electrical work, I am certain that your recommendation concerning my performance and work habits will help to establish my reliability.

I would appreciate your assistance, and I wish Southern Timber continued success.

Sincerely,

Charles H. Patterson

Home Address:

406 Joseph Avenue
Bainbridge, Georgia 31745

Letter Requesting a Recommendation: Example 10.B.2

May 7, 1990

Professor Mark Jones
Management Department
Tennessee Tech
Gatlin, Tennessee 40332

Dear Professor Jones:

As you will recall, I was in your Management 3023 class in the Spring of 1989. Now I am being considered for a Sales and Marketing position with IBM. Before my application file is complete, a letter of recommendation must be sent to them and I ask that you recommend me.

Some of the qualities that they are interested in are leadership ability, aptitude, attitude, and the ability to work with others. The intensive work we did in the team projects for your class should give you a basis for commenting on my work. In a summer job with Lancour I have already applied the market survey techniques we studied.

I ask that you send the recommendation by June 2, 1990, so that IBM can make a final decision on my application. I have enclosed a stamped, addressed envelope for you.

Sincerely,

Andrew A. Michael

Tennessee Tech Box 35757
Gatlin, Tennessee 30332
604-894-4623

B. LETTER REQUESTING A RECOMMENDATION

Construction

P.1. State the subject of your request in the first paragraph. "I am writing to ask that you recommend me for a job as a _____ with _____." "I would appreciate your writing a recommendation for me to _____."
- Second sentence: Mention the enclosed form to be filled out, or say that those to whom you are applying would like a letter.

P.2. Jog the memory of the teacher or employer. "You will recall that I was a student of yours in _____ during _____ quarter of _____." How much you say depends on how well you know the person and how long it has been since you were associated. This sentence can also lead into the emphasis you want in the recommendations.

A note received this week from one of my students reads, "Please note that these people are interested in those qualities that make for a good scientist or engineer; communication must certainly be one area." Such guidance to the writer is appropriate and most helpful, for you want a good balance from those who recommend you.

P.3. Give details about when it is due and enclose a stamped, addressed envelope. This is called "dated answer" and "easy answer" in letter writing. In the last sentence, express your appreciation; *not* "Thanking you in advance" but something more like "Your writing this will be a great help to me, and I really appreciate it."

Writing recommendations is part of any teacher's or employer's job. While most are happy to do that, be thoughtful of those who help you. Especially, give ample time for their efforts; procrastinate with your own time, if you will, but do not pressure someone of whom you are asking a favor.

It is especially gracious to write a note later to tell the writer the results of the application. *Always* thank people who help you; it is not only good manners but good business.

Thanks for a Plant Trip: Example 10.C.1

May 13, 1990

Ms. Sonya Norman
Personnel Director
West Point Pepperell, Inc.
Lindale Plant
Lindale, Tennessee 40147

Dear Ms. Norman:

This letter is to thank you and your personnel staff for the plant trip and luncheon on Thursday, May 12.

The tour was well planned and very informative, giving me valuable insight as to what actually takes place in a yarn manufacturing plant.

Sarah Spence of your chemical engineering staff was especially helpful in explaining your color matching process.

Again, thank you for the opportunity to visit and learn more about the Lindale plant.

Sincerely,

Lance Hilton

Individual Situation (notes):

Thank-You Letter: Example 10.C.2

May 5, 1990

Dr. Charles A. Branch
School of Civil Engineering
Pacific Northwest Tech
Portland, Oregon 90332

Dear Dr. Branch:

I am proud to have been chosen by the School of Civil Engineering to receive a graduate assistantship in Fluid Mechanics. The letter you wrote was instrumental in persuading the review board to accept my application and offer the award; for that help I am grateful.

The assistantship will allow me to continue my studies in Environmental Engineering and reach the professional goals I have set for myself.

Thank you again for your time and encouragement.

Sincerely,

Vaughn S. Davison
251 10th St., N.W. 16-148
Portland, Oregon 90318

Individual Situation (notes):

Thank-You Letter: Example 10.C.3

May 7, 1990

Mr. John Doe
Supervisor, College Recruiting
Major Industries, Inc.
2666 W. Their Ave.
Their Town, USA 30000

Dear Mr. Doe:

I would like to thank you for giving me the opportunity to visit your offices and plants in New York.

I have enjoyed talking with you and learning more about Major Industries.

I look forward to hearing from you in the near future.

Sincerely,

George P. Burdell
Ga. Tech Box 35747
Atlanta, Georgia 30332
404-984-2743

C. THANK-YOU LETTER

Construction

This should be a relatively brief letter. *DO* be specific.

If it is for a plant trip, cite things you found of special interest and remember and cite names of people you enjoyed talking with. *DON'T* be too effusive in your thanks; after all, however nice they were, it is a part of their job to try to interest prospective employees.

If it is for a letter of recommendation, mention to whom it was written, what position it recommended you for, and whether you got the job.

Letter Accepting a Job: Example 10.D

May 1, 1990

Mr. Edward R. Lee
Carter Advisor Services, Inc.
1200 Road, N.E.
Olney, Maryland 00123

Dear Mr. Lee:

I am pleased to accept your offer of a position as planning consultant at Carter Advisory Services, Inc. in Olney at $19,500.

I look forward to receiving further information about the company, as well as any necessary forms that I need to complete. I will be available to begin working June 23, 1990.

Again, thank you for your offer. I look forward to beginning my career at Carter Advisory Services, Inc.

Sincerely,

Aurora Leigh

D. LETTER ACCEPTING A JOB

Construction

This letter should be brief and courteous. Express your pleasure at receiving the offer. Also recognize that your relationship with the employer has changed. You are no longer the prospect being wooed; the advantage has shifted to the individual or company for whom you are going to work.

Establish your image as a good worker in two ways: 1) mention the paperwork involved in becoming employed (and complete and return it on time), 2) do not ask for special favors about when you will report to work or any such exceptions to company practices. Company personnel forms are not homework that you may or may not hand in on time, and the boss is not a teacher who may respond to a sad story about how you want two more days at the shore. You may negotiate a job on your terms, but you accept it on the employer's terms and act accordingly—beginning with your letter of acceptance.

Letter Refusing a Job: Exercises 10.1 and 10.2

June 1, 1990

Mr. Lou Sandor
Personnel Director
Archer Textiles
P. O. Box 6947
Edgefield, SC 40569

Dear Mr. Sandor:

Thank you for your job offer as quality control technician for the Clemson plant of Archer Textiles.

I enjoyed learning more about your growing company through the plant trip to Clemson and the literature you sent to me on the other facilities in North Carolina and Georgia. However I have decided that career goals are best served by continuing my education at the graduate school of North Carolina State University.

I will continue to watch with interest the growth of Archer Textiles and will look forward to talking with you again.

Sincerely,

Lanie Hilton

May 1, 1990

Mr. Edward R. Lee
Advisory Services, Inc.
1200 Carter Road, N.E.
Olney, MD 00123

Dear Mr. Lee:

Thank you for your offer of a position as planning consultant at Advisory Services, Inc. in Olney. I enjoyed meeting with you and discussing my future with your company.

However, because my wife and I have been offered jobs in Boston, Massachusetts, I am unable to accept.

Again, thank you for your offer. I regret being unable to accept, but am sure you will find my reason understandable.

Sincerely,

Aubrey Leigh

E. REFUSING A JOB

Construction

P.1. Thank the company for the offer. "However, I have decided to accept a job with _____."

P.2. Business writers are somewhat cagey at times about explaining. Give a good reason, but you do not have to give *every* reason. In a business letter, it is appropriate often times simply to assure the reader that his offer has been given careful and thorough consideration.

P.3. End with a brief courteous close. "Again, thank you for the opportunity to visit your plant and to discuss working with you."

In refusing a job, don't be so overconfident that you burn your bridges. You may one day need a job with this company you've refused. One young friend of mine who left a job to marry told her boss, "I'd dry up and blow away if I stayed in this little town." Three years later with divorce in hand she'd have jumped at the chance to be there.

Exercise: Employment Letters

Use the Letter Checklist in Appendix B to evaluate the employment letters which follow.

Letter of Application: Exercise 10.3

May 6, 1990

Mr. James Vincent
Vice President, Personnel
Maritime Systems, Inc.
Forfold, Virginia 18603

Dear Mr. Vincent,

With my college background, an insistence on perfection, and my experience as a leader I believe I can successfully and effectively fulfill the responsibilities of a design engineer for Maritime Systems, Inc.

I have prepared myself for this type of opportunity by specializing in vehicular technology. This course of study included 8 courses in the design and operation of mobile communication systems and related equipment. I maintained a consistant overall GPA of 3.0 at Western Tech while participating in several extra-curricular activities.

My participation in intercollegiate football while attending a junior college in Indiana has instilled in me the desire to do my very best and to settle for nothing less than perfection. This can only be achieved through hard work: hard work that I am willing to perform in the service of your company.

I can work well with people as well as take charge when the situation requires such a response. Making decisions under pressure is not new to me. While in junior college, I was the defensive signal caller and was elected team co-captain both years I was there. During the same period of time I was the junior varsity and assistant varsity wrestling coach at Lafayette High School.

My days in athletics are over, but the lessons I've learned will carry over into the business world.

The enclosed data sheet and inquiries to the references I've listed will probably give you all the information you need, but I will be glad to furnish any additional information you feel is necessary.

Mr. James Vincent
May 6, 1990
Page 2

I would appreciate an opportunity to meet with you and discuss a possible future I might have with your company.

Yours very truly,

Edward Smith

Home Address:

1008 Rover Street
Lafayette, Ind.

Letter of Application: Exercise 10.4

May 2, 1990

Mr. John Smith
Director of Personnel
Smith & Company
2802 Buffalo Road
New York, New York 54321

Dear Mr. Smith:

I am writing Smith & Company on the recommendation of one of my professors, Dr. Joseph Brown. You might recall that Dr. Brown was a consulting engineer with you during the summer of 1986. The purpose of this letter is to apply for a position as a junior engineer in research and development with Smith & Company.

I am currently a last-quarter senior in the School of Electrical Engineering at Eastern Institute of Technology. My course of study has oriented me toward the communications field. I have concentrated primarily on the design and functional aspects of systems applicable to signal processing. Prior to enrollment at Eastern Tech I was enlisted in the United States Army where I received training as a communication technician, training which offered me exposure to several state-of-the-art communications systems.

As is stated in the enclosed resume, in addition to the work experience I received in the Army, I have also worked in the medical electronics and sales fields while enrolled in college. Although this experience has been somewhat limited, I believe it gives me a good feel for the problems that an industry-related job entails, as well as a feel for the teamwork necessary in integrating the efforts of many people to get a job done.

I can be reached by phone at 608-652-7863 or at the address below.

>3008 Johnson St.
>Cleveland, TN 52311

Thank you for your attention and consideration. I hope I will hear from you soon.

Sincerely,

George P. Burdell

Letter of Application: Exercise 10.5

May 5, 1990

Mr. John Doe
Personnel Manager
General Electric Co.
Ogden, Utah 66666

Dear Mr. Doe:

I would like to apply for a test-department job with General Electric in Ogden. I heard of an opening through the Tennessee Tech Placement Center. I feel I am qualified for the job.

I will be graduating in June from Tennessee Institute of Technology with a Bachelor of Science degree in Electrical Engineering. I specialize in transformer design and feel I could contribute to the department. I am a graduate of Custer High in Smithville, where I have lived all my life.

My variety of job experience is limited. I started to work in the ninth grade with Carter's Bait Farm raising live bait and doing general farm work. In the eleventh grade I started to work for Western Wholesale in Smithville. I worked after school and in the summer up to the present filling sales orders and receiving freight.

Refusing a Job

I would appreciate being considered for the job. I will be happy to fill out any information which you will send me. I can be reached at my home address given below. References will be provided if needed.

Sincerely;

George P. Burdell

Home Address:

500 Elm Drive
Smithville, Utah

Letter of Application: Exercise 10.6

May 5, 1990

Dr. George P. Burdell
Professor, Electrical Engineering
Tennessee Institute of Technology
Cleveland, Tennessee

Dear Mr. Burdell:

I would like a letter of recommendation for a test-department job with Data-Pac in Salt Lake City. I need the letter received by them in five weeks.

I was in your transformer class last year. The job I am applying for involves testing transformers. I think the knowledge I acquired in your class will make me more than adequate for the job. I am enclosing a stamp addressed envelope for your conience. Please notify me if there are any problems.

I would greatly appreciate your help. I think your response could be a deciding factor in my getting the job. Thank you again for your cooperation.

Sincerely,

John H. Fell

School Address:

P. O. Box 36830
ext. 2743

Letter Requesting a Recommendation: Exercise 10.7

May 2, 1990

Mr. Ed David
Sales Manager, Southwestern Region
Electronics, Incorporated
2323 Sunset Drive
Austin, Texas 12345

Dear Mr. David,

I am writing this letter to ask you for a letter of recommendation.

I plan to continue my career in the sales field. Any favorable comments from you concerning my past experience and capabilities in sales will be greatly appreciated.

Thank you very much for your kind attention.

                              Sincerely,

                              George P. Burdell

Letter Accepting a Job: Exercise 10.8

May 5, 1990

Mr. John Doe
Personel Manager
General Electric Co.
Cedar Rapids, Iowa

Dear Mr. Doe:

Thank you for the test department job at your Cedar Rapids plant. The starting pay of twenty thousand is also appreciated.

I am really looking forward to working for you and General Electric. I have looked over and filled out the necessary information which I am enclosing. If there is anything else please notify me. If not I will see you two weeks after graduation ready for work.

Thank you again for the job. I hope to be a valuable asset to General Electric.

Sincerely,

George P. Burdell

Home Address:

500 Elm Drive
Rock Island, Illinois

Letter Accepting a Job: Exercise 10.9

May 2, 1990

Mr. John Jones
President
Jones and Company
398 Buffalo Road
New York, New York 54321

Dear Mr. Jones,

I am happy to accept Jones and Company's offer for employment as a Research and Development Engineer. Thank you very much for this opportunity to work with a progressive company in a position of practical importance.

I have enclosed a copy of my schedual through graduation. Hopefully it will help in planning the most convenient date for me to report to work.

It isn't often that a junior engineer is given an opportunity like this. I hope to fulfill the faith that Jones and Company has shown in me and I look forward to a challenging and successful future.

Thank you again.

Sincerely Yours,

Margie Taylor

Letter Accepting a Job: Exercise 10.10

U.S. Department of Housing and
    Urban Development
Attention: Harvey Jones
Personnel Staffing Specialist
98 South Street, S.W., Room 600
Kansas City, Missouri 54321

Dear Mr. Harvey:

Thank you for your offer as a Program Analyst GS-7 with the Office of Program Planning and Evaluation.

I appreciate receiving the new employee information package, and I will return promptly all the necessary forms to complete my personnel action folder.

In reference to our telephone conversation of May 1, I will be reporting to work on June 16th at 8:30 a.m.

Thanks again for your offer and I look forward to working with the Department of Housing and Urban Development during the summer.

Sincerely,

Mason Conner

CORRESPONDENCE SYSTEM FOR A LARGE TECHNICAL PROJECT

    Learning to write letters is a matter of writing one type of letter at a time. The actual practice of those skills in writing all the letters required in a large-scale project becomes a complex technical, managerial, and marketing job. The Introduction to this text discussed how many writers, operating on many levels, in various tasks, interact with each other in a system of communications. The Introduction also gave some indication of how a beginning technical writer might work in that system.

    This discussion of letters in a system goes into more detail about the various management and decision-making tasks which form the basis for letters. Management systems are weakened when technical professionals neglect to consider how important letters are as a component in a large project.

1. A correspondence system is an integral part of the systems design and management of any large technical project.
2. Writing is not a separate task in a large project but a logical extension of planning, drawing, budgeting, construction, and all other technical tasks associated with that work.
3. Clear and persuasive communications are necessary to communicate the solutions to technical problems. Technical work does not stand by itself; it must be supported by adequate communications.

Good correspondence systems are especially helpful to large-scale project managers, for they can help to:

1. Keep more adequate records.
2. Make writing a way of giving concentrated attention to stages of the work and, in that way, solve or avert problems.
3. Build good will with a client by good communications before later problems draw upon that fund of good will.

This system assumes four stages in any given project: the initial contact and proposal, the planning stage, the actual performance of the work, and the conclusion of the work.

Stage 1: Initial Contact and Proposal

The follow-up letter after the first meeting with a client should include three parts:

A. Paragraph 1, expressing your pleasure at meeting the client/prospect.
B. The body paragraphs, explaining your understanding of the problem and the kind of solution you have in mind. Especially if the client represents a lay audience or a small group like a church or a school, you can build credibility and good will by giving a clear explanation of what the problem is. If the client represents a larger technical audience, your credibility rests at this point upon a clear and detailed technical explanation of the problem.
C. The last paragraph, expressing your pleased expectation of working with these people on this job.

If a follow-up letter like the one described above will run more than two pages, write a short report with a cover letter. The cover letter should contain these features:

A. Paragraph 1, expressing your pleasure at meeting the client/prospect.
B. One body paragraph, which leads the reader into the short report and points out any special features to which you want to direct special attention.

C. A closing paragraph, expressing your pleased expectation of working with these people on this project.

The short report, which offers a preliminary discussion of the problem and a proposed solution, should be divided into parts, on broken pages with adequate white space. Use logical divisions, like these for a project involving construction:

- *Area* (Describe the area which you walked through.)
- *Discussion* (Discuss the design and engineering problems you were called in to study.)
- *Design* (Explain the proposed design necessary to solve the problem.)
- *Personnel* (How many workers with what skills necessary to do the job. Staff members you are assigning to the project. Especially in the case of a small group, give them some idea of the people they will be working with.)
- *Schedule* (If you can at this point, give some idea of when you can begin working and how long the work will take.)
- *Budget* (A schedule of fees for the various people who will work on the job, materials, etc.)

The short report provides an initial overview of the project. While the cover letter may be addressed to one person, the short report may be duplicated and distributed to a group of people involved in the decision-making process related to the project.

You gain these advantages by using a letter-short report combination:

1. The sales/PR content of the letter is addressed to one person and concentrated on one page.
2. The short report as a separate document for distribution to a group is easy to read, discuss, and refer to.
3. The logical division of the parts on the page gives evidence that you have done your homework in a very competent way.
4. The short report becomes an important part of your own records of a project separate from the sales/PR matter of the cover letter.

The writing in this initial stage of a project fulfills the three purposes of planning ahead: keeping an adequate record, using the writing process itself as a thinking process to organize your plans for the project, and building good will with good communication at an early stage.

The systematic approach for the first stage is perhaps best suited to a lay audience or a small group, but it can also function well with other professionals who work with you on a larger project. In this case, it is even more appropriate to use the short report to supply your colleague with the kind of technical data she needs in a well-organized and readable form.

Stage 2: Planning

The communications system here may require more writing than you customarily do, or may call for a different schedule of writing. However, a systematic approach accomplishes these purposes:

1. Concentrating time and attention on a project by replacing phone calls and many short letters with periodic progress reports.
2. Keeping good records as a way of resolving future problems.
3. Building good will to draw upon in case problems arise.

Concentrating Time

A given reading file may contain many very short letters to the same person about the same project. If at all possible, set aside a block of time to concentrate on answering several short letters and confirming several telephone calls with one long letter, or with a cover letter and short report. How often you do this depends, of course, upon the number of communications you have to answer.

The periodic progress letter should have the usual parts: opening, body, and closing.

A. Paragraph 1 of the letter should begin with something like one of these statements:

"For your records, I am sending a report on the progress of our work to date."

"To bring you up to date on our work thus far, I am writing to confirm our phone conversations since February 6 and to answer your recent letter."

B. The body of the letter or the short report should form a kind of communications log:

Telephone Conversations:

"To confirm and briefly review what we've discussed this week, my records show the following calls:

"February 1: Bob Toombs said that he cannot have the plans by March 1, but he assures me that he will have them the week afterward.

"February 9: We agreed to delay a final decision until we can get all the estimates from the electrical contractor, but we must finish this phase by March 15.

"If your recollections of these conversations do not agree with my understanding of the details we discussed, please let me know so that we can establish an accurate record."

Letters:

"To answer a series of questions posed by your letters since February 1, I have collected the following information:

"February 4: As you requested, we can change the location of the heating ducts.

"February 8: Changing the plan to move the north wall may lead us to structural problems which will add to the cost of the job; however, we can ask the architect to look at the plans again."

C. The final paragraph of this progress letter can include the sentence "If you have any further questions, please call." It is even better to use a sentence like this: "This letter should give you the information you need; however, if you have further questions, please call."

Concentrating Attention

It could be that concentrating time on the longer progress letter could serve the additional purpose of concentrating your attention on a given project. Writing a longer letter, rather than giving brief, situational replies with very short letters, could serve to relate several small parts to a whole project. A further possibility is that this concentrated attention might enable you to identify and avert problems as you periodically give a status report on a job. At the same time, your reader is likely to give more concentrated attention to reading a summary of calls and letters and, in turn, be able to raise a question that can be settled better at this stage than later on.

Even if you write a series of many short letters during the progress of a project, set aside time to review your file of letters from time to time, and in this way maintain an overview of communications. By reading your file you may discover that you've written three letters in three weeks about changing a particular feature of a design. Your correspondence file can alert you to a problem which needs attention.

Stage 3: Actual Performance

Keeping Good Records

Letters create a legal record, but they can also help to avoid problems, or to deal with them. When the written record is there, it is easier to set the record straight.

Letters which help to avoid problems: Problems are likely to arise if people misunderstand directions, if they delay in carrying out directions, or if workers on a project do not understand the terms of a compromise. Following are suggestions on how to avoid such problems:

Directions: Any directions you give can be no clearer than your own understanding of exactly what you want done. Your directions can be no more effective than your initial analysis of the audience who must understand and carry out the directions. Are you writing to another engineer, to a contractor, to an architect? Answering this question is an essential step. Break your directions into steps wherever you can. "Directions" does not mean a random list of fifteen items as they occur to you, but a logically-organized series of steps in a process. The clarity, completeness, and logical order of a good list of instructions convey the authenticity of your authority and credibility.

Motivations: In the E/T's concern for *how,* do not neglect to explain *why.* And rather than to give in to the pressures of time which make us want to say, "Because I *said so,*" take time in the early stages of a project to explain why. This is best done with *so that* motivation: explain that one thing must be completed by a certain date *so that* another part of the job can begin; explain the need for one item of information *so that* another part of the project can be planned. In that way, relate the specific action you request to another part of the project or to the project as a whole.

Using *so that* motivation may not always motivate the reader to the desired action. If, however, his delay causes a problem further along in the project, your "so that" letter forms a basis for later making a statement like this: "Our correspondence file shows that we wrote to you on February 1, requesting the revised drawings by February 14 *so that* we could complete the report by March 1. Since we have not yet received the drawings, we are two weeks behind schedule. Please let us hear from you *so that* we can avoid further delays."

The two assumptions which underlie the use of such motivation techniques are that people in business do not deliberately fail to do their jobs, and that a person will comply with a request more assiduously if you state your reason for making it or the significance of his action in a larger context.

Compromises: When it becomes necessary to yield a point and make a compromise in the planning and design of a project, a letter which states and discusses the terms of a compromise can be a valuable document if things do not work out well later on. For the record, review and discuss the options, state your willingness to cooperate and work out a compromise. Make clear at this point the client's role in shaping the decision: "Making this change will cut your initial costs, but in the long term, this HVAC system could increase your operating costs." Or "These, then, are the options. We are willing to support your decision based on your understanding of the situation and adopt option 2 in the plan."

If a problem arises as a result of the compromise, your record allows you to write in this vein: "You will recall that we agreed in January to ....", "Our initial design called for .... but the later revision we worked out with you...", or "My records show that when we agreed on this design in January, you did so with the understanding that adopting option 2 could lead to delays."

The clear record of the terms of a compromise becomes useful in future communications. The principle involved is that people like to avoid, if at all possible, assuming the responsibility for their choices. That is but one of the reasons why it is important to keep a clear record of any compromises which a project may require and to make a clear distinction of each party's role and responsibility in that decision.

Letters which deal with problems: Breaking bad news, announcing changes in schedules or costs, making apologies or explanations—having to write such letters makes E/Ts grow old before their time. There are, however, established techniques for dealing with such problems.

Bad News: Whenever possible, combine the positive with the negative. "Thus far the work has progressed on schedule, but an unforeseen delay has occurred in completing the drawings." "Usually we can provide an estimate on such a project within two weeks; in this case, however, the data gathering process has been delayed by...." The first sample sentence offsets bad news with good. The second is very useful in that it establishes this bad news as the exception in an established history of successful work. The assumption is that bad news *is* an exception; express it as such, as illustrated in the sample sentences.

Changes: In business writing, changes are usually described in these terms: "a *necessary* increase in cost" or a "*needed* change in schedule to allow time for...", "a *necessary* extension of the time usually required to.....". Such words, while almost subliminal, are still effective.

Apologies: Another accepted strategy in business writing is that one does not apologize unless specifically requested to do so. Instead, one "regrets any inconvenience caused by the delay" or "regrets the architect's decision to modify the design." One can quite sincerely regret a thing without taking on an unnecessary burden of responsibility by making an apology.

Explanations: Having to give an explanation when something on a project goes wrong raises a very valid question of how to explain and how much to say. It is a common practice in business writing that, if it is not possible or politic to give all the reasons which underlie a decision, it is acceptable to assure a client that the problem has had a thorough study and your full attention and consideration. Moreover, trying to re-trace every step in a problem and how it occurred and every option that you considered before making a decision may be more confusing than otherwise. An answer may best serve the client, not an elaborate reconstruction of the considerations involved in how you arrived at the answer.

Perspectives: When a serious problem arises with a client, the most useful method of dealing with that is to put the immediate problem into perspective. Careful attention to record-keeping and communications during the course of any job will enable you to do this. When you are under pressure from a client, having a good communications system becomes an advantage. When you can begin, "It may be helpful to review the entire question of the heating ducts to this point. My correspondence file contains a letter dated February 4 which . . . . . In a progress report to you on February 7, I agreed that . . . . , but to quote from that letter, 'In my judgment, if . . . . , then it is possible that . . . . will result,' " you avoid being on the defensive, and involve the client in the problem, rather than letting the client adopt a narrow, outside, blaming view.

Unfortunately, we don't always have the gift of foreknowledge which the example above reflects. However, putting a problem in a chronological perspective can serve to define exactly what the problem is, and is not, and where the division of responsibility lies.

If a chronological perspective seems ineffective, a second way to gain perspective is by comparing one job with another. This is particularly effective in dealing with colleagues; for no matter how much an architect may rage and stamp about, you can remind him of what he already knows: such problems are likely to arise, they have before, and they will again. Placing a problem in perspective in that way takes the pressure off the immediate point at issue.

When you deal with a lay client, you can accomplish the same thing by assuming the role of teacher and interpreter. For instance, a client who chooses custom work rather than commercial work automatically increases the risk of delays. When that delay occurs, the client can be reminded that such a problem lies at the very heart of the matter. An important part of your work thus becomes educating the client to what you are doing, a much easier task than placating a client.

Rhetorically this process is called "analysis for the root principle." The effect is to elevate the discussion from haggling over details to a more general area of understanding. The mural remains uncompleted and the painters are still working overtime; neither you nor the client can change that. But explaining the problem of delays inherent in special-order, customized items lends perspective which the client can understand.

Stage 4: Completing a Project

The visual and physical work of the E/T too often makes him rely too heavily upon the work itself. Indeed the work does stand as evidence of the worker's skill and creativity, but the client needs the communications link to teach him how to view that work. A second problem is that the worker's creative impulse is satisfied once the project has been completed, and he is busy looking ahead to the next project and/or simply sick of thinking about the completed project.

Good letters at this stage direct the client's view of the work, reinforce the client's pleasure in having the job done, and serve as a sales tool of inestimable importance.

Paragraph 1 of a final letter should state that the job has been completed. (If it is on schedule or ahead of schedule, note that. If there have been delays along the way, saying something like, "We are happy that all work is completed and inspected and all furnishings are in place and ready for your grand opening on August 14," tells the client that everything came out all right in the end even though he threatened you with court action two weeks ago.) Express your pleasure at the job and at working with the client.

Body Paragraphs should note in some way how the job originated from a client's need or from a problem which now has been solved. Briefly interpret how the problem has been solved. If the project was the satisfaction of a need, view the client somewhat as an artist views a person who says of a painting, "It's lovely. (pause) What's it supposed to *be*?" In the same way, a client may take enormous delight in viewing a re-designed space or a new heating system, and yet not know quite what to say about it. As a sales person or as an interpreter of your work, you can point out the special features or effect of the design, the economy to be derived from the solution of a technical problem, and especially how you fulfilled the original purpose of the project. Here again we see the role of the E/T as interpreter.

Final Paragraph expresses your pleasure in a job well done, and shares the client's pleasure. With this letter you are still selling the client on this job. Just as important, you are selling your company for the future to this client and, by his referrals, to other clients. Therefore, after the actual labor is completed on a job, communications is the final link. The last letter to the client is the finishing touch.

Conclusion

A good communications system made up of effective letters can be a pivotal point in managing a job. Such a system facilitates and unifies these work activities:

Initial contacts with clients

Planning

Record Keeping

Managing Time and Subordinates

Avoiding Problems

Dealing with Problems

Making every stage of work on a project a component in a sales system

part 3

Long-Term Writing Projects

11. Short Technical Reports
12. Case Studies
13. Lab Reports
14. Proposals
15. Technical Recommendations Reports
16. Technical Presentations
17. Professional Articles

Most academic disciplines require that students produce a lengthy formal document which reports the results of research. Professions which require technical writing extend that requirement into most major projects which engineers and technologists do. This section of the text is a catalogue of both academic and professional examples of long-term, or sustained, writing projects.

Short technical reports, lab reports, and case studies are a cluster of primarily academic forms—although each may be required of a practicing professional. As manuscripts, these are usually serially arranged rather than being divided into formal chapters. It is also unlikely that these will entail extensive documentation from other printed sources.

Proposals, technical recommendations reports, and technical presentations form a cluster of documents which usually make up the major communications forms in any large project. They are professional forms more than academic assignments. While the technical recommendations report may refer to other sources, it is not usually heavily documented from other sources unless done as an academic exercise in learning how to write such a sustained work.

The professional article represents the most professional of these documents. Either published in a journal or read at a professional meeting, this document is less prescribed in form than the others as a type of writing, and it is most closely tailored for a specific publication or professional audience.

References to proposals, short reports, technical recommendations reports, and the other long forms catalogued in this chapter can leave students confused about how each form relates to the others. Therefore, by way of further introduction to this section of the text, the charts which follow show the broad interrelationships of the forms.

The first chart gives a quick reference to the forms on the most traditional bases for considering any written work: audience, purpose, selecting a topic, narrowing a topic, and organizing the data. The forms are loosely grouped according to the setting where one is most likely to have to produce such a form: in an academic setting, in a primarily technical sense, and in a primarily professional sense for other professionals in that immediate setting. The last column on the chart shows how the academic application of all the forms differs somewhat from the actual practice of writing outside a technical writing course.

To further clarify what writing these forms entails, I have placed an * beside that component of audience, purpose, topic, or organization which I consider most essential for success in writing one of the areas outlined in the chart.

The second chart is a representative list of potential source materials to be used as data for the long forms. This chart is designed to give the writer a quick reference guide to the kind of data needed for a given form; again, the last column of the chart shows how students can adapt sources for the purposes of a course in technical writing. In general, the chart shows whether a given data source is common to a particular form, how that is used in the written

Long-Term Writing Projects

text, and how or whether the source is specifically cited in quotations, endnotes, and bibliography. Such a general representation of such a large and complex topic will not prove out exactly in every instance in research, but the chart should provide a comprehensive overview of how research sources and data are used in writing.

	ACADEMIC FORMS lab report case study	TECHNICAL FORMS short report, proposal, technical recommendations report	PROFESSIONAL FORMS professional article technical presentation	ACADEMIC EXERCISE
AUDIENCE	An identified and specific audience with the same level of interest and expertise in the subject.	An identified, specific audience characterized by different levels of expertise and different needs for reading, applying, and approving the document (client on management level in decision-making capacity, client with financial responsibility, a lay client, or a varied technical audience).	General for a published article and specifically identified for an oral presentation; both can be assumed to have a need to have the subject made interesting, understandable and technically accurate.	Primarily the person who will evaluate the work for a grade; other audience factors must be simulated.
PURPOSE	A specific technical purpose determined almost entirely by the prescribed procedure for the lab work or the case study.	*A specific purpose derived from an analysis and understanding of the identified audience, a technical and professional assessment of the audience's needs, and an understanding of the audience's request for proposals or commission to do a report with or without specific recommendations.	A specific purpose which derives in part from the writer's or speaker's assessment of how to adapt technical material to a highly specific situation where the audience has specific needs and interests.	*To present data according to one of the forms, but primarily to learn the parts, the construction, and the function of each of the forms as an academic exercise. The study of the form is as important as the practice of a given form.
SELECT TOPIC	The experiment itself for a lab report; an assigned case for the case study.	Determined both by the technical understanding of the actual work to be done and by a professional interpretation of what specific focus the project requires.	The oral presentation is most often a direct outgrowth of one of the technical forms; written articles are determined in part by the subject of study and by the publication.	Assigned or devised to meet the writing requirements of the assignment more than the technical content of the assignment.

NARROW TOPIC	The specifically assigned topic.	The process of selecting a topic most often also narrows that topic in professional writing situations.	*A very strict limitation based on the length of time allowed to present orally or by the length of article allowed in a given publication.
			According to the assignment: length requirements, time available, data available.
ORGANIZE DATA	*Derived 1) from the lab procedure and 2) from the assigned form for the lab report, determined by the assigned case study form.	*Formal*: prescribed parts for specific purposes in a prescribed order in a specific form (short report, proposal, technical recommendations report).	*Logical*: clear divisions of beginning, middle, end, and transitions but usually without formally prescribed parts.
			Stylistic: appropriate based on a careful assessment of the audience.
			Locate, collect, process, and then organize the data to fulfill the formal writing requirements of the course.

Data Sources	Lab Report	Case Study	Short Report	Technical Report	Proposal	Technical Presentation	Professional Article	Student Use
DICTIONARIES general specialized by discipline	refer to in text no bib or note, even for ""	only to establish basic terminology	refer to in text no bib. or note, even for ""	refer to in text no bib or note, even for ""	usually N/A	usually N/A except to establish basic terminology with the listeners	generally not referred to	to establish general command of subject
ENCYCLO-PEDIAS general specialized by discipline specialized handbooks	background, few or no citations	seldom or never used for such current data	usually only the most specialized to establish application of theory	background, few or no citations	only for the most general statement of the problem or for background	refer to only for the most general reference; to define a common body of terms	background, but generally not cited in text or bib.	to establish general command of subject
GOVERNMENT PUBLICATIONS laws and regulations agency regulations state and municipal codes	cite for prescribed procedures	cite especially for laws and regulations affecting business	refer to in text for specifications and procedures	cite in text general bib. to identify specs. and procedures	reference in text to state a problem to be addressed and solved	brief reference for background or to gain credibility with the listeners	usually has little bearing unless the article uses these as the subject of discussion	cite to identify procedures
books pamphlets	very rarely quotes, endnotes, and bib.	cite only if directly related to this case	rarely notes, quotes, or bib. in a short report	rarely notes, quotes, and bib. except for technical reference	identify as a resource for your work	quotes by author and title if directly applicable to subject	quotes, citations in the text, endnotes, and bib. as for any published source	same as for professional article
maps and charts	cite standard charts, tables	cite source for economic and financial data	cite as in a technical recommendations report	cite standard charts, tables; credit the source of all graphics	identify as a resource for the proposed work	use and refer to as graphic illustrations in the presentation	specialized used according to subject rather than a standard reference	same as for professional article

INDUSTRIAL PUBLICATIONS annual reports in-house documents	full technical reference within the text	a MAJOR source for case studies	cite in text; rarely note or bib.	cite in text, no note or bib. in most cases	general to state the problem	cite by author or title in introduction	cite in text, no note or bib. except for extensive use	cite in text, quote, end note, and bib. because of % secondary material
parts catalogues specifications	note and bib. for copyrighted material	general description in the text; no bib.	full technical reference in the text	full technical reference within the text	cite as a resource for work proposed	rarely applicable in the short space usually allowed	very rarely used unless the subject of study	same as for technical and lab reports
computer programs		note and bib. for any used	note and bib. if copyrighted material	note and bib. for copyrighted material	cite and discuss as a resource	refer to as the basis for statements	note and bib. for copyrighted material	same as for other forms
PROFESSIONAL PUBLICATIONS journals of research transactions conference proceedings	cite procedures	cite sources of information in text, notes, bib.	cite in text as in technical recommendations rep.	cite authorities, new techniques in text; no long bib.	usually N/A except to refer to an authority	refer to as sources of authority in making statements	cite in text, quote, end notes, bib. to give review of previous research	extensive use to form a data base for a paper, cite in text, quote, endnotes, bib.
text books by discipline	cite in text formulae, procedures; bib.	cite in text to establish terms or methods. general bib.	rarely or never used except to ref. theory or procedure	rarely or never used	almost always N/A and not on a professional level	rarely or never used here as a source referred to directly	rarely or never used as a source	background reading, definitions, formulae, authors cited in quotes, endnotes, bib.
GENERAL PUBLICATIONS books magazines general specialized newspapers	not a standard technical source	periodical reports a MAJOR source: notes, quotes, bib.	not a standard technical source	not a standard technical source	rarely except as a topical reference to describe a problem to be solved in the proposed work	rarely used except as a topical reference to dramatize a technical problem to be addressed and solved	used to a very limited extent except to define the existence of a problem or for use as a data base	potentially the major data base for a paper written as an academic exercise

LABORATORY RESEARCH experimental data photographs, drawings measurements calculations	major substance of the lab report	usually N/A	more specific and specialized than in many longer technical reports	discussed in text, graphics in text and appendices	general reference to procedures as a resource to do the work	may be the major basis of the presentation, but fully illustrate with graphics	depending on the subject and the publication discussed in text, graphic illustrations in article.	use depends upon the form being written and the assignment specs.
FIELD RESEARCH test data samples drawings photographs measurements	major substance of report if the procedure is a field procedure	interviews and data collection more by the methods of the social sciences	more specific and specialized in the text than in many longer technical reports	recorded and discussed in text, illustrated in text, fully recorded in appendices	general reference among general capabilities of the firm to do the proposed work	may be the major basis of the presentation but fully illustrate with graphics which are referred to and fully discussed	included if related to the subject for study	use depends upon the form being written, data base, assignment specs.
survey data	in the sense of a land survey	similar to professional article	in the sense of a land survey	in the sense of a site survey	an available skill	in both technical and social science sense	most common to the professional article	potential major source for an as signed paper
data from client company estimates, cost analyses	only if specifically required	a major source of data	discussed to state the problem	MAJOR feature of content, even if no citation	cite in text; cite understanding of the proposed work	specific and frequent use of client-listener	usually N/A unless the subject for study	usually N/A
computer data	as required, cite program	cite program and source	discuss, append as required, cite program	discuss, append as required, cite program	cite as available as a part of your capabilities	general reference or graphic illustration	discuss method, results, cite program	use depends upon the type of data and form

chapter 11

Short Technical Reports

The short report is useful because not all technological projects are of major scope. Some types of engineering or technological expertise lend themselves to work for clients with very specific problems: the roof on a warehouse is leaking, plastic panels in the light fixtures in a fast food chain have begun to crack and warp, a small city requires an engineering project, a private school needs a parking lot paved. E/Ts fan out into every area of our society to study such problems, devise solutions, and write short reports about their work.

Certain areas of technological expertise also lend themselves to narrowly defined projects. Computer companies, materials-testing firms, structural engineers, consulting firms—all of these may do small projects for individuals or organizations. In a given year, one company may do hundreds of such jobs. A narrow band of technical expertise may be covered by a short report within a much larger project. The short report is a primary medium for such situational or highly expert work.

The summary example which follows is designed to include a broad range of topics a short report may cover. Unlike the large-scale technical report which is formal in its arrangement of parts (Executive Summary, Table of Contents, Introduction, etc.), the short technical report is primarily logical in the selection and division of parts. That is, some short reports may require a description of a site and others will not. Some will include field procedures and others lab procedures. Some will suggest further needed work, and other projects will not call for recommendations at all.

The summary example, however, represents an effort to include most of the possible headings short reports cover in a variety of technical fields. The assumptions are that 1) you have data to report and know what to include and 2) you know basically how to arrange the data to give a logical presentation. Given that basic preparation, your major task is the actual writing, and the summary example is designed to be a thorough guide to writing the parts you need. Some of the material in the summary examples is repeated elsewhere in the text, but this section is designed to be a self-contained reference. If you need further help with overall organization in addition to the headings here, refer to the section on outlining.

Short Technical Report: Summary Example 11.1

SUBJECT: (This should contain all of the information usually contained in the subject heading: type of work, for whom, where. This may be distributed and filed separately, so give full information here. Depending on length, a formal title page with Abstract may be appropriate.)

JOB NUMBER: (The number of short reports generated often requires reference codes.)

1. Project Information: This introductory section serves to describe the project as the client described it to you. This establishes the client's problem/need in the client's terms. This paragraph has the dual function of: first, recalling to the client exactly what it was you were called upon to do; second, and perhaps most important, establishing how your task was initially described.

    1.1. Site Conditions: It is important, first of all, to describe the site where the work will take place. This paragraph has a dual function: (a) particularly for a large organization which may have many jobs under consideration, this description of the basic work area serves to define and describe the area about which decisions must be made and where the work must be done; (b) here you begin your professional interpretation; the client, like a patient, may describe a problem in layman's terms, while your job, like that of a physician, is to see and re-define the problem in technical and professional terms.

        Describing a Space: Three things are important to remember in describing a space:

Job number:
1. Like a camera panning an area, move logically from point to point. Set up a direction from one point to another along a vertical or horizontal line.
2. Again like a camera, pause to give concentrated attention to the pertinent points of the space under consideration. Selection of important detail is a part of any art, and a great failing of much space description is that it gives a one-dimensional view without giving the proper selection and emphasis to the most important parts of a given area under study.
3. Especially for lay audiences, reading a description of a space may be difficult. Given your orientation to space and objects within space, you probably have far more visual aptitude than your audience. Allow for that as you write by making sure that your description is logically organized.

1.2. <u>Purpose of the Report:</u> This paragraph describes in your more professional and technical terms what you were hired to do. A list at this point—as you see the three points listed above—is appropriate, for both logically and visually it sets up the parameters of your job.

1.3. <u>Scope of the Report:</u> Describing the purpose of the report may make a scope paragraph unnecessary. If, however, you want to make clear (a) how your job fits into a larger project (b) and what your job does <u>not</u> include, it is helpful to write a paragraph which defines the scope of your work. Should a problem or a question arise later, you can then refer to this scope paragraph.

2. <u>Investigative Procedures:</u> Before you begin the detailed account of what you actually did, it is appropriate here to give a brief introduction to how you went about doing your job. In classical rhetoric this is called the <u>exordium</u>, a figure from weaving which means "to lay out the threads." The modern formula for that is the first part of "Tell 'em what you're gonna tell 'em, tell 'em, tell 'em what you've told 'em." Either way, it is helpful for the reader of technical writing to have some idea of where he is going; this is not needless repetition, but very necessary reinforcement.

<u>Writing an Introduction:</u> Make a proper distinction between writing and reading; that is, don't feel that you have to write the introduction first because it appears first. Do it either first or last for these reasons:

Job number:
1. If writing an introduction will help you to organize your thoughts and set up the parameters for what you have to write, write it first as an aid for your writing.
2. Particularly if you're in a hurry, write the mass of technical matter you have to deal with. Then when you have a fresher, better overview of that mass of material, write an introduction for the reader. This takes less time than revising what you have written, and it provides the reader a guide through the technical matter. If you write the introduction last, simply cut-and-paste it at the head of the section.
   2.1. Field Procedures: Assuming that your introduction has given the necessary overview, it is necessary here only to:
      2.1.1. introduce this with a sentence like "A group/crew of (number) Stuart and Lee performed the following procedures under the direction/supervision of Edward R. Lee, Chief Engineer." (use of personnel information is optional)
      2.1.2. name the procedure: "Soil Sample," "Compression Test," etc.
      2.1.3. name the procedure next in logical order
      2.1.4. name the next procedure

   Break Each Page: As you establish a logical flow of the steps in the process involved in any field procedure, display each step to make it visually easy to read. This distributes a large part of the burden of writing and reading to "external" and "mechanical" devices. Assume also that the page will be re-read and actually used and worked with.

   2.2. Laboratory Procedures: The mechanics of putting this section of a report on the page are the same as for Field Procedures: write an introduction to serve as a guide, break the process into steps, arrange and place it on the page in such a way that it is easy to see and easy to work with.

   Process Writing: Describing a procedure can be deceptively simple because the process itself forms the outline for the writing. Process writing is more complex, however, than a one-dimensional, unselective account of everything that happened. A good process report

Job number:
includes three features, on three levels, for three purposes:
1. Write a technical description of the <u>process</u> itself for your own records and for the technical reader.
2. Explain the <u>principle</u> involved in each step and in the process as a whole. In a word, explain why and to what purpose the process is done. This lends perspective; it is especially helpful to a non-technical audience who can grasp the principle involved even if it does not understand every step in the technical description.
3. Relate the explanation of the process to your purpose in writing the report. Are there <u>implications</u> for reduced costs, increased efficiency, and the like? This level in process writing most often relates to sales and/or feasibility.

<u>Style in Process Writing</u>:   Since a process involves steps, lists are appropriate and can cover up a multitude of writing sins. Within the logical flow of a paragraph, use transitional words: <u>first, second, then, next, when, after</u>. Do not use <u>firstly, lastly</u>, or any <u>ly</u> word except <u>finally</u>.

Beware of <u>ing</u> words in process writing, for they can create an awkward rhythm. For <u>turning</u>, substitute <u>to turn</u>; for <u>using</u>, substitute <u>which uses</u>. If a passage sounds particularly awkward, check first for the "sprung" quality of too many <u>ing</u> words.

To introduce the principle in a process, use phrases like <u>so that, in order to, to determine how, to establish whether</u>.

<u>Defining Terms</u>:   Depending upon whether your reader is a technical or a non-technical audience, define terms which may be unfamiliar. Knowing when to do this without being condescending to the technical audience or too obscure for the non-technical audience is a matter of judgment.

1. For a <u>technical audience</u>, define a term primarily to assure that the two of you agree on what the term means. Distinguish your intended definition from any other possible definition to clarify your use of the term, and to stipulate the sense in which you use it.

Job number:
- 2. For a <u>non-technical audience,</u> give a brief definition the first time you use the term: "230 pounds per square inch (psi)..." Thereafter use psi. A word like <u>crazing</u>, or a word like <u>ambient</u> in the phrase "high ambient temperature" may need to be briefly defined in its technical sense.
- 3. For an <u>unusual word,</u> include a brief definition: "... Miocene Age marine deposit which is locally referred to as 'marl.' This material—described as a firm-to-very-dense, gray-green, silty sand—exhibited penetration resistances that..." (<u>Notes</u>: Once you have identified the word as an unusual word by using quotation marks around it, you need not use them thereafter. Note the use of commas in a definition which includes a series of terms; these are essential for any lay reader who needs to have an indication of the parts of the definition.)

3. <u>Discussion/Conclusion/Recommendations</u>: The purpose of this section of a report is to "tell 'em what you've told 'em." Up to this point the report is like the working-out of an equation; this section is the answer. In that your field and laboratory procedures investigated the causes of a problem, this section has to do with effects: what produced a given effect and what should be done to bring about a desired effect.

- 3.1 <u>Discussion</u>: Discussion should be clearly distinct from conclusion. Given your data from field procedures and laboratory investigations, a discussion of the data goes into the explanations of what you established by your study. In the Investigative Procedures section, you describe the collection of data. In the Discussion section, you deal with what those data mean.
- 3.2. <u>Conclusion</u>: "To conclude" means in the ancient sense "to shut up," as in shutting up a flock of sheep in a fold. Here you gather up the scattered points in your report into a final statement of opinion. This is what your study adds up to.
- 3.3. <u>Recommendation</u>: Most reports end in a recommendation for an action or a series of actions. The client has presented a problem for study with the question, either stated or implied, "What must be done about this problem?" Almost all reports end with a <u>list of proposed actions.</u>

    <u>Cause-and-Effect Writing</u>: Because determining cause and effect is a complex way of thinking, cause-and-effect writing should be carefully organized. Traditional

Job number:
> rhetoric divides cause and effect into these possible combinations:
> 1. single cause—single effect
> 2. single cause—multiple effects
> 3. multiple causes—single effect
> 4. multiple causes—multiple effects
>
> To clarify the writing process and the reading process, clearly divide causes and effects into logical parts.

Good cause-and-effect writing is similar to, and builds upon, good process writing. The difference is one of emphasis: whereas process aims at explaining how, cause and effect explains why; whereas the emphasis in process is to establish the steps, the emphasis in cause and effect is on the principle of operation and on the final outcome.

Technical Elements: (a) a clear statement of cause and effect on the mechanical or physical level defines a problem and establishes a basis for a solution, (b) a cause-and-effect explanation establishes a common ground of understanding with a technical audience, (c) being able to explain a cause-and-effect relationship to a lay audience establishes credibility: when a reader knows what has caused a problem and what will be required to solve it, he can make an informed decision.

Persuasive Elements: Cause and effect says, "If A happens, B will follow." Such a statement can be made with a high degree of certainty in a laboratory. Explaining to a client the ultimate effects of the recommendations in a report cannot be made with the same degree of certainty, for physical laws under laboratory conditions are not operative here. Therefore, be careful to distinguish between what can be proved technically and what can only be recommended. The technical is the present object; the recommendation is the future possibility—however well grounded in technical data.

Style: Use adequate transitions which express causation: because, since, thus, therefore, for that reason, with the result that, when...then. Do not use as for because in cause-and-effect writing. This misuse of as can be particularly confusing to a non-technical reader who reads as correctly as a word which relates to time and motion, not to causation.

Job number:

4. Further Studies: If your report is a part of a larger project, it is appropriate to recommend further studies and/or to give directions as to how one or more of your recommendations is to be carried out.

> Giving Directions: Suggestions for further study may involve giving instructions on how this recommendation is to be carried out. Writing directions is a form of process writing: the operation is divided into steps; explaining the principle is very important because it helps the reader to understand both why and how a process is carried out; no directive process can be clearer than the writer's understanding of what must be done and how; therefore, clarity is nowhere more important than in the writing of directions.

Short Report to a Small Non-Technical Audience: Summary Example 11.2

SUBJECT:

JOB NUMBER:

1. STATEMENT OF THE PROBLEM: Assuming that the description of the problem has come from an individual—and even by verbal description—this may be rather informal for a small audience with the use of you and we.

2. PURPOSE OF THE REPORT: This can be a very short paragraph of no more than two sentences. The purpose here is as much to define exactly what your job is as to communicate to the reader. Record-keeping is the primary function.

3. INVESTIGATION. The investigative procedure may be quite routine in comparison to the jobs you usually do. For the sake of the non-technical reader, however, make sure that the investigative procedure is logically and clearly written.

> Citing Figures: Especially for a non-technical reader it is important to do more than say, "See Figure 5." Make sure (a) that the text reference refers to the page so that it will be easy to find (b) that the figure itself is fully identified (c) that, if need be, a caption on the figure explains what it is and what it means. A common failing

Job number:

of engineering writing is to append a few pages of text to many pages of figures—and, in effect, leave the bewildered reader to "write" the report as he reads it.

4. <u>CONCLUSION(S)</u>: For a lay audience, the conclusions are less formal than for a larger job. They may, in fact, take the form of an expression and a discussion of a professional opinion.

5. <u>RECOMMENDATIONS</u>: These take the form of recommended action to solve the problem which you have defined and discussed. Suggestions for further study are also appropriate here.

<u>Note:</u> "Write to communicate, not to impress." This simple axiom of technical writing is nowhere more important than in writing to a small non-technical audience. The ability to reduce a complex problem to clear and simplified terms for a lay reader involves not only writing skill, but also a profound understanding of the problem. The small job, like the largest project, deserves careful and thorough attention.

The evaluation checklist for short technical reports can be found in Appendix B.

Short Technical Report with Transmittal Letter: Exercise 11.1
Cover Letter

May 30, 1990

Mr. Henry Jones
Energy Corporation, Inc.
P. O. Box 8310
Orlando, Florida 32160

Re: Feasibility Study for a Solar-Powered Air Conditioner

Dear Mr. Jones:

I am pleased to submit the final report on the computer simulation of the Energy Corporation, Inc., solar-powered air conditioner. The design and operating specifications are being sent under separate cover.

You will be especially interested to see that the proposed system can obtain a CoP (coefficient of performance) of 0.435 using a standard 20°F approach temperature in the vapor generator. This CoP is enough to warrant further study along with the construction of a working model.

An existing routine was used to simulate the thermal compressor; therefore, the final cost was reduced to $3200.

Since Com Sim, Inc., has set up the simulation and shown it to be accurate, any further trials can be executed at a cost of only $1000.

When Energy Corporation, Inc., needs further simulation work I hope Com Sim, Inc., can be of service. We always appreciate the opportunity to work on projects of this type.

Sincerely,

COM SIM, INC.

Georgette P. Burdell

Report

FINAL REPORT:	Process Simulation
SUBJECT:	Feasibility Study for a Solar-Powered Air Conditioner
JOB NUMBER:	1001-1

1. SUMMARY: The evaluation of a solar powered air conditioner was done by comparison of its thermal compression refrigeration cycle to a mechanical compression refrigeration cycle. The simulation of a 4-ton refrigeration system, 48M BTUs, was performed, using a pump, solar-heated hot water at 190°F, and two thermal compressors. The resulting coefficient of performance (CoP) was 0.435. This CoP is almost twice the CoP obtainable from standard mechanical refrigeration systems.

2. INTRODUCTION: Air conditioning costs are rising rapidly with the cost of electrical energy. In an effort to reduce this high cost, Energy Corporation, Inc., examined the possibility of using a thermal compressor combined with solar-heated water to provide

Job Number: 1001-1
Page 2

air conditioning. The largest energy user in a conventional air conditioning system is the refrigerant compressor; therefore, in the system under study the compressor would be replaced by a pump, two thermal compressors, and solar-heated water.

2.1. Purpose of the Report: Com Sim, Inc., was to evaluate the proposed system for engineering feasibility.

2.2. Scope of the Report: Com Sim, Inc., was to provide the necessary information to design and operate either a scale or a full-sized model of the proposed solar-powered air conditioning system. This information is being sent under separate cover.

3. INVESTIGATIVE PROCEDURE: Most of the equipment used in the proposed air-conditioning system had been simulated in prior jobs; however, the thermal compressors had not. It was necessary first to develop a routine to simulate the thermal compressor, and then to combine the thermal compressor routine with other routines to provide a total system simulation. FLOWTRAK was chosen as the best computer simulation program to use as a base upon which to build the required simulation routines.

3.1. Thermal-Compressor Simulation Routine: The thermal compressor can be simulated with a simple flash-drum simulation routine. This routine takes two streams and adds them together: then it flashes the combined stream to a desired pressure. From an enthalpy balance around the flash drum the condition of the outlet stream can be determined.

3.1.1. Enthalpy: Enthalpy (H) is an arbitrary scale for quantizing the heat value of a material. Since only the enthalpy difference (ΔH) of the streams is used, the

Job Number: 1001-1
Page 3

temperature and pressure chosen as a reference are of no consequence.

3.1.2. Condition: The condition of a stream is the weight fraction of the stream that is vapor. If the condition of a stream is 80%, the stream is a mixture of 80% vapor and 20% liquid by weight.

3.2. Process Simulation: The air-conditioning system was simulated by optimization of the CoP. This was done by using a parametic study, with the pump-pressure rise as the independent variable. With each different pump-pressure rise, the simulation would find the optimum operating state; using the operating state, a CoP was then calculated. After several different simulations were performed at different pump-pressure rises, a plot of CoP vs. pump-pressure rise was made. From this plot the optimum CoP could be found along with the design and operating conditions for the optimum system.

3.2.1. Parametric Study: a parametric study is done on a system by holding all except one variable constant and evaluating the results in comparison with that variable.

3.2.2. Pump Pressure Rise: The pressure difference between the streams entering and leaving the pump is the pump-pressure rise.

4. DISCUSSION: Several areas of the simulation presented problems. Each problem along with its solution follows:

4.1. Stream condition leaving the thermal compressors: When the condition of a stream is 100%, the stream is all vapor,

Job Number: 1001-1
Page 4

or a saturated vapor. This is the desired result for a stream leaving a thermal compressor. However, when the system was simulated, the stream leaving the thermal compressor would pass the condition of saturated vapor and become superheated vapor, or a vapor which contains more energy than necessary to maintain all vapor. If a stream leaving the thermal compressor is a superheated vapor, the system is not operating at total efficiency. To insure that the stream leaving the thermal compressors would be saturated vapor, the exit streams were controlled at a condition of 99.99%.

4.2. <u>Stream Split Ratios</u>: There are two places in the system where one stream is split into two streams. The ratio of one exit-stream flow to the other exit-stream flow had to be controlled at the optimum; otherwise, variable-changing in the parametric study would not have been limited to one variable. This ratio was achieved by controlling one exit-flow rate and by using controller routines, coupled with stream multipliers, to adjust the other exit stream to the optimum split ratio automatically.

4.3. <u>Heat Exchanger Approach Temperature</u>: Proper operation for heat exchangers requires a finite difference between the temperatures of the hot and cold streams in the heat exchanger. The standard of a 20°F approach was chosen. The <u>approach</u> is the minimum temperature difference allowed between the hot and cold stream at any point in the heat exchanger.

5. <u>CONCLUSIONS</u>: The use of solar power to drive a thermal-compression refrigeration cycle to be used in air conditioning is

Job Number: 1001-1
Page 5

technologically feasible. The hot water requirements for the 4 tons of refrigeration are achievable with current solar water heaters. The CoP obtained is almost twice the CoP of current air-conditioning systems. The approximate cost of operating the solar-powered air conditioning system would be ⅓ of the cost of the conventional mechanical-compression system.

6. RECOMMENDATIONS: The unit studied would provide enough air conditioning for the average-sized home. The areas that suggest further study are:

1. Industrial-sized units
2. Automotive-sized units
3. Energy storage for night use.

Com Sim, Inc. has the capabilities to perform the necessary studies to evaluate all such systems.

Short Technical Report: Exercise 11.2

TITLE: City of Augeas, Ohio, Municipal Waste-Transfer Station

SUBJECT: A report from the Public Works Department to the City Manager concerning the planning and construction of a municipal waste-transfer station for the City of Augeas, Ohio.

SUMMARY: The design, construction and start-up of the municipal waste-transfer station for the City of Augeas has been completed and carried through within the allowable budget. Start-up was delayed, and full capacity operation was not achieved until March 5, 1991, 33 days beyond the target date of February 1. While additional costs were accumulated during this delay, they were not sufficient to cause a budget deficit.

PURPOSE: The purpose of this report is to follow up the Public Works Department's proposal of May 21, 1990, with a record of the work done following authorization of that proposal.

DISCUSSION: The following sections describe and discuss the major features of the project.

1. Project Information: On May 7, 1990, the City Council passed a resolution calling for all departments of the city government to investigate ways and means of reducing the city's budget, with emphasis on both the short and the long term. The response of the Public Works Department was to introduce the idea for a municipal waste-transfer station. The incorporation of such a station into the city's waste-disposal system was seen by the P.W.D. as a logical way to upgrade and make more efficient the city's waste-handling capabilities. Costs would be reduced immediately and over the long term; furthermore, the city would

Public Works Report
April 21, 1991
Page 2

be better prepared to handle the substantial increase in solid-waste volume projected for the coming years. A proposal for the project was submitted to the City Manager on May 21; authorization to begin work was approved by City Council and City Manager on June 17, 1990. All phases of the project were within the jurisdiction of the P.W.D., with the Engineering Section, under the direction of E. R. Lee, responsible for actual design and implementation.

1.1. Site Conditions: The transfer station is located on 5 acres of land situated 3.5 miles from the city limits on U.S. 481 North. The land is 700 feet due west of the highway and is accessed by entry and exit roads which serve to keep ingoing and outgoing traffic separated and flowing smoothly. An industrial security fence surrounds the entire property and includes gates which can be locked at the entrance and exit. The soil at the site is homogeneous, or uniform in nature, throughout the site and down to a depth of 20 feet where bedrock is encountered. The residual clay found here is ideal for foundation because very little settling is expected over the life of the site.

1.2. Construction: The floor of the station, 2000 square feet in area, is formed of steel-reinforced concrete with special additives to resist corrosion. Three waste-receiving bays were built on the floor to accommodate present and future waste volume. Only one is now in operation, with conveyor and compactor installed, handling an average of 20 tons of waste daily. Each bay is designed for a capacity of 33 tons per day,

Public Works Report
April 21, 1991
Page 3

but plans are to bring a second bay into operation when the daily volume per bay reaches 25 tons per day. This volume is expected to be reached in approximately 3 years. Walls and roof of the structure are of weather-resistant corrugated aluminum with an expected life of 20 years. Roof supports are of the simple but sturdy K-truss variety.

2. Investigative Procedures: The development process for design and construction of the project was divided into 4 parts: research, evaluation of existing conditions, calculations, and construction. Although some planning and work was done in all four areas immediately, completion of the project on time and within budget depended on a thorough investigation of each step in its turn.

2.1. Research: Cities comparable to Augeas, Ohio, in size and population, with similar projects completed, were sought out for comparison. Two were located: Elis, Iowa, and Hercules, Texas. Both cities volunteered statistics and cost data that were invaluable to us in terms of providing guidelines and time-saving ideas for our work.

2.2. Evaluation of Existing Conditions: Augeas' waste-disposal system was studied first in terms of efficiency. Graphical comparisons were made to determine the breakeven point for the present hauled system vs. the same system with a transfer station (see Appendix). Using this method helped to locate an optimum location for the transfer station. Soil tests were performed on the several available sites; 60% tested satisfactorily. This fortunate availability of desirable

Public Works Report
April 21, 1991
Page 4

soil allowed us to choose a site very near the optimum location.

2.3. Calculations: While some of the cost analysis was done by hand, the reiterative or complex calculations were performed by computer. The Farmers and Mechanics Bank of Augeas sold this office the necessary computer time.

2.4. Construction: Construction was performed by Craft Construc-tion Company of Augeas, Ohio, the low bidder of seven competing companies. Construction standards were maintained by the Engineering Section through on-site evaluation of work in progress.

3. Financial: The project was completed with 8.7% of the allocated $347,000 remaining, which was diverted to the general fund. The following is a breakdown of project costs:

Engineering	$ 37,000.00
Computer Time	875.20
Land	7,000.00
Soils Analysis	446.00
Equipment	162,000.00
Contracting	108,489.80
	$315,811.00

The efficiency of the transfer station allowed the retirement of 8 compacting trucks. Of these, 4 were sold and 4 were kept for replacements during breakdowns. The personnel from the retired

Public Works Report
April 21, 1991
Page 5

compacting vehicles were assigned to work at the transfer station, eliminating the need for additional hirings.

4. Recommendations for Further Development: Besides being designed for efficiency and convenience, the transfer station was also laid out to provide for possible recycling activities. Space was allocated so that separation devices, e.g., magnetic separators, could be installed with a minimum of difficulty. We strongly suggest that the city actively pursue recycling as a means of solid waste reduction and as a possible source of revenue.

CONCLUSION: With the incorporation of the waste-transfer station, the efficiency of the city's waste-disposal system has been greatly improved. The benefits accruing from this greater efficiency will become more evident as Augeas grows and produces an ever-increasing volume of solid waste that must be disposed of routinely. We believe that the project has been a success and will benefit the people of Augeas in the years to come by allowing more efficient allocation of their tax monies.

APPENDIX

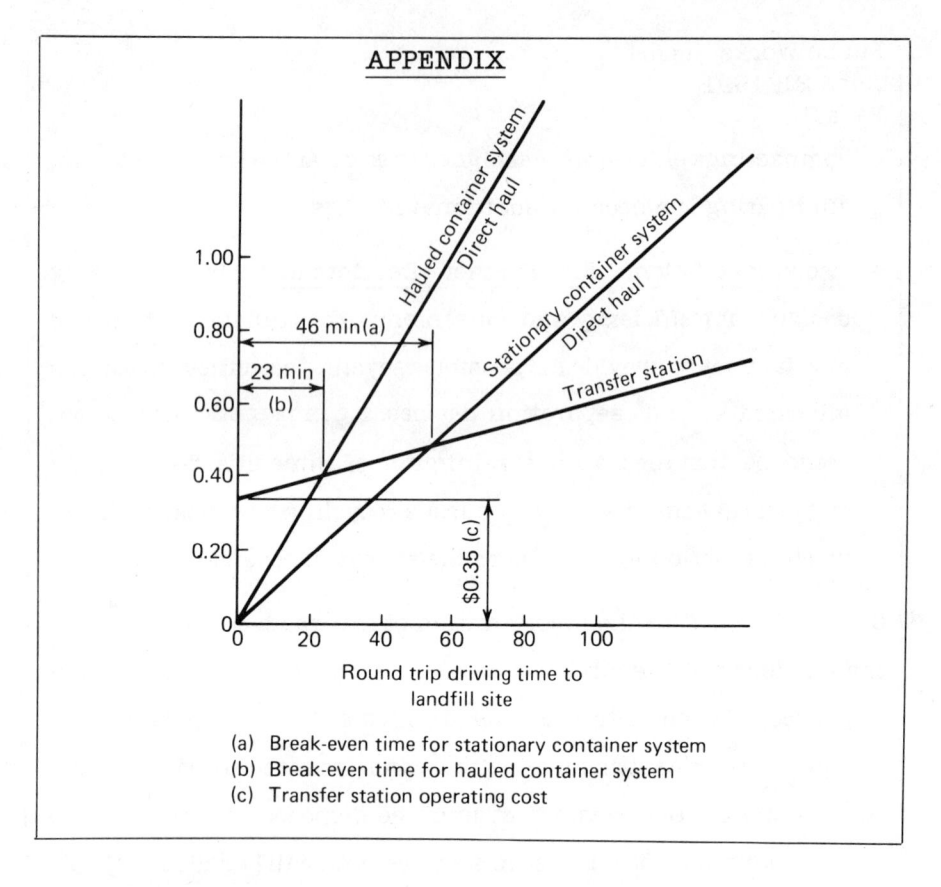

(a) Break-even time for stationary container system
(b) Break-even time for hauled container system
(c) Transfer station operating cost

Manuscript specifications and typing instructions appear in Appendix A, pages 475–481. An Evaluation Checklist appears in Appendix B, pages 491–492.

chapter 12

Case Studies

The case study is a teaching technique which facilitates the student's understanding of realistic business problems. It requires the student to do an in-depth analysis of a firm and an industry so that he may devise alternative strategies for the firm he is studying. Although these strategies may not be of the high quality required for use by a large corporation, the development of the strategies helps the student gain an appreciation and understanding of the time and effort that must go into any business decision. The case study requires the student to analyze industrial problems, decide upon a course of action, and present analysis and recommendations in good written form. It is an invaluable teaching aid which requires outside research to familiarize the student with the business world.

Like the short technical report, a case study may or may not make recommendations, depending upon what the task requires. One approach to case studies asks students to make recommendations and to offer constructive criticism of past company practices. The second form requires only that the student analyze the historical data; he is dissuaded from offering criticisms or recommendations. The following Summary Example is written with recommendations, while the subsequent Example is written without recommendations. The differences in the two approaches are not as evident as one might think because they are both written in the same basic format.

Case Study: Summary Example

Format

The case study should conform to the following basic guidelines:

1. Margins should be used: at least one inch at the top and at the bottom of the page, and at least 1½ inches on the left side to make the paper readable after it is stapled or bound.
2. Exhibits can be placed as appendices to the paper. These could include such things as xeroxed graphs, reprints, or supporting articles.
3. When you quote figures or present arguments taken from other sources, cite the reference in a footnote. Give sufficient information for the reader to locate the reference, but not necessarily in formal footnote form. Keep it simple but be consistent.
4. Divide your report into main issues, or most important topics. Avoid dividing into purely functional areas, for instance, Production, Marketing, and Finance. Identify major headings with Roman numeral items.
5. The first page should contain a list of the issues covered in your report. The reader should be able to scan the list and get a good synopsis of your report.
6. Maximum length is 10 double-spaced pages, exclusive of exhibits in the appendix.

Summary of Analysis

This is the section of the case study that the reader will read first, and the section that you will write last. Everything that you have said in the rest of the report must be reduced to two paragraphs. Nothing should be brought up in the summary of analysis that is not discussed in the body of your report. Do not be concerned that you are being repetitious. The person who reads the report needs the repetition for reinforcement. In this summary, make sure you convince the reader of your understanding of the case study's topic.

Major Issues

Provide a very complete listing of what you perceived as problems/issues in the case. Develop your listing so that the analysis which follows will be a logical outgrowth of the problems/issues listed. This list of major issues will be the closest thing to an outline the reader will have, so make it an accurate reflection of the contents of your report.

Analysis

The analysis is the backbone of the case study. In this section, a complete study of the firm and its environment is done, in order to fully develop alternative solutions and an overall strategy for the firm. There is no set number of sub-headings for this section; the only requirement is that the material be covered sufficiently to enable you to support a master strategy.

Industry Analysis:

1. The political, social, and economic dimensions of the industry should be covered, including international as well as national economic policies and any government regulation which might have an effect on the industry.
2. The market, including desire for product, availability of substitutes, and prevailing distribution system is another dimension of the industry that warrants study.
3. Product and technological dimensions of the industry should also be given attention. The resulting cost predictions and the industry's ability to fill demand are important determinants of a successful strategy.

Competitive Dimension:

1. The size, strength, and attitude of other companies is essential knowledge.
2. Major competitors' strategies should be analyzed to predict their future actions.
3. Whether or not strong trade associations exist could be important.

4. After studying the competition, the next step is to recognize marketing opportunities and threats.

Current Trends:

This section of the report is necessary if you are to succeed in developing any type of strategy for the firm. You could learn everything there is to know about the history and current state of the firm and its industry; if you failed to take tomorrow into account, your analysis and consequent strategy would be worthless. A master strategy must include trends and predictions to be useful.

Competitive Resource Strengths and Weaknesses:

1. What are the crucial factors for success in the industry? These might include research and development, low costs, adaptability to local needs, large financial resources, or creative imagination.
2. What are the company's strengths and weaknesses within the market? The answer to this question should include reference to the firm's distribution system, and consideration of the reputation of the company's products.
3. How does the service capability of the firm compare with that of its competitors? Also, are the firm's physical facilities and available labor conducive to performance?
4. What are the financial and management capabilities of the firm? This analysis is important in gaining an understanding of the firm's resource limitations and areas that need improving.

Alternative Solutions:

A study of past experiences, along with the creative thought process, should be used to come up with different ways in which resources may be committed. The consequences of these alternatives should be studied, including brief statements of the pros and cons of each possible solution.

Strategy and Policies:

This is the final and most important section of the analysis. It should focus on developing a master strategy while following these guidelines:

1. Be as explicit as possible in the development of policies, procedures, rules, and methods to support the master strategy.
2. Suggest and define the scope of major programs, projects, and budgets which will be needed to implement the master strategy.
3. Suggest possible sequences and timing of changes. Consider the right time to act in terms of the receptivity of the environment.
4. Consider coordination requirements.
5. Be aware of resource limitations.

Case Study: I. Summary of Analysis

The bearing industry is a highly capital-intensive industry characterized by keen, oligopolistic competition. A recent boost in sales has been counterbalanced by depressed prices; the result has been a low profit margin for the industry as a whole. Because of favorable government trust rulings, and high performance by the smaller bearing companies, there was a trend toward acquisition in the industry during the 1970s. More recently, bearing distributors have become more diversified, and have integrated into different markets, to decrease their single-product dependency and to realize additional profits outside the actual bearing market.

Since its change in ownership, Wabash Bearing has become a conglomeration of functional departments rather than a company dominated by one individual. The decentralization has developed into staff management and the more efficient use of the decision-makers within the company. That is, delegation of authority has enabled issues of secondary importance to be resolved at lower levels in the organization. Wabash Bearing is depending on quality control and improved customer service to obtain a greater share of the replacement-bearing market. Also, Wabash Bearing is attempting to secure a niche in the original-equipment market by offering specialized engineering work to firms with unique specification needs in the bearing market.

Construction

How

1. Write the Summary of Analysis last, when the report is so clearly defined in your mind that you can sum it up in a page.
2. The first paragraph of the Summary of Analysis should be a brief summary of the pertinent industry, including the current trends of the industry. The second paragraph should summarize your findings on the firm under study, including the firm's overall strategy and policies.

Why

As in the Executive Summary, *why* is really more important than *how*:

1. For the reader under a time constraint, the Summary of Analysis must be a good enough sales job to warrant the reading of the entire report, or its routing to someone else.
2. For the reader of the entire case study, the Summary should provide a very basic outline of what the report will cover and the conclusions to which that coverage will lead.

II. Major Issues

> 1. What is the background of the bearing industry, and what are the present conditions and trends of the industry?
> 2. What are Wabash Bearing's competitive-resource strengths and weaknesses?
> 3. What are Wabash Bearing's priorities, and what is its overall strategy?

Construction

How

1. List the major issues or problems that your report will cover, as the first step in writing a case study. Your initial identification of the issues and symptoms of the firm will help you in laying the groundwork for the Analysis section of your case study.

2. The list should be complete, but not limiting. That is, do not attempt to present a detailed outline for the reader; let the list of issues serve as basic questions that will be answered in the Analysis.

Why

The Major Issues section should allow the reader to gain an understanding of the direction your report will take, and should provide a logical outgrowth for the Analysis which follows.

III. Analysis

A. The Background of the Bearing Industry and the Competition Faced by the Wabash Bearing Company

The bearing industry is a highly capital-intensive industry characterized by relatively stable technological developments and very keep competition. The industry is best described as an oligopoly, because four firms produce over 85% of the total bearing output, but it is unique in how many smaller firms account for the other 15% of American production. Bearings have uses in a wide variety of markets, including the automobile industry, consumer durables, machine tools, farm equipment, aircraft engines, other aerospace applications, computers—in fact, in every industry which produces motor-driven products. There are three distinctively different bearing types with corresponding unique applications:
1. Ball bearings—used in small machines and motors.
2. Tapered roller bearings—used primarily in automobiles.
3. Cylindrical roller bearings—used in construction equipment. (Wabash Bearing concentrates on these.)

Bearing distributors are placed into one of three categories for the sake of simplicity, and as a guideline for the federal government in anti-trust cases:

> III. ANALYSIS (cont.)
>
> 1. Small distributors—those with less than $4.2 million in sales.
> 2. Medium distributors—those with from $4.2 million to $10 million in sales.
> 3. Large distributors—those with $10 million or more in sales.
>
> Wabash Bearing would be classified as a medium-sized distributor, along with many other bearing companies. There are four American bearing companies considered to be large distributors:
>
> - Timken
> - General Motors division of New Department—Hyatt
> - Fafnir
> - Ingersoll-Rand's Torrington division
>
> The biggest problem Wasbash Bearing has with its competitors, and especially with the major four, is its lack of share in the original-equipment market. That is, instead of having contracts with original-equipment producers, Wabash Bearing must rely primarily on supplying replacement bearings. The problem with that market emphasis is that most bearing users request the same brand of bearing as the original, which many times leaves Wabash Bearing as a second-choice bearing supplier.

Construction

How

1. The first section of the Analysis should define the industry and environment which are relevant to the case study.
2. Besides an analysis of the industry in this first section, attention should be given to the competition of the firm under study.
3. The basic thrust of this section should be to identify threats and opportunities.

Why

1. The background of the firm's industry is knowledge essential to understanding the market and product dimensions of the firm.
2. The size, strength, and attitude of other companies is important information if the firm is going to realize its marketing opportunities.

B. Current Trends in the Bearing Industry

The bearing industry has experienced a boom in sales since 1979, but because of suppressed prices, profit margins have remained relatively low. The increased demand for bearings is largely attributable to the present emphasis, found in all industries, on minimizing energy wastage. Automobile manufacturers are increasingly investing in tapered roller bearings in an attempt to save energy by reducing friction. The economic situation of "stagflation" has influenced companies to hold onto equipment longer and hence resulted in an upsurge in the replacement-bearing market. An acquisition period in the industry occurred in the early 1970s, and then again in the late 1970s. The trend was for the large distributors to acquire the small and medium-sized distributors, a trend facilitated by favorable government trust rulings. One reason for the acquisitions may have been the consistent tendencies of the smaller operations to outperform the larger concerns in both return on sales and return on assets. Another trend in the industry was the virtual extinction of the pure bearing distributors. That is, the bearing companies have become more integrated and diversified by dealing in and selling more than just bearings.

Foreign bearing firms continue to increase their sales in the American market by selling their product at a lower price. The most notable competition comes from the following:

1. SKF, a Swedish bearing firm, now accounts for over 20% of total world production.
2. Several Japanese bearing firms are threatening to acquire a larger share of the original-equipment bearing market in the United States.
3. In the 1980s, Eastern European bearing firms have entered the

> B. Current Trends in the Bearing Industry (cont.)
>
> American market by surpassing domestic firms in certain technological breakthroughs.
>
> If the current industry growth rate of 2.6% slows, there will likely be cutthroat price competition as domestic and foreign firms battle to maintain their current shares of the market.

Construction

How

   Current trends of the industry should include product, market, economic, and competitive trends.

Why

1. Any strategy that the firm decides to enact will involve a time lag. Therefore, so that at the time of implementation the strategy will still be applicable, your case study should consider the future of the industry.
2. Cost predictions and technological developments could have a profound effect on the direction the firm might take.

C. The Wabash Bearing Company's Competitive-Resource Strengths and Weaknesses

Wabash Bearing's biggest competitive advantage is its ability to provide engineering help on small-volume problems. The combination of custom engineering and production strength gives Wabash Bearing an advantage over the larger bearing firms in small orders. That is, Wabash Bearing is better equipped to make short production runs than are the larger concerns, because of the general setup of its plants. While these short runs have been used to produce replacement bearings, Wabash Bearing has recently expanded into two original-equipment markets: snowmobiles and printing-press equipment. By entering these new markets, Wabash Bearing has decreased its dependency on the provision of replacement bearings for construction equipment.

Since Firth Investment purchased Wabash Bearing, several areas have attracted a lot of attention:

1. Direct labor costs have been noticeably reduced through a reevaluation of line personnel.
2. A large capital outlay was made to introduce a quality-control system. This new control system should eliminate the drop-off in quality which occurred with increased production runs.
3. Added emphasis was placed on the customer-service segment of the business. Wabash Bearing realizes that to make up for its concentration in the replacement-bearing market it will have to provide such good service that customers will ask for Wabash by name. This is because firms have a tendency to request the same brand of bearing as the original. Wabash Bearing's biggest drawback is its lack of leverage in the bearing market. Because customers have a tendency to ask for the same brand of bearing as was in their original equipment, while Wabash Bearing makes very

> C. The Wabash Bearing Company's Competitive-Resource Strengths and Weaknesses (cont.)
>
> few original-equipment bearings, Wabash finds itself at an immediate disadvantage.
>
> When Wabash Bearing changed ownership the first time, there was an immediate increase in staff personnel, and hence salaries and fixed expenses. The tremendous overhead which was created when Firth Investment Fund took ownership is now affecting the gross profit margins, which have steadily decreased over the past five years. Firth Investment Fund made some large investments in plants and equipment, which immediately increased liabilities as the result of the large bank loan obtained. This bank loan now looms larger as Wabash Bearing's profit margins continue to decline.

Construction

How

1. The firm's management, service, and financial capabilities should be analyzed in this section of your report.
2. After reviewing this section, your reader should have an understanding of the crucial factors for success in the industry.

Why

A study of the firm's resource strengths and weaknesses is crucial in making an accurate appraisal of the firm's competitive position within the industry. This appraisal is necessary to continue into the next section, covering alternative solutions.

> ### D. Possible New Directions for the Wabash Bearing Company
>
> Wabash Bearing would like to increase its sales to correct the deteriorating profit margins. The firm has several alternatives which would enable it to do this:
>
> 1. Taking over a smaller company with either:
>   a. a geographical advantage, or
>   b. a strong market position in a special type of bearing
> 2. Gaining a foothold in an original-equipment market, such as:
>   a. snowmobiles
>   b. printing presses
> 3. Specializing in small-volume problems:
>   Contracting out its engineering capabilities to firms with a specific need that is not well handled by the general bearing market.
> 4. Merger:
>   Before considering merger, Wabash Bearing would have to make an all-out effort to improve its poor bargaining position so that Firth Investment Fund would not suffer a severe loss.

Construction

How

1. In this section of your report, you should develop alternative solutions or strategies.
2. These alternatives should be ways in which the firm's resources may be committed.
3. The basis for these alternatives includes:
   a. study of past experiences
   b. creative thought-process

Why

1. The alternative-solutions section of the case study should be a synthesis of all your prior work on the case.
2. This section is the "bottom line" of all your work; it should be an accurate appraisal of the options the firm has.

E. Wabash Bearing's Overall Strategy and Policy

Wabash Bearing is in a transition stage in even more ways than the change in ownership and basic organization would imply. From the time Wabash Bearing was a small, budding enterprise until the time it was sold, the company was led by one dominant individual, Angus McCabe, who did both the long-range planning and the day-to-day managing. When the irreplaceable Mr. McCabe died and Firth Investment Fund took over, Wabash Bearing was no longer run by one dynamic manager; instead, responsibility and authority were delegated to functional departments within the firm. Firth Investment implemented a systematic approach to building the needed corps of executives by studying the open positions and the available personnel for those positions.

Wabash Bearing distinguished its service unit from its operating unit by placing added emphasis on customer service. By decentralizing its authority structure, the company created staff management which facilitated the handling of matters of secondary importance. Wabash Bearing's strategy has involved improved quality control and customer service to differentiate Wabash Bearing's product from the original-equipment bearings on the market. By expanding into other markets and helping customers with specific design problems, Wabash Bearing hopes to gain a unique market segment for itself.

Construction

How

1. This section should have a coordinating effect on all your previous research and conclusions regarding this case.
2. These two paragraphs should make it clear where the firm stands, and where the firm wants to go.

Why

1. The strategy and policy section should tell the reader what you have already told him, but in a more concise form.
2. Since that is the last section the reader will read, it is imperative that you make a strong impression.
3. Use forceful, positive language so that the reader is struck by your confidence and by your knowledge of the material.

Note: Thomas Q. Langstaff, a graduating senior in the College of Management, Georgia Institute of Technology, supplied the text for this chapter on the Case Study.

chapter 13

Laboratory Reports

Unlike a short technical report, a lab report does not vary in form according to audience and subject. Instead, it has prescribed parts which appear in a prescribed order. Aside from that difference from the short technical report, the lab report can be defined and discussed on the basis of factors common to all technical writing.

COMPONENTS OF THE LABORATORY REPORT

Subject

The subject for a lab report is determined by the lab procedure it describes and the problem which that procedure addresses and solves; thus a built-in feature of subject automatically narrows the topic.

Purpose

The purpose is also determined by the problem the laboratory procedure addresses: is the lab work done to determine the rate of flow across an airfoil?

to find an unknown substance in a chemical solution? to test the strength of a material? to test the durability of a metal subjected to stress or to corrosion?

Audience

Relationship of writer to audience is perhaps less direct in a lab report than in any other form of technical writing; the aim is to write *about* a technical subject exclusively, not *to* someone. The lab report of the experiment and the results exists in and of itself as a self-contained entity, whether an audience with a compelling need or interest accesses the report or not.

Design

Like other formats in technical writing, a lab report has a specific design format; it also contains specified parts which must be written to satisfy a specific purpose within the document as a whole.

Style

This section on the lab report calls for a special note on style. Nowhere else in this text will you find a style so formal and so markedly characterized by use of the passive voice. The absence of a direct connection with a reading audience accounts for the style. Conventions of usage in math, science, and engineering also govern the selection and arrangement of words in many statements, especially those which contain calculations. Students who read text books in math and science will be familiar with these conventional modes of expression.

As the discussion of the parts of the lab report shows, many such features of the lab report must be modified to meet the conditions of writing in a profession for a paying client.

Following is an example of an assignment for a lab report made by Professor John Harper of the Aerospace Department at Georgia Tech. Note how prescriptive the instructions are. Much of the work of writing is actually to meet the specifications of form and content. Therefore, the ability to do the technical work itself and the ability to write to specifications are far more important in doing a lab report than skill or talent in writing. Being able to master this form, however, is an excellent introduction to writing a long technical recommendations report or even a professional article; for a writer must first of all be aware of design and specification in any written form.

The evaluation checklist for the lab report can be found in Appendix B.

Lab Report Assignment: Example 13.1

A.E. 3001

LABORATORY 1

Purpose

To test the integral momentum equation for steady flow experimentally.

Content

The forces on a flat plate and on a cascade of turbine blades mounted in a jet flow will be experimentally determined and compared with the values predicted by applying the integral momentum equation.

Part I. Flat Plate Tests and Analysis

Experimental Apparatus

A uniform jet flow is produced by ducting regulated air from a high-pressure air-storage tank through a 6" diameter pipe with a converging nozzle which discharges the air to the atmosphere. The pipe and nozzle configuration are shown in Figure 13.1. The nozzle exit area is $A_j = \pi$ sq. in. (i.e., the exit area is equivalent to that of a circular nozzle with a 1" radius). The pipe area is $A_1 = 9\pi$ sq. in.

The jet flow will impinge on a flat plate mounted normal to the jet axis as shown in Figure 13.1.2. The plate surface area is much larger than the jet cross-sectional area. Consequently, the plate will turn the jet flow through 90° (i.e., the jet flow will be discharged parallel to the plate surface around the outer edges of the plate). The plate will be supported by a "strain-gage balance" as illustrated in Figure 13.1.2.

A schematic of the strain-gage balance is shown in Figure 13.1.3. This balance has two strain gages which are used in a Wheatstone bridge to measure lift, L, and drag, D. The output will be measured

Components of the Laboratory Report

in "lift counts" (LC) and "drag counts" (DC) which are directly proportional to the voltage output. Previous calibrations have yielded

$$\frac{\Delta(LC)}{\Delta L} = COUNTS/\#_F$$

$$\frac{\Delta(DC)}{\Delta D} = COUNTS/\#_F$$

Test Procedures

Before measuring the force on the plate due to the jet impingement, the drag calibration of the strain-gage balance will be checked. First, the zero load drag counts (ODC) will be measured. Then, a known drag load, D, will be applied and the drag counts will be measured. These results will yield the calibration check.

Next, the drag force due to jet impingement will be measured. Before the jet flow is started the zero load drag count will again be measured and recorded. Then the jet flow will be started and the upstream control valve set so that initially $(P_1 - P_a) \approx 19"H_2O$. The pressure difference $(P_1 - P_a)$ where P_a is the atmospheric pressure will be measured with a water manometer. Because this test uses a blow-down system $(P_1 - P_a)$ will decrease slowly with time. When $(P_1 - P_a) = 18"H_2O$ the drag counts will be measured and recorded. Finally, the drag force will be computed.

The required data and calculation sheets are attached.

Analysis. **The integral momentum equation is**

$$\iint_{c.s.} \zeta \bar{v}(\bar{v} \cdot \bar{n}) dS = \bar{F} \qquad (13.1.1)$$

where $\bar{F}$ represents the sum of the forces acting on the fluid passing through the control volume. The x-component of Equation (1) is

$$F_x = \iint_{c.s.} \zeta u(\bar{v} \cdot \bar{n}) dS \qquad (13.1.2)$$

If we make the reasonable assumption that the jet flow is uniform with absolute velocity V_j, and use the control surface and coordinates shown in Figure 13.6 below, Equation (2) becomes

$$F_x = \zeta_j V_j (-V_j) A_j = -\zeta_j V_j^2 A_j$$

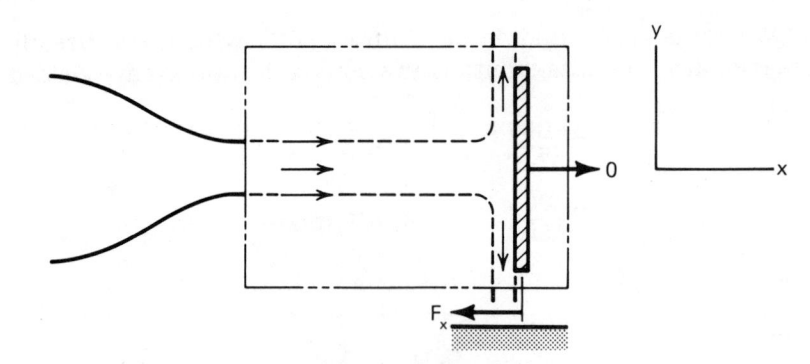

Figure 13.6

The drag force D (force on the plate) is opposite that on the fluid F_x. Thus

$$D = -F_x = \zeta_j V_j^2 A_j \qquad (13.1.3)$$

Equation (3) states that the drag force is equal to the jet momentum flow rate. It is convenient to define $J = \zeta_j V_j^2 A_j$ so that

$$D = J = \zeta_j V_j^2 A_j \qquad (13.1.3a)$$

J can be evaluated from the measured pressure difference ($P_1 - P_a$). Since the flow velocities are low, it is acceptable to assume incompressible flow with $\zeta_j = \zeta$ = constant. Furthermore, it is reasonable to assume that the flow is reversible from station 1 to the jet exit plane and that $P_j = P_a$.

Bernoulli's equation then gives

$$\frac{\zeta V_j^2}{2} - \frac{\zeta V_1^2}{2} = P_1 - P_j = P_1 - P_a \qquad (13.1.4)$$

In addition the continuity equation gives

$$V_j A_j = V_1 A_1 \qquad (13.1.5)$$

Using Equations (4) and (5) in Equation (3) gives

$$J = 2(P_1 - P_a)A_j / \left[1 - \left(\frac{A_j}{A_1} \right)^2 \right] \qquad (13.1.6)$$

Part II. Cascade Tests and Analysis

Experimental Apparatus

The experimental apparatus for these tests is the same as that for Part I except that the flat plate is replaced by a cascade of airfoils as shown in Figure 13.4. The airfoils have a trailing edge angle (mean camber line) of 45°. The cascade will turn the flow through an angle Θ slightly less than 45°. Due to turning the jet, the cascade will experience lift and drag forces.

Test Procedures

The zero lift and drag counts (OLC and ODC) will be measured before the jet flow is started. Then, following the procedure of Part I, the flow will be started and set for an initial reading of $(P_1 - P_a) \approx 19"H_2O$. The drag counts will be read when $(P_1 - P_a)$ decreases to $18" H_2O$. This will be repeated to measure the lift counts with $(P_1 - P_a) = 18" H_2O$. These results will be used to determine D and L.

The required data and calculation sheets are attached.

Analysis. The x- and y- components of the integral momentum equation are

$$F_x = \iint_{c.s.} \zeta u\, (\bar{v}\cdot\bar{n})ds \tag{13.1.7}$$

are

$$F_y = \iint_{c.s.} \zeta \mu (\bar{v}\cdot\bar{n})ds \tag{13.1.8}$$

It is a reasonably good approximation to assume reversible flow through the cascade. Therefore, using the control surface shown in Figure 13.7 below, and uniform flow at absolute velocity V_j both upstream and downstream of the cascade, Equations (7) and (8) become

$$F_x = \zeta V_j(-V_j)\, A_j + \zeta V_j \cos\Theta\, (V_j)A_j \tag{13.1.9}$$

and

$$F_y = \zeta(-V_j \sin\Theta)(V_j)A_j \tag{13.1.10}$$

or

$$F_x = -J(1-\cos\Theta) \tag{13.1.11}$$
$$F_y = -J \sin\Theta \tag{13.1.12}$$

Since $D = -F_x$ and $L = -F_y$ these give

$$D = J(1-\cos\Theta) \tag{13.1.13}$$
$$L = J \sin\Theta \tag{13.1.14}$$

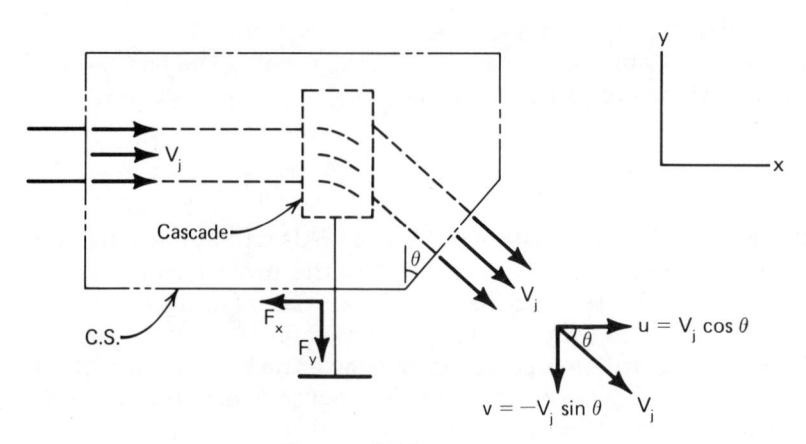

Figure 13.1

Preliminary Homework

1. Assuming that $(P_1 - P_a) = 18''$ H$_2$O calculate the jet momentum J where
$$J = \zeta_j V_j^2 A_j = \zeta V_j^2 A_j$$
Answer: J=4.135 #F.

2. Using J from problem 1 and assuming that the cascade of airfoils turns the flow through $\Theta = 45°$ calculate the lift L and the drag D on the cascade.
Answer: L = 2.924 #F; D = 1.211 #F

3. To show that V_j is relatively low and, thus, that the assumption of incompressible flow is reasonable, calculate V_j assuming $P_j = 14.7$ psia and $T_j = 70°F$.
Answer: $V_j = 285$ ft/sec.

Components of the Laboratory Report

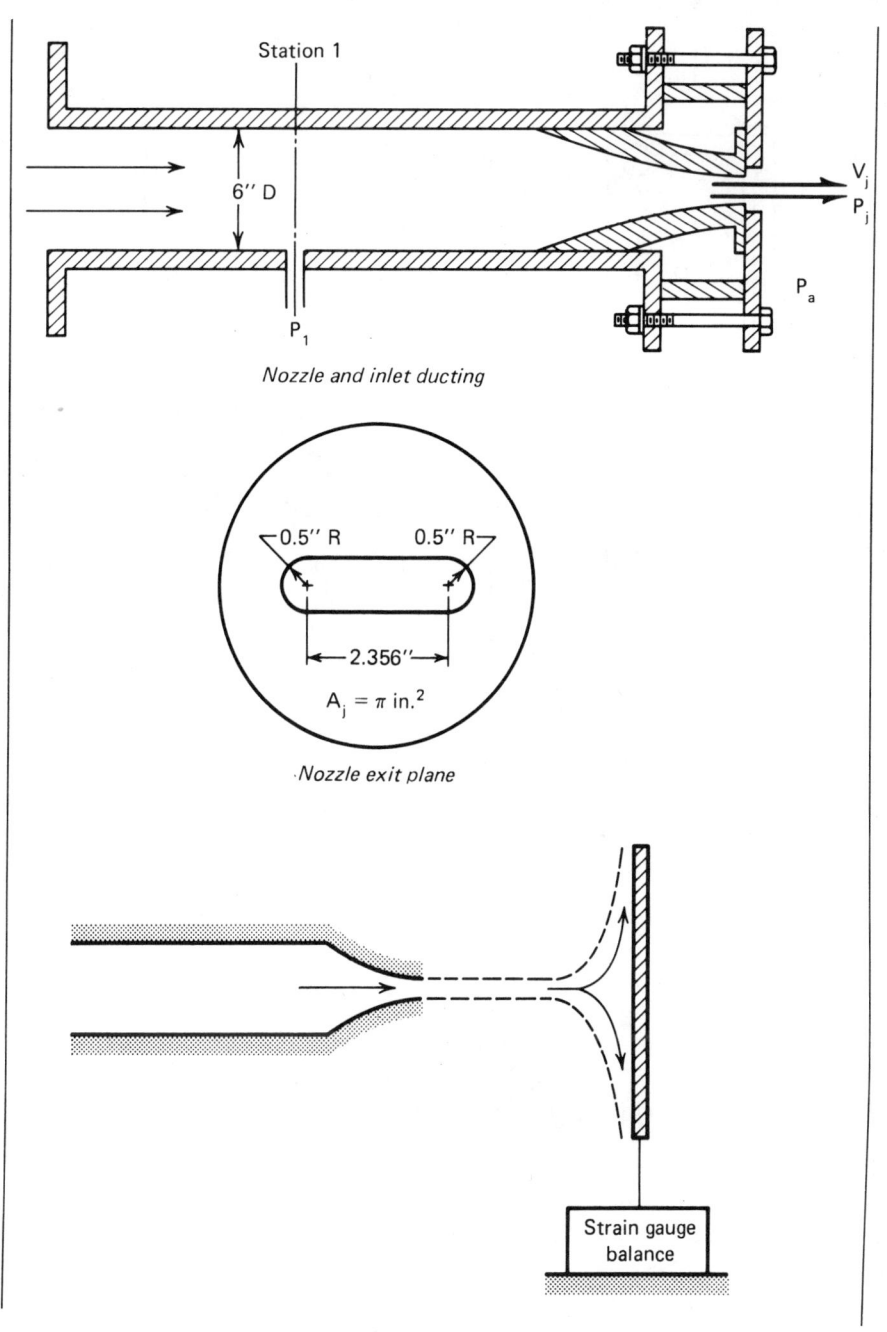

Nozzle and inlet ducting

Nozzle exit plane

$A_j = \pi$ in.^{2}

Figure 13.2

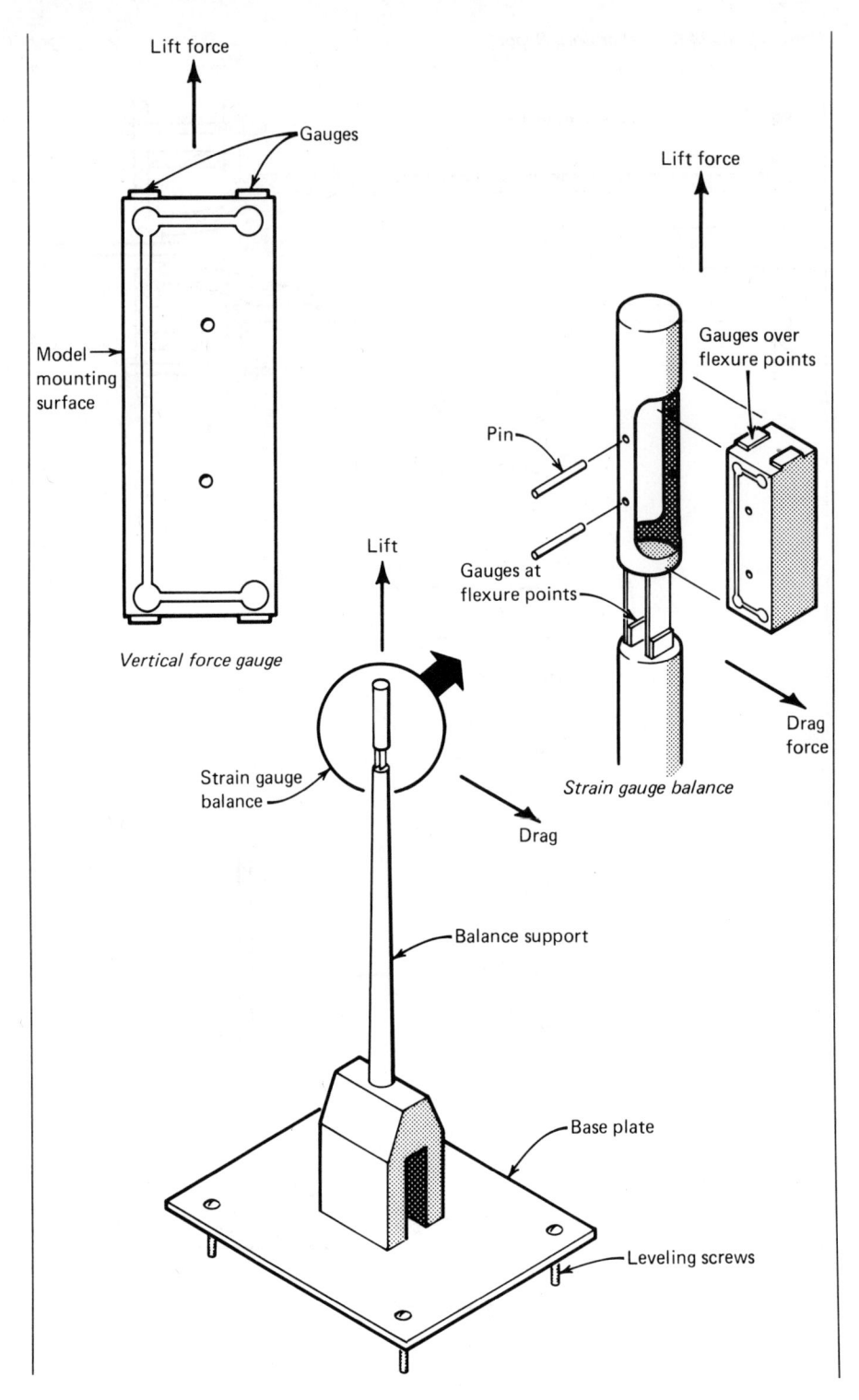

Figure 13.3. Strain-gage balance.

Components of the Laboratory Report

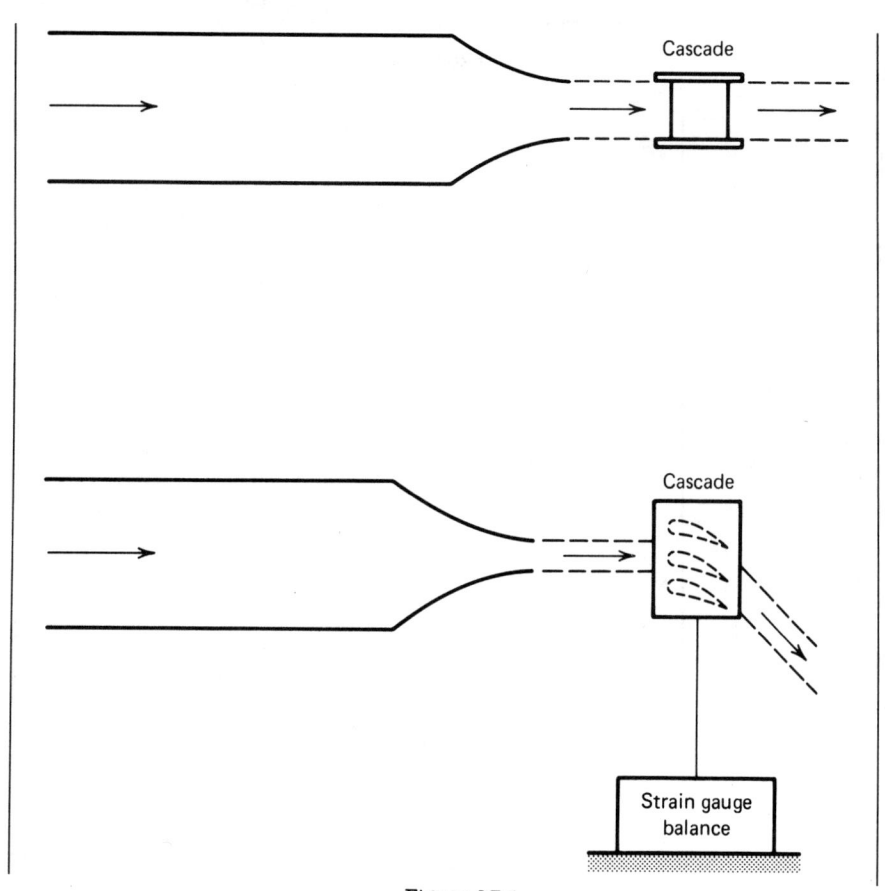

Figure 13.4

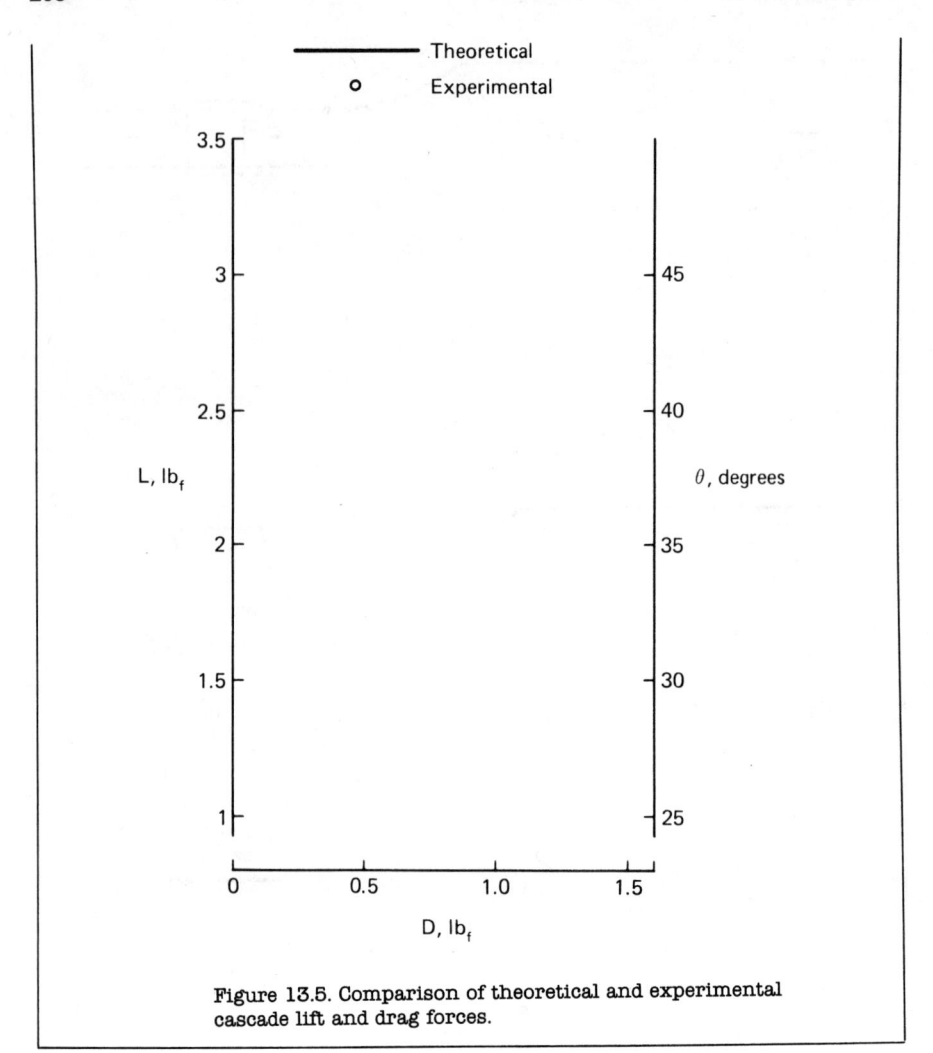

Figure 13.5. Comparison of theoretical and experimental cascade lift and drag forces.

Results

PART I. FLAT PLATE

1. Calibration of Strain-Gage Balance

    ODC = (　　　) counts
    D = (　　　) #F
    DC = (　　　) counts

    $$\frac{\Delta(DC)}{\Delta D} = \frac{DC - ODC}{D} = \frac{(\quad\quad) - (\quad\quad)}{(\quad\quad\quad)} = (\quad\quad) \frac{\text{counts}}{\#F}$$

2. Experimental Data

    ODC = (　　　) counts
    $P_1 - P_a$ = (　　　) in. H_2O = (　　　) psi
    DC = (　　　) counts

3. Experimental drag.

    $\Delta(DC)$ = DC − ODC = (　　　) counts

    $$D = \frac{\Delta(DC)}{(\Delta(DC)/\Delta D)} = \frac{(\quad\quad)}{(\quad\quad)} = (\quad\quad)\ \#F$$

4. Theoretical drag.

    $$J = \frac{2(P_1 - P_2)A_1}{1 - (A_j/A_1)^2} = \frac{2(\quad\quad)(\quad\quad)}{1 - (\quad\quad)^2}$$

    J = (　　　) #F

PART II. CASCADE

1. Experimental data

    ODC = (　　　) counts
    OLC = (　　　) counts
    $P_1 - P_a$ = (　　　) in. H_2O = (　　　) psi
    DC = (　　　) counts
    LC = (　　　) counts

2. Experimental drag and lift

    $$D = \frac{\Delta(DC)}{(\Delta(DC)/\Delta D)} = \frac{DC - ODC}{(\Delta(DC)/\Delta D)} = \frac{(\quad) - (\quad)}{(\quad\quad)}$$

    D = (　　　) #F

$$L = \frac{\Delta(LC)}{(\Delta(LC)/\Delta L)} = \frac{LC - OLC}{(\Delta(LC)/\Delta L)} = \frac{(\quad\quad) - (\quad\quad)}{(\quad\quad)}$$

$L = (\quad\quad) \# F$

Plot these results on Figure 13.5.

3. Theoretical drag and lift

$$D = J(1 - \cos\Theta) = (\quad\quad)(1 - \cos\Theta) \quad (13.9)$$
$$L = J(\sin\Theta) = (\quad\quad)(\sin\Theta) \quad (13.10)$$

Complete the following table and plot the results on Figure 5.

Θ	$D_1 \# F$	$L_1 \# F$
25°		
30°		
35°		
40°		
45°		

Annotated Lab Report: Example 13.2

EXPERIMENTAL VERIFICATION

of the

INTEGRAL MOMENTUM EQUATION

by

George P. Burdell III

for

AE 3001/ENGL 3023

October 16, 1990

SCHOOL OF AEROSPACE ENGINEERING

GEORGIA INSTITUTE OF TECHNOLOGY

SYMBOLS

A_j	Area of Jet Nozzle
A_1	Area of 6" dia. Pipe
D	Drag Force
F_x	Force in x direction
F_y	Force in y direction
J	Jet Momentum Flow Rate
L	Lift Force
P_a	Atmospheric Pressure
P_j	Pressure at Nozzle Exit Plane
P_1	Pressure in 6" dia. Pipe
V_j	Velocity of Jet Flow
V_1	Velocity of 6" dia. Pipe
P	Density
P_1	Density in 6" dia. Pipe
P_j	Density in Jet Flow
θ	Angle Between Mean Chord Line and Jet Flow

Construction

An alphabetized list of symbols serves as a guide to the reader of a laboratory report. Even more important, the symbols create an exact, uniform reference to constant elements in the calculations which all such reports contain.

Summary

Experimental values were obtained for the lift and drag forces created by a uniform jet impingement. These results were then compared to predicted results calculated from the integral momentum equation (IME). The resulting percentage of disagreement was found to be within explainable limits.

Construction

Most technical documents begin with a summary—or the answer to the problem which the document investigates. Perhaps the clearest analogy to how a summary in a lab report operates is this: some math teachers require students to place the *answer* to a problem first; then the working out of the problem follows. The grader can then look at only the answer to the problem, or choose to study each step of the demonstration.

The extensive use of abstracts in technical documentation also reflects the function of a summary. It is the abstract of results—or the applicable solution—not the tedious record of how the results were derived which is more useful to the reader.

Introduction

> Ducting air from a high-pressure storage tank through a nozzle produced a jet flow. The resulting flow impinged on a cascade of 3 airfoils; in the second part of the experiment, it impinged on a flat plate, set normal to the flow, instead. Strain gages measured the lift and drag forces created by the jet flow against the plate or through the cascade. The measurements were compared with values calculated using the IME:
>
> $$\Sigma F = \iiint_{cv} p\vec{v}\,d\vec{v} + \iint_{cs} p\vec{v}(\vec{v}\cdot\vec{n})\,ds \qquad (13.2.1)$$
>
> It should be noted that for steady flow the first term of the IME is equal to zero. There also are no terms which evaluate pressure and viscous effects; thus the IME assumes incompressible and frictionless flow.

Construction

"Statement of the Problem" is an almost univeral way of describing the subject of a technical document. Some writers who practice engineering as a business object to the connotations of the word *problem*, thinking that such usage will create a negative impression in the client's mind. Students sometimes question this usage because of the same negative connotation.

There is really no adequate substitute, however, for the phrase, "statement of the problem" to express in technical and scientific terms the ideas: "this is the subject the experiment investigates," "this is the situation which the report discusses," "this is the defined area of study which the scope of the following document (memo, letter, report) covers." It is very important, then, to be aware of the function of "statement of the problem" as a standard term in technical documents.

Apparatus

The source of the jet flow was a high-pressure storage tank. The air in this tank was ducted through a 6" diameter pipe and finally a nozzle of cross sectional area $A_j = \pi \text{in}^2$. A pressure port was installed in the 6" dia. pipe and attached to a water manometer. This allowed measurement of the pressure within the pipe in relation to atmosphere pressure $(P_1 - P_a)$. A blow-down valve was installed upstream of the nozzle to enable control of the volume of flow to a magnitude of $(P_1 - P_a)$. Because a blow down system was used, the value of $(P_1 - P_a)$ decreased with time. Figure 13.2.1 illustrates the configuration of the pipe and nozzle assembly.

For the first part of the experiment a cascade of three airfoils was fixed in front of the jet flow. The width and height of the cascade was such that it may be assumed that the entire flow was deflected. Figure 13.2.2 illustrates the position of the cascade and also the trailing edge angle (mean camber line) of 45°.

During the second part of the experiment a flat plate was fixed normal to the jet flow. As shown in Figure 13.2.3, the height and width of the plate were such that the flow was deflected through an angle of 90°.

A strain-gage balance held both the cascade and the plate in place. The balance consisted of two strain gages used in a Wheatstone

> balance—one gage to measure lift and the other to measure drag. The gages were electronically connected to an SR-4 readout which displayed lift counts and drag counts; the number of counts was directly proportional to the voltage received from the gages.

Construction

A verbal report of a lab experiment cannot be separated from the equipment used to derive and test the results. The apparatus which the experimenter uses, and how the apparatus is set up, can be the most significant factors in how the problem is solved or in what results the experiment produces. A careful description of the apparatus and how it is used can also be a guide to other researchers who wish to perform the same experiment.

To describe an apparatus and its operation can be a real test of skill in spatial, process, and cause-and-effect writing. Discussions of those patterns appear in the writing labs on paragraphs and outlines. Describing an apparatus also illustrates the interplay between verbal description and graphic illustration. Use of each of these techniques in the shorter, more formal lab report is a forerunner to the extended use of both techniques in the longer technical report and professional article.

Procedure

The strain gage and SR-4 assembly were first calibrated. A description of the procedure used for this calibration can be found in the Appendix. Tables 13.2.I and 13.2.II re-word the results and form the graphs found in Figures 13.2.4, 13.2.5, and 13.2.6. These graphs make it possible to transform lift-count and drag-count readings to numerical values of lift and drag.

After calibration, the apparatus was set up for the cascade portion of the experiment. The blow-down valve was opened and adjusted so that $(P_1 - P_a) \approx 19"H_2O$. The pressure then began to decrease. Lift-count and drag-count readings were recorded when $(P_1 - P_a) = 18"H_2O$. These readings were then transformed into actual lift and drag loads using Figures 13.2.5, 13.2.6, and 13.2.7. Table III records the results.

Next the flat plate was put in position and the procedure was repeated. This time, however, the valve was opened until $(P_1 - P_a) \approx 14"H_2O$ and the drag reading was taken at $13"\ H_2O$ instead of $19"\ H_2O$. This procedure prevented unnecessary stress on the strain-gage balance. The lift reading was not taken for the flat plate because, as a result of its orientation to the flow, it experienced no lift. Table IV shows the results of the flat-plate portion of the experiment.

That step concludes the experimental part of the lab exercise. The next step was to evaluate the IME to obtain theoretical values for the drag of the plate and the lift and drag of the cascade.

Theoretical Calculations

The x-component of the IME is

$$F_x = \iint_{cs} \rho u(\bar{v}\cdot\bar{n})ds \qquad (13.2.2)$$

on the assumption that the velocity of the jet was constant. The control surfaces and coordinates in Figure 13.2.1 were used; equation 13.2.2 then became

$$F_x = P_j V_j(-V_j)A_j = P_j V_j^2 A_j \qquad (13.2.2a)$$

The drag force D on the plate is opposite to that on the fluid; thus

$$D = -F_x = P_j V_j^2 A_j \qquad (13.2.3)$$

This says that the drag force is equal to the jet momentum flow rate J.

$$D = J = P_j V_j^2 A_j \qquad (13.2.3a)$$

Since the flow velocities were low, the flow was assumed to be incompressible with constant density

$$P_j = P_1 = \text{constant} \qquad (13.2.4)$$

It was also assumed that the flow was reversible and thus the pressure at the jet exit plane, P_j, was equal to atmospheric pressure, P_a; that is, $P_j = P_a$ \qquad (13.2.5)

Bernoulli's equation (see Appendix Section 2) then gave

$$\frac{P_j V_j^2}{2} - \frac{P_1 V_1^2}{2} = P_1 - P_j \qquad (13.2.6)$$

substituting from (13.2.5) yields

$$\frac{P_j V_j^2}{2} - \frac{P_1 V_1^2}{2} = P_1 - P_a \qquad (13.2.6a)$$

The continuity equation (see Appendix Section 3) gives

$$V_j A_j = V_1 A_1 \qquad (13.2.7)$$

Substituting equations (13.2.6a) and (13.2.7) in equation (13.2.a) gives

$$J = 2(P_1 - P_a)A_j / \left[1 - \left(\frac{A_j}{A_1}\right)^2\right] \qquad (13.2.8)$$

Evaluating equation (13.2.8) with

$$(P_1 - P_a) = 13'' H_2O = 67.58 \, \text{lb/ft}^2$$

$$A_j = \pi J \mu^2 = 2.18 \times 10^{-2} \text{ft}^2$$

$$A_1 = \frac{\pi 6^2}{4} \times \frac{1}{144} = 1.96 \times 10^{-1} \text{ft}^2$$

gives

$$J = 2.983 \text{ lbs.}$$

and hence by equation (13.2.3a)

$$D = 2.983 \text{ lbs.}$$

for the flat plate. Table IV shows this result.

For the cascade the x and y components of the IME were

$$F_x = \iint_{cs} \rho u (\vec{v} \cdot \vec{n}) ds \qquad (13.2.9)$$

and

$$F_y = \iint_{cs} \rho v (\vec{v} \cdot \vec{n}) ds \qquad (13.2.10)$$

Reversible flow was assumed and the control surface shown in figure 13.2.2 was used. These equations then give

$$F_x = \rho V_j(-V_j)A_j + \rho V_j \cos\theta (V_j) A_j \qquad (13.2.9a)$$

and

$$F_y = \rho(-V_j \sin\theta)(V_j) A_j \qquad (13.2.10a)$$

Using equation (3a) yields

$$F_x = -J(1 - \cos\theta) \qquad (13.2.9b)$$

and

$$F_y = -J \sin\theta \qquad (13.2.10b)$$

The drag $D = -F_x$ and the lift $L = -F_y$, so

$$D = J(1 - \cos\theta) \qquad (13.2.11)$$

and

$$L - J \sin\Theta \qquad (13.2.12)$$

Substituting the known values

$$A_j = 2.18 \times 10^{-2} \text{ft}^2$$
$$A. = 1.96 \times 10^{-1} \text{ft}^2$$
$$P_1 - P_a = 18"H_2O = 93.58 \text{ lb/ft}^3$$
$$\theta = 45°$$

results in
$$D = 1.210 \text{ lbs.}$$
and
$$L = 2.92 \text{ lbs.}$$
Table III shows these results.

Finally, the experimental and theoretical results were compared. Tables III and IV show how the results were compared.

Construction

How an experiment proceeds once the problem has been stated and the apparatus has been assembled is a key factor in a lab report. Here the lab report moves out of an almost exclusively verbal mode and into a mathematical mode. A student with a problem in geometry or calculus is limited to numerals and symbols; a lab report expands that method by interspersing words as well. The demonstration becomes a narrative of how calculations operated at various points in the experiment; these pages can be called an annotated math problem.

The procedure section of a lab report is also a forerunner of the kind of writing required in a technical report. Since a lab report addresses a subject for itself more than for an audience, relying mainly upon calculations in a kind of running narrative is appropriate. In contrast, a technical report is addressed to an audience which includes, in addition to technical readers who could easily read a lab report, readers from management, financial officers, lay clients, and a whole host of readers who may be unable to understand the procedures expressed in mathematical terms.

The technical report writer, therefore, must apply the axiom "never express mathematically what can be expressed verbally." While calculations form the major substance of a lab report, a verbal narrative of how the experiment was performed and what results it yielded would appear in the body of the report, and a procedures chapter like this one would most often be relegated to an appendix. That shift of style from addressing the technical subject as a self-contained exercise to communicating the same kind of technical data to an audience marks a significant difference between being a student who must write for a class and a professional who writes for a client.

Results and Discussion

> The results recorded in Tables II and III show that both the lift and drag values obtained for the cascade through experiment were somewhat lower than the predicted values. One explanation of this difference is that the jet flow was not turned through an angle of 45°. Figure 7 shows calculated lift and drag values for various angles of Θ. The experimental values are also plotted on this graph. These data suggest that θ was approximately equal to 40° rather than 45°.
>
> This hypothesis is further supported by the small percentage disagreement in the flat-plate results. As Table IV shows, this disagreement was only 2 percent. Such a small deviation is probably due to human error and the limitations of the equipment used. It is, however, reasonably safe to assume the jet flow was turned a full 90° to impinge on the flat plate. All other assumptions made were the same for both tests.
>
> Using θ = 40°
>
> and
>
> error for lift = 1.7%
>
> these results are well within experimental limits.

Construction

Whereas the summary gave the answer to the problem, the results and discussion section expands the answer with discussion of variables which the writer considered just before reaching a final answer. Most experimental work deals in exact quantities, but most careful researchers also qualify their findings. To perform an experiment at all assumes that some question exists about the exactitude of the subject under study, or it assumes that some exact answer can be derived. The answer may lie within reasonable limits—or "within experimental limits" as the writer expresses. To describe what those limits are, even in one-hundredths of a percentage point, can be very useful to researchers who study the lab report.

Beginning students sometimes question any answer that is not exact. The more any researcher matures in his work, however, the more he comes to appreciate that the most important data may lie in the qualifiers which define the measures of inexactitude in experimental results. Therefore, to discuss variations or to qualify results can be the mark of the careful researcher, not the mark of someone who has failed to produce an absolutely exact answer.

Conclusion

> This lab exercise demonstrates experimentally that the IME is an excellent approximation of real fluid flows. The results also suggest that an airfoil does not turn the flow over its surface through an angle equal to its mean camber angle. Instead the flow is turned through a slightly smaller angle.

Construction

Again unlike the summary which introduces a report, a conclusion is a more general discussion of the experimental results. More than any other part of a formal technical document, the conclusion calls upon the researcher to draw together data to form a personal judgment. The conclusion is not stated in the first person (*I, me, my*), but it nevertheless represents how all the separate factors are drawn together to render a conclusive opinion.

The conclusion which the student writer reaches in the lab report is once more a forerunner of how a technical-report writer must operate as a professional. Suppose, for instance, that this experiment on lift and drag had been performed at a major aircraft manufacturer's lab. The results of the experiment might determine a major design change in the wing of a cargo plane. This, in turn, could improve the efficiency of the plane, reduce fuel consumption, increase the range of flight, and increase the payload. To achieve such an improvement in design could place the company in a more competitive position to win a government contract or to apply the design to commercial airline use. Billions of dollars could ride upon a conclusion in that instance.

The researcher's job after completing the lab experiment would be first to sell the new design to company management. This would be a major work of communications far beyond writing a lab report. If the company decided to make the changes necessary to produce the newly designed wing, the next task would be to market the design to the Department of Defense and to major airlines. That also would require a communications effort of a magnitude undreamed of by most undergraduates in an aerospace-engineering lab.

Simply reporting the results of a lab experiment, then, is merely a first step toward further demands upon a student writer's ability to communicate.

References

AE Fluid Mechanics, Junior Level Laboratories

Keuthe, Arnold M., and Chuen-Yen Chow. Foundations of Aerodynamics. New York: John Wiley and Sons, 1976.

Graphics

A student lab report to an audience of one professor does not call for the production requirements demanded by a technical report for a client. Therefore, it is sufficient here to append all figures. When a student carries the same method of appending graphics into a professional situation, that produces one of the major weaknesses of technical reports: the writer writes five pages of discussion, appends twenty pages of figures, and leaves to the reader, in effect, the work of making sense of the report.

Therefore, pay very careful attention to the discussion of graphics in section four of the text.

1. Graphics must be used within the text of a technical report to illustrate a logical point, to summarize a long discussion of several elements, or to give a readily understandable arrangement of the classification and division of data. In a report discussion, graphics become a rhetorical tool, not exclusively a technical record of experimental results.
2. Carefully prepared graphics can have a major sales impact because the force and effectiveness of good graphics can sometimes do more than words.
3. A report writer must develop a sensitivity to an audience's needs for illustration and, in turn, develop judgment skills about when to include graphics as a rhetorical aid in the text, and when to append graphics to give technical illustration and support for points which appear in the discussion.
4. Strong transitions within the written text and adequate identification of illustrations are also very important to meet the needs of a reader. A professor can immediately identify a graph or a table of figures in a lab report; a lay reader with a million dollars to spend may not be able to make that association. Therefore, the careful technical writer allows for the needs of the audience in placing graphics, by giving transitional directions about how to find them, and by providing adequate identification.

Appendix

SECTION 1
CALIBRATION

The SR-4 was calibrated by taking readings at known values and then recording this information on the graphs of figures 13.2.4, 13.2.5, and 13.2.6. Because of the configuration of the strain gages, lift counts are registered as a result of a drag force. In order to obtain a corrected value for the lift, use the equation

Lift Counts = Lift Counts Uncorrected − Life due to Drag or

$$L = L_U - L_D$$

The value of L_D for any value of drag can be found from Figure 13.2.6.

TABLE 13.2.I

APPLIED LIFT	*LIFT COUNTS
0	31,000
1 lb	30,803
2 lb	30,608
3 lb	30,413

*A negative lift count value signifies a positive lift

TABLE 13.2.II

APPLIED DRAG	‡DRAG COUNTS	*LIFT COUNTS
0	31,000	31,000
1 lb	32,228	31,297
2 lb	33,483	31,603
3 lb	34,727	31,767

*A negative lift count value signifies a positive lift
‡A positive drag count value signifies a positive drag

TABLE III

RESULTS FOR CASCADE

TABLE 13.2.IIIA

Drag Counts	1,169
Drag (From Figure 4)	95 lbs.
Theoretical Drag	1.210 lbs.
% Error	21.5%

TABLE 13.2.IIIB

Lift Counts	−208
Lift in Drag (From Figure 6)	285
*Actual Lift Counts	−493
Lift (From Figure 5)	2.55 lbs.
Theoretical Lift	2.92 lbs.
% Error	12.7%

*Actual Lift Counts = Recorded Counts − Lift in Drag

TABLE IV

RESULTS FOR FLAT PLATE

Drag Counts	3,765
Drag (From Figure 4)	3.05 lbs.
Theoretical Drag	2.983 lbs.
% Error	2%

TABLE V

DRAG & LIFT VS. Θ

	Drag (lbs.)	Lift (lbs.)
25°	.387	1.746
30°	.553	2.066
35°	.747	2.370
40°	.966	2.655
45°	1.210	2.921

Drag = $J(1 - \cos \Theta)$ $J = 4.13$ AT
Lift = $J(\sin \Theta)$ $(P_1 - P_A) = 18''\ H_2O$

SECTION II
BERNOULLI'S EQUATION

Bernoulli's equation states that

$$P_1 + \tfrac{1}{2}P_1V_1^2 = P_2 + \tfrac{1}{2}P_2V_2^2$$

SECTION III

CONTINUITY EQUATION

The continuity equation states that

$$P_1V_1A_1 = P_2V_2A_2$$

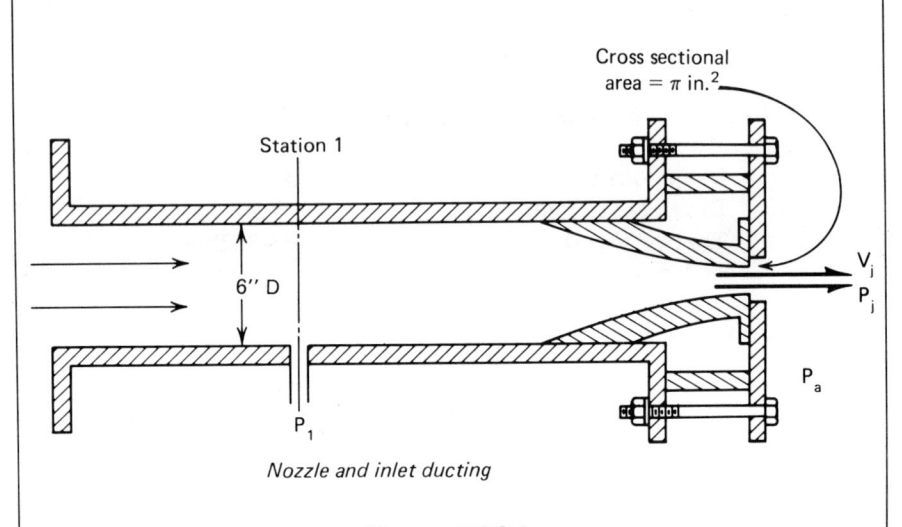

Nozzle and inlet ducting

Figure 13.2.1

Components of the Laboratory Report

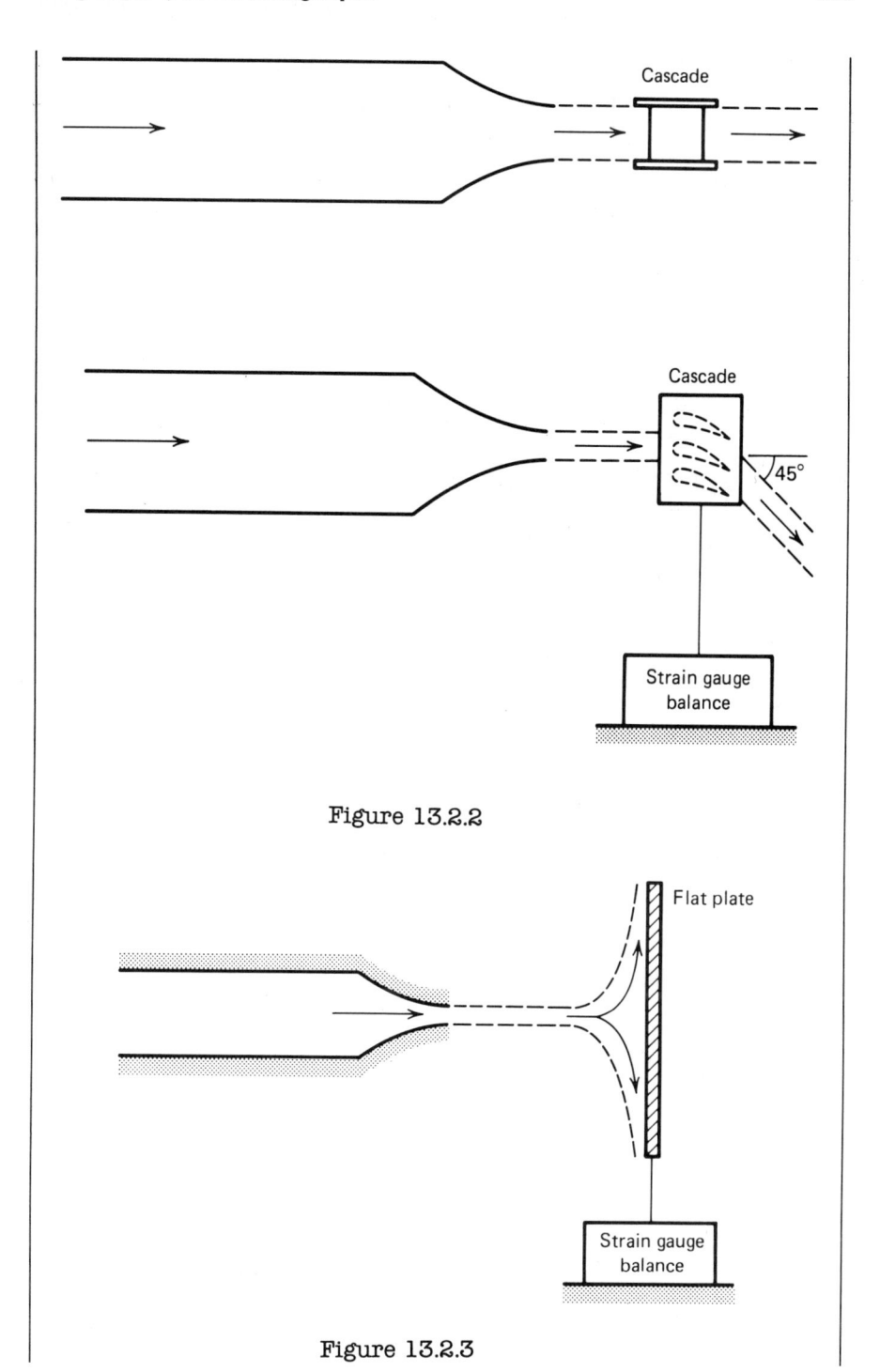

Figure 13.2.2

Figure 13.2.3

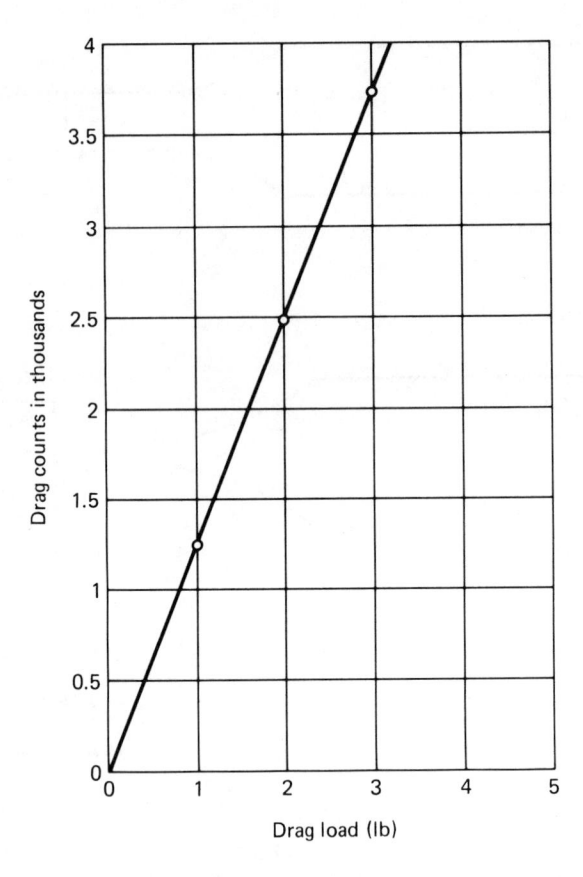

Figure 13.2.4

Components of the Laboratory Report 227

Figure 13.2.5

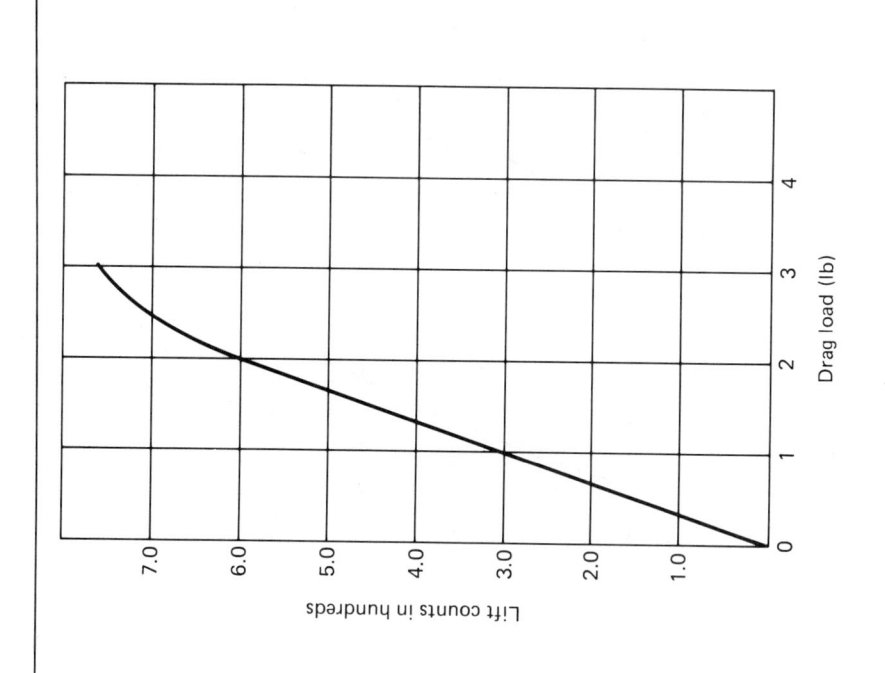

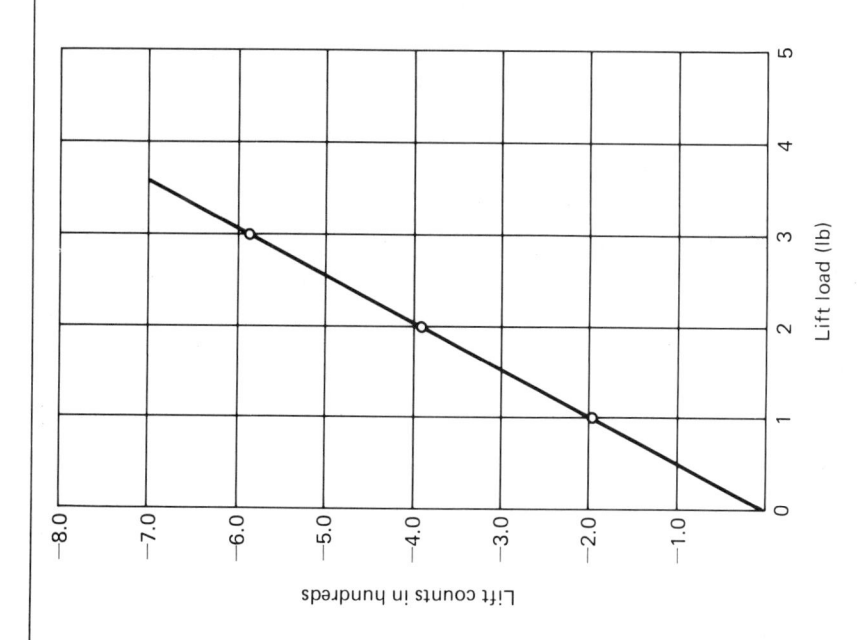

Figure 13.2.6

Figure 13.2.7

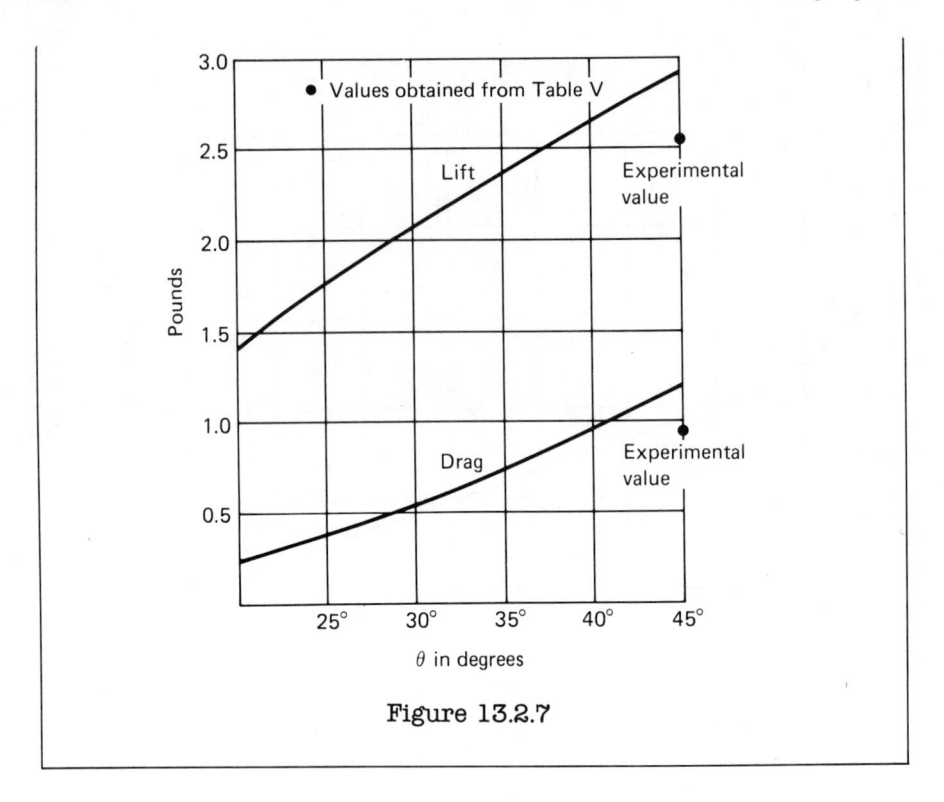

An Evaluation Checklist for Lab Reports appears in Appendix B, pages 492–493. Manuscript specifications appear in Appendix A, pages 474, 475–481.

chapter 14

Proposals

We live in a world where we are constantly bombarded with commercials. The "hot media" of television commercials, the ads on radio, the attention-getting sight of huge billboards, slick and colorful magazine layouts, the somewhat "cool media" of understated sales pitches which could pass for news stories or editorials: each of these media bids for our attention and our consumer dollars at almost every turn in an average day.

Engineers and technologists do advertising too. We have already established that, in one way or another, *every* letter is a sales letter. The primary mode of advertising for an E/T, however, is the proposal. It is through the medium of the proposal that the vast majority of projects become paying jobs for E/Ts. As an advertising medium which sells services and solutions for engineering and technological problems, the proposal is among the most demanding forms of writing an E/T does.

Let us review the forms already covered. The memorandum is somewhat cut and dried in its form and the audience is usually in-house. The letter requires a wider range of writing and management skills. The short report places even more demands upon the writer's ability to communicate. This sequence of increasing demands upon a writer's judgment and abilities is analogous to the work of a juggler: the performer continually moves from relatively simple tricks to increasingly more complex tricks in increments of difficulty.

The proposal extends the sequence, for there are several elements, some of them new to our discussion, which the proposal-writer must juggle:

1. *audience:* The proposal-writer relates to the reading audience in a direct sales transaction. It is in this document that the E/T makes a direct bid for the reader's business. For that reason, the sales/PR element is not as subtly underlying a factor as it would be in a letter. Often, too, a prospective client who publishes a "Request for Proposals" (RFP) will set up strict specifications about what to include.

2. *subject:* There are two principal difficulties with subject: a) descriptions and explanations must be as technically accurate as they would be in a technical report, but the *purpose* in writing them is sales, not technical reporting; b) the proposal describes what will, can, and may be done, not what has already happened. Thus the E/T is using experience and judgment to explain technical solutions for the future.

3. *sales:* Taken together, the audience and subject of a proposal bind the E/T proposal writer within very strict parameters. A professional selling technological services simply does not have the range of marketing capabilities that a vacuum cleaner salesman or an automobile salesperson has. Nowhere is "argument through exposition"—or depending upon clear technical explanations to carry the persuasive force—more important than in a proposal.

4. *production:* A major way of investing resources within the design parameters for a proposal is to make the document itself as attractive as possible. Huge sums of money may be poured into graphics for the cover, design for the individual pages, and very fine paper. None of these, however, can substitute for logical organization and clear exposition.

A special note to student writers: Perhaps the most difficult feature to understand about the proposal is this: a proposal *projects what will happen in the future*, whereas students are more accustomed to *reporting what has already occurred*. Most student writing is done for the purpose of saying, "This is the work I have completed; what is my grade to indicate how well I have done the work?"

The proposal perhaps best represents what any professional must learn to do: project out of a combination of experience, judgment, and study what she can do in the future. That marks a critical shift in your intellectual development, and you must accomplish that change of mind-set to fulfill any class assignment on proposal writing.

For a professional, the "grade" in a real sense, comes when the client contracts for his services, possibly before any actual work has begun beyond writing the proposal. Keep in mind this shift of focus when you write any proposal.

The checklist for evaluating proposals is in Appendix B.

Proposals   231

Summary Example 14.0

PROPOSAL:

SUBJECT: (All necessary title information to identify and file)

JOB NUMBER: (if applicable)

> Format: Depending upon the length of the proposal and upon whether it is likely to be duplicated for distribution to a multiple audience:
> 1. present it as a separate proposal with a cover letter if it will be more than 3 pages; or
> 2. present it as part of a letter (It should form the body portion of a letter which contains all other features of the letter of transmittal in the opening and closing sections.)

1. PROJECT INFORMATION: Be aware from the outset of writing a proposal that, in constrast to writing a report, you are more closely in communication with the client as you negotiate for the job. For that reason, this first section on project information relies heavily upon the client's description of the job and your understanding of the job as the client describes it. If your proposal does not accord exactly with the client's idea of what he needs, presumably there is the possibility of further negotiation. In a report, the burden of proof for this paragraph rests upon the engineer-report writer. In the proposal, it rests at least partially with the client as well. Make sure to establish an understanding here.

2. PROPOSAL: First be specific in selecting a heading: SCOPE OF SERVICES, ENGINEERING SERVICES, PROPOSED INVESTIGATION or the like. A report of work that has actually been done can contain a high degree of accuracy and exactness. No matter how firmly based upon experience with similar situations, a proposal nevertheless remains a projection of what can happen in the future. In a report, the work more nearly sells itself than in a proposal, where the dynamics of communication are more strongly bound up with the client. This diagram illustrates the difference:

   REPORT: work ⟶ writer ⟶ client
   PROPOSAL: writer ⟶ client ⟶ work

In the first, the writer is a medium; in the other the writer is more nearly a causative factor.

Job Number

> What the proposal-writer does: The essential task of the proposal-writer is to define a problem and propose a solution. Thus, like a report, a proposal is grounded in engineering. The rhetorical task is to describe the process which is necessary to solve the problem or satisfy the need expressed by the client. The proposal writer should rely heavily upon explanation of the principles involved in the proposed process; not only how a proposed test works, but what it is designed to accomplish.
>
> How the proposal writer involves the client: Poets, novelists, and English teachers may fancy that only they have imagination. Audience analysis in engineering writing also calls for imagination of a very high order. That is, it requires the ability to project what is necessary to solve a given problem, to predict the outcome of mechanical and physical processes, and to appeal to the reader in such a way that she will buy the engineer's services. The work of the proposal-writer, especially, goes beyond dealing only with the engineering problem to these further considerations:
>
> 1. the conditions of the engineering work: Will it be necessary to gain access for your equipment through adjacent property? What other conditions are likely to occur which must be accounted for in the initial proposal?
>
> 2. the cooperation of the client: What conditions will require it? To make your needs clear without seeming to make undue demands requires diplomacy on the part of the proposal writer.
>
> 3. the concerns of the client: What about damage to man-made objects; what about the condition of the site after the field investigations? Predicting such worries can require more imagination than is required of a novelist who creates a setting and characters out of thin air.
>
> All these considerations require of the writer a general awareness of the implications of the actual engineering work for the client.
>
> Qualities the printed proposal should have:
>
> 1. Logical Development: A report describes things which exist; a proposal projects what can exist in the future. Since it lacks an objective referent, a proposal must depend heavily upon the logic of its development: careful division of a process into its steps, or careful description of the characteristics of a site, or careful explanation of what a test is

Job Number

designed to do—in other words, careful analysis of any applicable elements.

The report writer describes how logical order has been imposed on a project; the proposal writer wins credibility by perceiving order in what may appear to be chaotic to the client. At the proposal stage of a project, the only order may exist in the definition and solution of the problem as it is described and explained in the proposal.

2. <u>Clear Visual Design</u>: Like a report, a proposal should be divided into parts which are clearly identified on the page. Imagine a committee with copies of the proposal referring to and discussing various features of the proposal. Imagine a phone call about "subsection 1 under section 2" rather than "on the third page about 10 lines down in that paragraph about subsurface investigation."

3. <u>Ease of Physical Handling</u>: A proposal looks like a book with paragraph-after-paragraph and page-after-page of type, but it is not read like a book. Always assume that an engineering document is a working document, that it may be passed through many hands, read in parts, re-read, read during constant interruptions, and that two people may stand (even ankle-deep in mud) looking at the written page. Here again, imagination comes into play.

3. <u>FEES</u>: Of course a client is interested in how much any proposed work will cost. Any statement of estimates should be accurate; at the same time, however, the writer should make clear that an estimate may vary under actual working conditions. Cite experience ("Based on our experience with similar projects, we estimate..."). If you use computers to project time and costs, by all means use that as a selling point to distinguish your estimate from a "guesstimate." If you have a good record for avoiding cost over-runs, say so.

In my opinion, it is better to enclose a schedule of fees to support any necessary discussion of estimates. This gives the reader a more secure sense of standardized fees charged by the company on a given job. If the schedule of fees is included in the body of the proposal, identify them as standard to serve as a constant and operate with the variable of the estimate.

4. <u>SCHEDULE</u>: Give some idea of how long a job of this type usually takes, or at least discuss any factors which are likely to influence the scheduling of the job and the time required to complete it.

Job Number

> 5. PERSONNEL: The general experience and capabilities of the staff are appropriate for mention, as is an indication of how many people will be required to do such a job. In having a proposal accepted, much depends upon the professional credibility of the firm itself; citing individuals serves to reinforce that credibility as a selling point.

GENERAL QUALITIES OF A PROPOSAL

The subject of a report and a proposal may be the same, and the substance may be highly similar. The chief difference lies in the nature and purpose of a proposal:

1. *Professionally Credible:* As a bid for professional services, a proposal is necessarily a sales document. Any sales appeal, however, must be subtle; a proposal is, after all, not a pitch for a hair tonic, but marketing of complex and expensive expertise.

2. *Technically Sound:* Ultimately the quality of the engineering work sells the proposal. The technical substance does not, however—and, especially with a lay audience, cannot—stand by itself. A proposal which is clearly and logically written, visually easy to see and refer to, and physically easy to handle forms a strong foundation for the technical matter itself.

3. *At Least Partially Theoretical:* The immediate task is to win a specific contract. The larger task is to interpret your engineering discipline as a problem-solving activity, and to market your work.

An Evaluation Checklist for Proposals appears in Appendix B on page 494. Manuscript specifications appear in Appendix A, pages 474, 475–481.

General Qualities of a Proposal

Proposal: Example 14.1

TITLE: General Outline for Project

SUBJECT: (All necessary information to identify and file)

JOB NUMBER:

1. PROJECT INFORMATION: This is based upon your understanding of the client's description and explanation of the project. Gain a clear understanding of what the project entails and express that clearly in this paragraph to form the basis for the entire proposal—and, in turn, for the entire project.

2. SITE CONDITIONS: Like the paragraph on project information, this description of the site lays a foundation for the proposal and the project. Taken together, these two sections have three functions:

    2.1. to render your understanding of the client's description and explanation of the project.

    2.2. to communicate to the client your grasp of the basic problem at hand.

    2.3. to serve as a record of the basis on which the project began and from which the work progressed.

3. PROPOSED CONSTRUCTION: Divide the project into its logical parts and relate your services to each part.

    3.1. Phase 1: Explain your understanding of the client's description of what will be done during Phase 1.

    3.2. Proposed Services: Explain how your work relates to this phase. Further sub-divide your services:

        3.2.1. Testing

        3.2.2. Construction Evaluation

    3.3 Phase 2

        3.3.1. Testing

        3.3.2. Type of Work Proposed

    In the substance of the actual proposal of work on a large project, it is useful to divide your understanding of the project as it stands from your proposed future work. This method distributes the burden of proof between the client's description on the one hand and your proposal on the other. Such a division may also allow

Job Number

leeway to negotiate a part without having the client reject the entire proposal.

4. *ESTIMATED COST:*

5.
   .
   .
   . (As deemed appropriate, add sections SCHEDULE, PERSONNEL, AUTHORIZATION.)

Proposal: Example 14.2

TITLE: General Outline to Describe Services

SUBJECT: (All necessary information to identify and file)

JOB NUMBER:

1. PROJECT INFORMATION: This is based upon your understanding of the client's description and explanation of the project. Gain credibility by rendering a clear understanding of what the project entails.

2. CAPABILITIES OF YOUR FIRM: Don't write a long, dense paragraph; instead make sure that the various functions are set off and highlighted on the page.

    2.1. Testing

    2.2. Construction Evaluation

   Rather than needing to re-invent the wheel each time you have to describe these capabilities, it is well to have a paragraph bank so that you can save time by plugging in the needed exposition. Such a resource is especially easy to have in our new age of word-processing equipment.

3. SCOPE OF SERVICES: Match the order of parts in section 3 to those in section 2. Concentrate the work of writing, not upon repeating the matter usually contained in section 2, but on explaining how each of those services can be specifically adapted to the proposed project. While you may be repeating what you know only too well, this is needed reinforcement for the reader.

    3.1. Testing: Relate the established capability of your firm to this particular problem.

    3.2. Construction Evaluation

General Qualities of a Proposal 237

> 4. ESTIMATED COST
> 5. SCHEDULE
> 6. PERSONNEL
> 7. AUTHORIZATION: Explain what the authorization commits the client to, how to prepare it, and how to return it.

A proposal is a form of selling, selling is a form of persuasion, but the marketing of professional services should be both subtle and tasteful. "Argument by exposition" is the traditional form of argument/persuasion which works this way: its basic aim is to teach, explain, and interpret. The assumption underlying this technique is that if the reader knows, he will then believe, and, in the case of a proposal, buy. For a competent professional with an established reputation, this is a valid approach. The strength of the proposal lies in the clarity of the technical explanation; this strength is in itself persuasive, and should not need "hype" to support it.

Letter of Transmittal: Proposal to a Large Technical Audience

> Inside Address and Salutation: Wherever possible address this is to an individual; then type a "direct line of communication" without the "Attention" buffer to a direct address:
> Name
> Address
> Dear (Name):
>
> P.1: The first paragraph is always the subject paragraph, addressed to the need of a busy reader to know what your business with him is. My term for this in a cover letter for a proposal is "grabber paragraph." It at least should engage the reader's attention:
>
> - Stuart and Lee is pleased to submit this proposal for (cite the kind of work, for whom, where).
> - In response to your call for proposals, Stuart and Lee is pleased to submit a description of our capabilities in (e.g., construction engineering) as they relate to your plans for (cite the kind of job, place).
>
> P.2: This paragraph directs the reader's attention to the enclosed proposal. Hence my term for it: "reader's guide paragraph," in

that it draws the reader into the reading of the proposal and gives the desired emphasis to his reading:

- A feature of the proposal which (should be of particular interest to you/is particularly addressed to the _____ problem/is specifically designed to address your concern for _____) is contained in Section 2. (describe briefly)
- The (cite the kind of engineering work) which you plan accords with (our experience/our record of success) in similar (jobs/projects)/in (addressing/solving) problems of this type/in working under the conditions you describe.

  Because such emphasis is often inappropriate in the engineering writing of the proposal itself, here is where you can highlight an important feature of your proposal and give it added sales value.

P.3: Acknowledgements serve here to build good will:
- We would like to thank John Pelham of your staff for his help in supplying the topographic plan and boundary survey maps needed for our work.
- We would like to acknowledge the cooperation of (name/group) at Hood Engineering for (cite their contribution).

P.4: The first sentence should cite any enclosed documents (authorization, general conditions), how they relate to the proposal, and how to return them if necessary.

P.5: This need be no more than a sentence, but make it count:
- We appreciate the opportunity to submit this proposal and look forward to (working with you/receiving your reply).

Letter of Transmittal: Example 14.3

May 1, 1990

Mr. Avery Jones
Energy Corporation, Inc.
P.O. Box 8310
Orlando, Florida 32160

Subject: Feasibility Study of a Solar-Powered Air Conditioner

Dear Mr. Jones:

Com Sim, Inc., is pleased to submit this proposal for a computer feasibility study of your solar-powered air conditioning system.

Because of our recent assistance to the Monrovia Company, we were able to purchase the rights to their computer simulation system FLOTRAK. Thus we are the only company allowed to offer this simulation system to companies that wish to have proposed processes simulated.

Section 3 of the proposal shows how important this system is to your study. It will explain how we can give you enough information to design, construct, and cost-analyze your solar-powered air conditioner.

We would like to thank your engineer Ms. Belle Boyd for her help in analyzing the exact system to be simulated.

With this proposal we hope to establish a productive working relationship with Energy Corporation, Inc.

Sincerely,

COM SIM, INC.

Georgette P. Burdell, P.E.
Vice President

Proposal: Exercise 14.1

TITLE:	Process Simulation
SUBJECT:	Feasibility Study of a Solar-Powered Air Conditioner
JOB NUMBER:	1001-1

1. PROJECT INFORMATION: The rising cost of electricity has caused air-conditioning costs to rise sharply. In the conventional A/C, the mechanical compressor uses the largest portion of the electricity. However, a new type of air-conditioning has been proposed. In the new system the mechanical compressor is replaced by a pump, two thermal compressors, and solar-heated water. A computer simulation will provide the necessary information to evaluate the proposed system. From this evaluation a decision can be made either to build a working model of the system or to report that the system is not feasible. The simulation will take place on one of our computers and all results will be checked for accuracy.

2. CAPABILITIES OF COM SIM, INC. Com Sim, Inc., maintains a large variety of computer simulation routines. The one best suited to your needs is FLOTRAK. FLOTRAK is a computer simulation program developed by the Monrovia Company to provide the ability to evaluate a proposed chemical or thermodynamic process before a company commits large sums of money to the design and construction of a scale model.

    2.1. Operating State: With only minimal input information, FLOTRAK employs standard short-cut calculation methods to provide all operating conditions.

    2.2. Design Specifications: FLOTRAK provides all necessary information to design a working model.

Job Number

    2.3. <u>Accuracy</u>: FLOTRAK has proven to be 95% accurate in previous simulations.

3. SCOPE OF SERVICES: Using FLOTRACK, Com Sim, Inc., will provide all necessary information for the design and operation of the solar-powered air-conditioning system.

    3.1. <u>Operating State:</u> Few variables are needed for the FLOTRAK simulation:

    the amount of refrigeration required

    the temperature of the solar-heated water

    the highest expected outside air temperature

Using this information FLOTRAK will calculate the optimum operating conditions. This will include all flow rates, temperatures, and pressures.

    3.2. <u>Design Specifications</u>: The simulation will provide the heat duty of all heat exchangers, the pump horsepower requirements, and the quantity of solar-heated water needed. The design of the thermal compressors can be accomplished with the flow rates, temperatures, and pressures that are obtained from the FLOTRAK output.

    3.3. <u>Accuracy</u>: All necessary routines except the one for the thermal compressors have been used in prior simulations. Should the modified routine for the thermal compressors fail to meet the criterion of less than 5% error, either an existing routine will be modified or a new routine will be written to simulate the thermal compressors. After the simulation is performed, the critical points of operation will be checked to assure that the simulation is correct.

4. ESTIMATED COST: The total cost of services should be $7000. This estimate is based on the expected time required to develop

Job Number

the simulation routine, computer time, and the time required to verify the accuracy. If an established routine can be used to simulate the thermal compressors, the cost will be sharply reduced.

5. SCHEDULE: The work should be completed 10 days after commencement. This will involve 3 days to prepare the simulation, 2 days to run the simulation, and 5 days to verify the accuracy.

6. PERSONNEL: The entire operation requires two engineers. One engineer prepares and runs the simulation and the other verifies the accuracy of the results.

7. AUTHORIZATION: Work can begin as soon as we verify your authorization. Should you decide to cancel the work, there will be no obligation on your part provided cancellation is received before work is initiated; however, if cancellation is received after work has begun, you will forfeit half of the estimated cost, and all results will become the property of Com Sim, Inc.

Proposal: Exercise 14.2

TITLE: Nurse-Staffing Study

SUBJECT: Development of a Nurse-Staffing Model for the Emergency Department of St. Helena's Hospital

1. PROJECT INFORMATION: This proposal addresses the problem of developing a nurse-staffing model for the Emergency Department of St. Helena's Hospital. Two objectives will be met through this proposal. It will evaluate the adequacy of nurse staffing in the Emergency Department. In addition, the data obtained in the examination of nurse staffing can be formulated into a workload model capable of predicting future staff planning and evaluation.

   In order to perform a more thorough job, Health, Inc., will also attempt to meet two sub-objectives. The present clerical staffing in the Emergency Department will be examined to see how much time the nurses spend in doing clerical activities. Our staff of engineers will also examine reasons for the delays associated with processing a patient through the ER.

2. WORK CONDITIONS: There will be no disruption of the work patterns of the St. Helena's staff. In fact, it is our hope that conditions will be as close to normal as possible to make a truly accurate staffing analysis. An observer will visit the ER at random times to conduct a work-sampling study.

3. PROPOSED STAFFING MODEL: The goals of developing a nurse-staffing model in the ER can be met through the analysis of staff utilization, patient delays, and historical data. Finally, the data gathered from these three phases of activities can be summarized and developed into a nurse-staffing model.

Job Number

3.1 Phase 1: Phase 1 will consist of an analysis of staff utilization. Work sampling has proven to be one of the most reliable methods for measuring work, especially non-repetitive hospital activities. Work sampling is the estimation of the proportion of time devoted to a given type of activity over a certain period of time by means of intermittent, randomly spaced observations. Specifically, our team of professionally trained management consultants will measure the proportion of time each employee is productively utilized.

3.2 Proposed Services: The first step in the design of the work-sampling study will be to specify the categories of activities to be observed. This involves more than simply a productive/nonproductive dichotomy. Categories such as patient assistance time, direct semi-professional patient care, and indirect professional patient care are utilized. These categories are based on those used in similar studies and can be modified to reflect the particular needs of St. Helena's.

    3.2.1. Testing: Design of the sampling procedure involves estimation of the sample size and selection of the period of time over which the study is to be conducted. It also includes specification of the schedule of observations and design of the observation sheet.

    Determination of the sample size will represent a trade-off between cost (time) and statistical accuracy. In order to determine an optimal time limit, our consultants will base their decision on a statistical formula which relates sample size, confidence interval (the value in which the staff utilization lies with some

General Qualities of a Proposal 245

Job Number

    given probability), and confidence level (the probability desired by the experimenter).

   3.2.2. Schedule: A period of two months has been found to work out well to generate a representative sample of shifts. The schedule of observations can be determined using random numbers. This should be done so that observer-chosen times will not bias the results.

3.3 Phase 2: The second phase of the proposal is an analysis of patient delays. However, prior to analyzing the delays within the department, it will be necessary to gain an understanding of the various patient flows within the department.

3.4 Proposed Services: A flow chart will be constructed which includes all possible patient flows connected with an emergency. In addition, subsystems which are perceived to have significant delays will be flow-charted in detail.

3.5 Phase 3: In order to gain a perspective on the present situation in the ER, it will be necessary to take a historical look at the utilization data. Patient-volume data from the past year will be used; then patient census by day and by shift will be graphed in order to characterize variations in patient load and to identify developing trends.

3.6 Phase 4: The final phase of the proposal is the development of the nurse-staffing model. This phase of the project will express the relationship among work-sampling results and will form the basis of a longer-term staff-utilization model. This is based on the assumption that staff utilization is dependent upon workload.

Job Number

3.7 <u>Proposed Services</u>: Linear regression models can express this relationship between workload and staff utilization. Linear regression is especially useful as a statistical technique which expresses a dependent variable as a linear function of an independent variable. Since staffing is planned by shift, a separate regression model will be developed for each shift of the day. One of the benefits of linear regression models is that not only can they be used to analyze the present utilization-workload relationship, but they can also be used to help develop future forecasting trends.

4. <u>ESTIMATED COST</u>: Based on previous studies on hospitals of St. Helena's size, the estimated cost of Health, Inc.'s work will total approximately $26,000. This also includes follow-up data, as well as computerized printouts of all the pertinent data collected.

5. <u>SCHEDULE</u>: The length of time this project will require depends on the sample size. In order to complete the study with an optimal cost/benefit ratio, we recommend a two-month study.

6. <u>PERSONNEL</u>: To do the best job possible, Health, Inc., will require a temporary office for a staff of four management consultants as well as facilities to set up two computer terminals. If we can do the majority of our work on-site, it will expedite our job and result in greater cost savings to St. Helena's.

7. <u>AUTHORIZATION</u>: The signing and returning of this proposal will indicate to Health, Inc., that we may begin work. If there is any part of this proposal which needs further discussion, feel free to contact Health, Inc. We will be happy to clarify any questions you may have.

Proposal: Exercise 14.3

SUBJECT: General Design of an Electronic Digital Sports Scoreboard

1. PROJECT INFORMATION: Electronics Corp. is searching for a replacement for the existing electro-mechanical scoreboard which they now produce. Because this design is no longer attractive, they want an integrated digital system. The proposed system should respond to an operator's touch within one second and should contain features to prevent mistakes due to human error. It should be completely weatherproof and lightning-protected. The design should adapt to different sports—including football, baseball, and basketball—with only minor modifications in the electronic circuitry.

2. PROPOSAL: The rapid development of integrated circuits during recent years is drastically driving down the price of electronic equipment. The microprocessor is now the major tool for correlating input equipment (keyboard) and output equipment (scoreboard). Stuart and Lee Engineering has the necessary personnel and up-to-date expertise to design microprocessors and other digital circuitry; we believe that we can design an efficient system at a relatively low production cost by using the following major components:

   2.1. Keyboard: The keyboard, the only source of input to the system, will connect directly to the microprocessor.

   2.2. Microprocessor: The microprocessor will digest the input by using a preset program stored in permanent memory. It will then output the proper responses to an amplifier for transmission to the scoreboard. Both the microprocessor and the keyboard are to be placed inside the game official's booth.

Job Number

2.3. Amplifier: The output signal of the microprocessor will not be large enough to travel the long distance between the booth and the scoreboard; therefore, an amplifier will be added to boost the signal.

2.4. Decoder: This component will decode the signal at the scoreboard and distribute it to the appropriate lights.

2.5. Scoreboard Lights: The scoreboard lights will be either the incandescent lamps currently used by Electronics Corp., or large LED's like those used in calculators.

3. CAPABILITIES OF STUART AND LEE: Below is a list of the facilities Stuart and Lee has at its disposal.

3.1. Readily Available Outside Resources: Stuart and Lee employs 34 electrical engineers who keep abreast with current developments in digital systems. The firm also has access to a Control Data corporation Cyber 74 computer capable of microprocessor simulations.

3.2. Development and Testing: As part of the project development, a test system will be constructed, using one of the microprocessors at Stuart and Lee, and a scoreboard and keyboard built by the staff. This model will be available for demonstration to representatives of Electronics Corp. approximately 8 weeks after project initiation.

4. ESTIMATED SYSTEM COST: The components, including the microprocessor board, amplifier, scoreboard, keyboard, lamps, power supplies, and various wiring and P.C. boards, are expected to cost approximately $252 per scoreboard. This cost does not

Job Number
include development, production, and packaging costs for the system. It also does not include fluctuations in price due to economic pressures.

5. SCHEDULE: A timetable for each stage of the project is shown below.

   5.1. Development: (Includes design and construction of prototype) 6 weeks.
   5.2. Testing: 2 weeks.
   5.3. Demonstration: 1 day.
   5.4. Changes: 1 week.
   5.5. Report: 1 week.
   5.6. Total Estimated Time: 10 weeks.

6. FEES: A fee of $5,546 will cover the development of the initial model. Further development will cost $90 per day in addition to any material costs incurred during this stage.

7. PERSONNEL: Two engineers will be assigned to Electronics Corp.'s project: one director and one assistant. The director, J. B. Stuart, has 10 years' experience with digital systems. Mr. Stuart, and the team which will assist in this project, developed the Digital Instrumentation for Radar Tracking, a project for the U.S. Air Force. The assistant engineer, C. J. Polk, has worked with Mr. Stuart for 3 years. The technician, David Marsh, has worked with the team for 2 years.

8. AUTHORIZATION: If Electronics Corp. chooses to accept this proposal, a letter of authorization will be necessary to initiate contract proceedings. This letter should include a tentative agree-

Job Number

ment to pay the costs described above, with an initial payment of 25% of the total estimated cost to be paid before the project is begun, and the balance to be paid upon completion of the project. The letter should be addressed to:

>Stuart and Lee
>165 Battery Avenue
>Charleston, West Virginia 18603

chapter 15

Technical Recommendations Reports

One of the most familiar and useful forms of technical writing to the engineer, the technician, the manager, and many other practicing professionals is the technical recommendations report. This document underlies most major projects in the business, industrial, and governmental world. To the general reader, this form of writing is probably most familiar from news reports such as those which mentioned "The Sibley Report" and "The Pullen Report"; both of those reports commented at length upon the state of education in Georgia and in Atlanta and made recommendations for change. A citizens committee or a "blue-ribbon panel" or a team of visiting experts will frequently make a study of an organization or a governmental body and publish the results of the study as a recommendations report. Colleges will at times call in an evaluation team to study all or a segment of the institution; the team will then produce a recommendations report. Those examples from education establish the form as a useful, and indeed indispensable, document for leading the way toward change.

The recommendations report is indispensable to the practicing technological professional as well. A road is planned through a residential area; a recommendations report explains where, how, and even whether. Energy-use reduction is suggested for a large office complex; a recommendations report tells where, how, when, and at what projected cost saving. A fire has destroyed a hotel and before renovation the owners need to know whether the existing stone arches can be used in the reconstruction; a recommendations report answers their question. A cotton mill is discharging wastes into a nearby river and is under federal order to deal more safety with chemical wastes; a technical recommendations report tells how and at what cost. A large hospital needs to cut costs for nursing care; a recommendations report tells how to accomplish that. Management in any kind of organization may sense and understand a need for change; to determine how, why, and whether changes must be made, they most often call for a recommendations report.

Those introductory paragraphs describe the *purpose* and *audience* of the technical recommendations report. The instructions which follow on how to write a technical recommendations report are divided into two parts. This division is necessary because a report can be a massive management project as well as a writing project; the production of the report, namely the planning, the physical writing, and the assembling of the final product, is a project of considerable scope.

The first section of the instructions is a kind of formula to guide a team of writers or a single writer through all the necessary steps in assembling the data for a report. Since so many texts stress the use of the library, it is necessary to point out here that these instructions do not assume a library of printed sources as the primary data source for most technical reports. As the chart on sources in section four shows, secondary sources necessarily contain valuable aids in establishing the technological methods which a report writer uses; however, the purpose in report writing is not to advance the state of the art through linking the study to the research which came before it. In contrast to a history or to a research article, the technical report is focused outward, first upon the specific needs of an identified audience, and second upon practical workable solutions to some technical problem. Technical education, an enormous range of technical resources, and the body of knowledge in technology underlie the report, but they are not visible to the reader through extensive documentation—just as those who enter a building are not aware of the vast substructure which supports the floor of an inviting lobby and the first-floor offices stretching beyond.

The second section of the instructions treats each part of the report in turn, as it should appear in the finished draft, and gives specific instructions about how to assemble the report. These instructions should be especially helpful because a work of such scope is not written at one sitting, nor is it written in serial order. Many a report writer stalls, bites his pen, and tears his hair, because he begins by writing the Executive Summary on the grounds that it appears first in the finished document. The instructions in both parts of the guide to writing give a strict order to lead writers to a better designed management system for writing.

The instructions by-pass the invention step in writing and deal primarily with the arrangement steps of assembling an existing body of data. In recognition of the special needs of students who must write a report as an academic exercise for a technical writing class, a section for students is contained in section four. This guide should lead students to ideas about how to generate data for the report.

Aside from these introductory notes about the technical recommendations report, the most significant skill required is not writing skill, but the skill and the discipline to manage time and resources and thus to manage the entire process of producing a manuscript to specification on or before the assigned deadline. Hence this assignment will help you build one of the job skills most significant for any working engineer, technologist, or other professional—management of your time and personal resources.

GUIDE FOR WRITING THE FIRST DRAFT OF A TECHNICAL RECOMMENDATIONS REPORT

Writing a technical recommendations report can be a major long-term project. Therefore, a project of such scope requires breaking the process into its components and giving directions on how to perform each part of the task.

The section which follows is designed to lead a writer (or a team of writers) through the report-writing process which occurs before the production of the final draft. Ideally, you should set up a system in a looseleaf notebook somewhat like an engineer's field book. Then you should write a part of the report as you complete each stage of the work. The work of writing can then parallel and interact with the technical labor at every point of the process.

Since the majority of long reports are written by teams of E/Ts, this method of writing can be particularly useful. A meeting of team members to determine an overall outline of the report and to assign writing tasks can be like the design conference which occurs in the planning stages of every technical project. With an outline established and with the writing tasks divided among the team members, the writers can then use the "field book" to contain a working draft of the report as it takes shape and before a designated E/T, a project manager, or a professional technical editor or writer performs the writing and editing of the final draft for production.

Who

Reports begin and end with WHO. Someone has a problem; someone reads your recommendations report; someone decides whether or not to hire your firm to solve the problem. For those reasons, your first job in report writing is to know "whom you're talking to." Determine your levels of communication. Before you begin to write, answer these questions:

1. Who will read the report in order to make the final decision to spend money for the project? This reader will most probably be a manager, not an engineer. Even if an engineer, the reader may not be fully informed about the technical implications of the problem you have been hired to deal with.
2. Who will read the report in order to do the actual work required by the project? This person will probably be an engineer or a technician. The technical details she needs to know are different from the strategic overview the manager needs.
3. Will anyone else read the report? For instance, nontechnical people may have a part in deciding whether to adopt the recommendations in the report: the board of directors for a hospital or a church group or a panel of legislators, perhaps.
4. Will a specific individual read the report? Do you know this person well? Are there certain emphases in the report which should be made to reach this particular reader? View the report as a sales document: is there a friend of yours within the audience who will personally influence the adoption of your recommendations?

Structuring the Discussion

Putting together the major section of the report (the *body* or the *discussion*) is mainly a matter of arranging the parts you have written during the solving of the practical problem itself. Write the answers to WHAT, WHERE, HOW, WHY and HOW MUCH as you go; then put them together to form the report.

1. WHAT: After you have decided what the problem is, use your statement of the problem as the first section of the discussion segment of the report. It can be headed *"Statement of the Problem."* This is the place, too, to define any overall terms, such as "intermediate-care facility" or "maximum security," according to audience need.
2. WHERE: This is a part of your general description: where the problem exists. This forms a necessary context for describing the problem.
3. HOW/WHY: The answer to these questions represents the major body of technical data and your interpretation of it. If time estimates are made, *how long* also belongs here.
4. HOW MUCH: These are the figures on what the job will cost.

The aim here is to divide and conquer. The writing of the report should not be a tiresome task to be done once you have completed the project. The writing, if it is divided into parts, can fully coincide with the working out of the practical problem itself. Writing, then, is not a separate task, but a part of the technical work.

WHAT (stating the problem)

Any researcher in any field begins by *stating the problem*. No solution is possible until you first know what the problem is.

1. What is the problem *as your client described it?*
2. What questions do you need to answer to derive your own working definition of the problem? List these questions before you make an on-site investigation. List any research you need to do in printed sources. What other data do you need and what support personnel?
3. State the problem to be solved. Your rhetorical problem in doing this will usually be cause-and-effect analysis. Select the rhetorical pattern which applies:
   - single cause—single effect
   - single cause—multiple effects
   - multiple causes—single effect
   - multiple causes—multiple effects

   (list multiple causes or multiple effects in order of importance)

WHAT (defining for the reader)

Allowing for levels of readership (management, engineering, technical, nontechnical), be sure that you define terms in order to make the report accessible to all levels.

Define in these ways:

1. By using parentheses:

   "The Department of Offender Rehabilitation (DOR)"
   Thereafter save time by using DOR
   "5000 BTUs (British Therman Units)"

2. By using parenthetical expressions:

   "5000 BTUs, the standard unit of measurement for heat, are ..."

Determine what your audience's needs for definition are. If brief parenthetical definitions will do, use them. Remember, too, that even experts in the same field may need a brief re-statement of a definition to clarify the sense in which the term is being used in the report and to ensure that the reader is reading the term in the sense intended. This technique of definition creates concentric circles to include the technical colleague, the informed reader, and the person new to the subject. Your task is to determine how far to go; you cannot sit within the narrow circle of your specialty pouting, "If you were an engineer, you'd understand."

If you have a nontechnical audience, consider a glossary of terms to be included at the end of the report. Remember, too, that figurative language can be appropriate in a report. Example: "The artificial knee prosthesis compares to a natural knee as the two hulls of an out-rigger compare to the single hull of common boats."

WHERE

How long and detailed the answer to the question WHERE will be depends on your audience. If this is a local church group, you need only describe the site and the necessary features of the site to provide a context for the work you will do. If this is a large company, make sure that management in one city has a clear idea of where the work will be done in another city.

1. Describe the site. Like a camera that pans slowly over a space, do this in an orderly way: left to right, top to bottom, and so on. Remember that the seemingly simple task of describing a space can be difficult. Give particular care to sentences; avoid long awkward constructions.
2. Focus on the part(s) of the site that will be altered by the work you propose to do or the problem you are identifying. Enable the reader to see through your expertly trained eyes.

HOW

Once you have stated the problem, HOW details the steps in the solution of the problem. This usually forms the body of the report. This is where the management of large amounts of material becomes a problem. Here are several ways of managing the material.

1. If you have conducted a series of tests which each lead to a set of conclusions and recommendations, consider this pattern:
   - I. A. Test I (Your subhead will be more descriptive.)
      B. Results
   - II. A. Test II
      B. Results
2. If you are presenting a single process as a solution to the problem, consider this pattern.
   - I. Step I (Your subhead would name the step)

Guide for Writing the First Draft

II. Step II (Again, name the step)

Describe what is involved in each step: the work that will be necessary, and the principle involved (e.g., time saved, money saved, motion saved).

3. If you are presenting alternative processes, simply divide them and treat them as single processes as described in 2.

Remember that a report is an argument in which winning means winning a contract; remember that how you arrange the parts within the body is important. What is your strategy?

1. Do you want your best solution to appear first, so that those that follow are obviously inferior?
2. Do you want to save the best solution for last to point out its obvious superiority?

In making these decisions, think of how the report will strike the reader(s) who will determine whether you win or lose the argument/contract.

WHY (in the body of the report)

WHY cannot be entirely separated from HOW in the body of the report. As you work out *how* a problem is solved, *why* usually emerges as a part of the process of figuring out *how*. One of the major weaknesses of a report can be making your reader follow you through all the *how* steps before you divulge the *why*. Consider the following ways of dealing with that problem; remembering that your job is not only to give data but to *interpret* it.

1. Write the *how* process as you normally would, and summarize the section by saying *why* this works. Then, simply take that last paragaph and, in your final draft, place it first in that section. (If you're saying, "What final draft?," simply cut and paste the last paragraph. When you reach the last paragraph, now that the whole subject is clear in your mind, write it as an *introduction* to the section.) This is necessary because the reader may know *why* without following you through the working out of it. In addition, if he knows *why* at the outset, *how* will be clearer to him as he reads along.
2. Consider your sales strategy. Usually you tell the reader why and then how, as described above. The assumption is that the reader who needs to know why in order to make a decision to spend the money will read the first paragraph of the section and leave the working-out of the problem to the people on a more technical level of the firm. If, however, you need to lead the reader step-by-step to your conclusion, put him through the how process before you tell why finally. In this case, write the steps in the process which will solve the problem as steps in an argument for the adoption of your solution and for winning the contract to do the work.

HOW MUCH

Where to answer the HOW MUCH question in the report is a matter of judgment. If you are offering three alternative solutions, you can do this in one of three ways:

1. Give the cost analysis of each solution as you propose it.
2. Give a comparative cost analysis as a separate chapter once you have proposed alternative solutions.
3. Do (1), then write a summary of the relative costs, then place it in an introduction position.

What you do depends upon the strategy you adopt:

- Will it be better to give a series of smaller figures within each part of the report?
- Will it be better to delay cost considerations until you have argued your case on each solution?

Whatever your decision, costs should be presented in an orderly and readable way. Do not neglect the importance of what tech writers call "white space" in detailing costs on a page.

A principle in general sales writing is that it is effective always to express the cost per unit. The reason for this is that the reader can conceive of the smallest unit of cost—and is more receptive to it—more easily than he can conceive of the total cost of a number of units combined.

Appendices

When to include an illustration within a report or when to append it is a matter of judgment. You should base your judgment on these principles:

1. If the illustration is needed to aid the reader's *understanding* of your argument or explanation as you are giving it, use it as an element which advances the reader's understanding of that point before you lead him to the next point.
2. If the illustration shows how you arrive at your conclusion, if it is the background which underpins what you are saying, put it in an appendix. The appendix is for matter that is *not immediately needed* in understanding a page or a paragraph. It is, instead, the *underlying* structure of what you have said.

Thinking of the reader and what the reader needs helps to make this judgment. Especially, do not write five pages of recommendations, append all your data, and in effect ask the reader to "write" the report by figuring out the steps of your reasoning from text to appendix.

Conclusion (the final WHY)

At this point, the reader has read perhaps a dazzling, but one would hope instead an illuminating, array of definitions, solutions, and costs. Your job here is to briefly refresh his mind on WHAT, HOW, and WHY. This section need be no more than a page or a page and a half, typed and double-spaced. It is a "breather" in which you help collect the reader's throughts before you present your recommendations.

If you fear that you are being too repetitious—a common phobia among report writers—remember these pointers:

- The manager may not have read the body; he may simply be flipping over to the conclusion in a cursory review of the report.
- Technical matter is difficult to read and assimilate; your reader probably *needs* this reinforcement.
- Good thesis writing is based, in any case, on three steps:
  1. Tell 'em what you're gonna tell 'em.
  2. Tell 'em.
  3. Tell 'em what you've told 'em.

Recommendations (the final HOW)

Technical writing is defined as "writing that makes things happen." Thus the recommendations section lists the actions that should be taken. Remember these things:

1. List recommendations on a page which has plenty of white space.
2. If the list is long, subdivide the recommendations:

    Renovation of Existing Structures:
    1.
    2.

    Construction of New Facilities:
    1.
    2.

3. State recommendations very clearly and simply.

Introduction

Now that you have prepared the report, you want to lead the reader into it in an effective way. This will tell the reader what you have done, and where he can expect to go in reading the report. To accomplish this write these four parts:

> *Subject:* What problem does this report investigate, where, and for whom?
>
> *Purpose:* What was your job? To make a study? To make recommendations?
>
> *Scope:* This should give a precise statement of the limits of your assignment: heating only, site preparation only. If there are other related problems which you were not hired to consider, mention their existence and specify that these do not lie within the scope of this report.
>
> This is a very important step. It tells what the limits of your job are. It also covers you if questions are raised about your not doing tasks that lie outside the scope of this report.
>
> *Development:* Very briefly this should give the reader an idea of the direction the report will take. Will it be a report of a series of tests and results? Will you present three possible solutions and your assessment of each? This is a statement which will keep the reader oriented to the thrust of the report when the item-by-item discussion in the body of the report becomes confusing.

The Introduction does not need to be more than four paragraphs as described above. To save the work of writing transitions, you may use the subheads *"Subject:," "Purpose:,"* etc., as shown.

Executive Summary

This summary is the last thing you should write before you do the Abstract for the title page. By this time the report is one clearly refined idea. You should be able to state in a short series of clear sentences the subject, site, reasoning, alternative methods, conclusions, and recommendations.

In considering how to write this, knowing who will read it is all-important.

1. Your reader may not read the entire report.
2. She may need to know only:
   a. where to route it within the organization
   b. what her advisors are asking her to spend money for
   c. whether it will serve as a record of how the company reached its decision, when the final review of the project occurs.
3. Your reader may have two dozen other reports which she is considering along with yours.

If in the process of writing your introduction you have reduced the burden of the report to four paragraphs, you should be able to reduce the introductory summary to eight or ten sentences. It is appropriate also to list your recommendations here.

GUIDE FOR CONSTRUCTING THE FINISHED DRAFT FOR A TECHNICAL RECOMMENDATIONS REPORT

The section which follows divides a technical recommendations report into its formal parts and provides instructions on how to produce each part. The task of the writer is twofold: to write the content as prescribed by the form, and to design the specific parts of the document according to specifications.

The instructions first display an example of a finished page and then explain how and why a report writer produces such a design.

See Appendix A, pages 475–481, for manuscript specifications.

Table of Contents

1. Executive Summary 3
2. Introduction 4
3. Building Description 6
   - 3.1. Location 6
   - 3.2. Construction 7
   - 3.3. Occupancy 9
4. Existing Energy Systems 11
   - 4.1. Power Distribution 13
   - 4.2. Natural Gas Distribution 15
   - 4.3. Lighting 16
     - 4.3.1. Exterior 17
     - 4.3.2. Interior 19
5. Annual Energy Consumption 20
   - 5.1. Historical Energy Consumption 21
   - 5.2. Actual Energy Consumption 24
6. Conclusion and Recommendations 26

7. Appendix 28
7.1. Floor Plans 29
7.2. Heat Transmission Coefficients 34
7.3. Computer Input Narrative Description 39
7.4. HVAC Equipment 45

Construction

How

    The Table of Contents is your finished outline in readable form.

- Use Arabic numerals.
- Use leaders (........) to page numbers.
- Break subjects into sub-divisions, though usually not beyond the third division (3.2.4.).

Why

- *Verbally* the words and phrases of an outline can give a reader an initial overall understanding of what the report covers.
- *Visually* the Table of Contents should be *readable* for the reader who needs quick reference to parts of the report.
- *Physically* the Table of Contents should make the report easy to handle for a reader who needs quick and easy reference to its parts.

    The Table of Contents allows for the fact that a report is a document which is not read through once by a reader who reads as though he were sitting down with a good book. It is a *working* document. Some readers will read all of it, others will only read parts of it, some parts of it will have to be used and referred to. The Table of Contents facilitates that activity.

1. Executive Summary

> The Energy Reduction Study of the Davis Building in Richmond, Virginia, shows that in 1986 and 1987 this unit of the State of Virginia consumed a monthly average of 1,213,875 KWH and a monthly average of 9605 Therms. The computer model utilized in this study assessed building construction, occupancy, and area-by-area consumption of gas and electricity in this facility. This computer model projects that energy consumption can be reduced to a yearly average of 1,029,403 KWH and 7,775 Therms, a reduction which represents a considerable saving in utility bills to the State of Virginia. Later phases of this study will show how altered designs in the existing energy-consuming systems can bring about these savings.

Construction

How

1. Write the Executive Summary last, when the report is so clearly defined in your mind you can sum it up in a page.
2. A handy way to sum it up is to reduce each of the four paragraphs in the Introduction to a sentence, reduce your conclusion to a few sentences, and list your recommendations.

Why

Why is really more important than *how* for the Executive Summary. As much as or more than any other part of the report, it is directed toward a particular type of reader with a particular type of need:

1. An executive may read this summary in order to route the report to someone in the organization who will deal with the specifics.
2. An executive who will receive verbal briefings from subordinates on the substance of the report may read only the Summary—and perhaps the Introduction.

Executive Summary and the Writer

It is said that a writer cannot write a successful novel unless he can reduce that to a sentence. To use a mathematical analogy, the Executive Summary is the statement of an equation and a highly condensed explanation of what it signifies.

Executive Summary and the Reader

This Summary may be the only part of the report which someone with decision-making power reads, so make every word count.

Again, do not be concerned that you are being overly repetitious. If you have devised some especially effective phrase in the Introduction, the Conclusion, or the body of the report, by all means repeat it here. The person who reads the entire report needs the repetition for reinforcement, and the person who reads the Executive Summary may not read any further.

2. Introduction

> This report presents a study of alterations, modifications, and improvements which could feasibly be made for the Davis Building in Richmond, Virginia. It is the first stage of a three-stage Energy Reduction Study to determine how best to reduce the overall building energy consumption and how to lower energy demand.
>
> The purpose of this report is to establish an analytical computer model of the energy consumption of energy-using equipment in the building: electric, gas, heating, air-conditioning, and ventilating. Both historical energy consumption and operations under existinng conditions will be subjected to computer analysis.
>
> The scope of this report takes in past and present energy consumption for the Davis Building. In subsequent phases of the study, the same computer model will be used to evaluate energy-reducing modifications and improvements. This report, then, will serve as a basis for later study.

> The report will describe and evaluate the construction of the building, the occupancy area by area, and the existing energy systems. The discussion of the last includes an area-by-area assessment of power distribution, natural gas use, and use of electricity. These assessments include references to function (lighting, heating, cooling, ventilation), equipment (furnaces, water heaters, water coolers, fans), and design. Building plans, equipment shop drawings, schematic control diagrams, and conversations with the building engineer underlie this assessment. Finally, the report will provide a computer model based on this input.

Construction

How

1. After you write the working outline, you should next write the Introduction. It will initially serve you as a statement of the parameters within which you will work as you write the body of the report. Like the first rough outline, the Introduction takes shape as the report develops. You should rewrite it in final form either just before or just after you write the Conclusion and Recommendations, to align these three sections.
2. The first paragraph of the Introduction states the subject. It should tell *what, where, when,* and *for whom.*
3. The second paragraph should tell the purpose of the report: *why.*
4. The third paragraph should tell the scope of the report: what it will do, and what it will not do.
5. The fourth paragraph of the Introduction sets forth the pattern of development. It is a summary of the major points in the outline which states the sequence of subjects which the report will cover.

Why

1. The literal meaning of "introduce" is "lead into." For that reason, the introduction is about the report itself, not about the subject of the report. You are letting the reader know what to expect in the actual reading of the report, offering a guide which leads into it.
2. The *Subject* paragraph tells what the report is about. Is it about energy, about corrections institutions, about an electrical system? What kind of subject, what kind of place, what kind of problem will the report concern itself with?
3. The *Purpose* paragraph tells why this report was written. What is the problem this report is designed to address? What is the nature of the problem, what is the need it is designed to satisfy? The overall purpose of writing a report, any report, is to solve a problem; why is this particular report being written?
4. The *Scope* paragraph defines what this report covers. It not only tells the reader what to expect, it tells the reader/client what *not* to expect. This is important, because if the limits of the report are clearly defined, you are protected. When the client says, "But you didn't ... ," you cite the scope paragraph and reply, "That did not lie within the scope of the report."
5. The *Development* paragraph is a map of what course the report will take. It is the first step in the three steps of thesis writing: (1) tell 'em what you gonna tell 'em; (2) tell 'em; (3) tell 'em what you told 'em. This first step maps out where the report will lead, and your outline is a model for what it should include.

The classical term for this is *exordium*, a term from weaving which means literally "to lay out the threads." The threads are the major subjects of the report in the order that they will appear. The writing of the report itself weaves those together, and the conclusion draws them together into final recommendations.

To use another analogy, the outline is the skeleton of the report, this paragraph of the introduction begins to flesh out the skeleton, and the full development of the subject will be the body of the report.

Introduction and the Writer

Writing the Introduction of the report after you have written the finished outline becomes your own introduction to the actual work of writing the report. The time which writing it consumes is well spent, because you define your limits and establish a direction for yourself. You will work better if you take the time at the early stages of the report to determine exactly what you are doing, why, within what scope, and in what direction.

Introduction and the Reader

After the Introduction has served its purpose as a guide for you as writer, you should revise it in the final stages of writing to align it with the Conclusions and Recommendations. Given the dynamics of the research, investigation, and design process, it may not remain rigid, especially in the parts on scope and development. This revision makes it a guide for the reader.

"But I'm being repetitious" is often heard from the technical writer. Not so. Generations of English teachers who have taught us to search for synonyms and to strive to surprise and delight the reader are talking about creative writing. Technical writing is difficult matter to read; busy readers are often distracted and inattentive, and such readers need all the help they can get. The repetition of setting forth a method of development, developing the report in that pattern, and re-stating that in the conclusion is not dull repetition, but reinforcement which the reader very much needs.

The Introduction is thus a pivotal point, as an aid both to you as writer and to your reader.

3. Beginning a Chapter

COMPRESSED-AIR DRYERS AND BOILERS

The compressed-air dryers and the boilers are two parts of the existing system which require redesign. This chapter describes their present condition, the design needed to bring them to the proper operational level, and both the system operation and the economics involved in addressing the problem.

    3.1. Existing Compressed-Air Dryer Equipment: The compressed-air dryers currently in use are of the heatless-desiccant, moisture-absorption type. Their function in the total process of . . . .

Construction

How

1. The Development paragraph of the Introduction mapped out for the reader the major points to be covered in the report. On the chapter level, the introduction performs the same function of letting the reader know the main points covered in the chapter.
2. Do not number a chapter title (3.) and then number the first paragraph of the chapter as a sub-section (3.1.). The very existence of a sub-section suggests the existence of a larger part.
3. As you begin a chapter, the substance that should be in the introduction may not be clearly worked out in your mind. For that reason, do not delay too long over this part; first go on with writing the chapter itself; then abstract your development of the chapter afterward, and reposition it as an introduction.

Why

1. Create strong transitions from one chapter to another in a long report; remember that a reader may not read a report serially, and that even a reader who is reading the report through may need the repetition and reinforcement.
2. Just as the Introduction to the entire report is a reader's guide to the course which the report will follow, the introduction to each chapter should clearly map out the direction which the reader must take in reading.
3. Unlike an essay, which plays directly upon the reader's reactions with first-person pronouns, feelings, and opinions, a report does not give you direct access to the reader. To compensate for the necessary separation from the reader which technical writing requires, talk about the report and how a chapter functions. This objective, directive use of strong transitions can guide, hold, and influence the reader.

Conclusions and Recommendations

This study has established the need for improved security control at the Braxton Bragg Dam. Most visitors were unaware of what constitutes a hydroelectric and flood control facility and thus view the entire installation simply as "the dam." As we have shown, this level of understanding encourages unintentional trespass which could easily result in serious injury to the curious but uninformed visitor who approaches a dangerous situation. Theft and vandalism, too, are problems, given the remote location of the facility and the small number of staffers on duty.

To meet the need for improved security, we recommend adequate signage to provide visitors with the following information:

- clear definition of public and restricted areas;
- hours of operation when visitors are permitted;
- directional instruction;
- explanation of warning signs.

To meet the need to secure the facility against theft and vandalism, we recommend the installation of the following items:

1. Heavy duty metal doors and locksets at personnel entry doors and at access to the exterior transformer platforms.
2. Security bars on windows and on air-intake openings at the rear of the powerhouse.
3. An infrared intrusion-detection system in all lobby areas and in the storage buildings.

Construction

How

1. *Conclude* means literally "to shut up"—in the sense of gathering all the sheep into the fold and shutting them up. To translate that ancient rhetorical figure into your terms, you gather together the points you have made, neatly tie together the end of the threads you have woven.
2. Draw upon the Introduction, the Table of Contents, and the introductions to major chapters of the report, both for the substance of the Conclusion and for phrases and sentences.
3. Re-state the problem. State that it has been solved.
4. Briefly review how you solved the problem; if it is significant, give emphasis to the method you used. Do not neglect to note any innovative and creative solution, for they sell your work.
5. *Recommend:* this is where the buck stops. Doing this is how you earn the megabucks. What is the answer? What is the solution to the problem? 1, 2, 3, 4,—what must the client *do*?

Why

1. Again, do not fear that you are being repetitious. You may be sick of saying a certain thing by the time you get to this point in the writing, but your reader needs the repetition. If she has read the entire report, she needs the lengthy series of major sections drawn together and re-stated in shorter form. You are about to tell her the answer; once again, tell her how you have arrived at this point—a point where you may be asking her to spend $$$$$$$$$.
2. Note the use of the pronoun *we* in the Conclusions and Recommendations. Banned until recently from report writing, this is now being used. *We* enables you to avoid the passive voice in describing an action you have taken; you can also express your recommendations in the active voice. Action is important, and for sales value, *we* expresses that you have done the job that you can be justly proud of.
3. Clearly express what your recommendations are. If you are offering a series of options from which the client must choose, make that clear. If you offer a series of options and your job is to recommend which one(s) to adopt, make that clear also. In the Conclusion, there is also the sense of the word *conclusive*.
4. If this report represents but one stage in a series of studies, make clear what point you have achieved in that progression; then clearly state what problems remain to be solved or what action will be necessary to complete the entire project.

5. Some research arrives only at the statement of a problem. In that case, your job is not to recommend, but to state clearly what the problem is and, in some cases, to project in more general terms what measures might be taken to address the problem.

Whether you write this section according to 3, 4, or 5, above, depends upon your purpose in writing the report: final recommendations, report on one stage of work, or defining a problem, respectively.

READING AND EVALUATING TECHNICAL REPORTS

Learning to read precedes learning how to write. Therefore it is most important to look carefully at how all the components of a report work together to form a comprehensive whole.

The reports which follow are selected to illustrate the point that audience analysis is the essential determinant in what a technical writer says. Each report deals with some aspect of warehousing, yet each one displays a selection of detail based upon the specified needs of an identified audience.

If you will recall the dialogue with the young technologist in the chapter on How Technical Professionals Write for an Audience, the job is not simply to tell all you know about a given subject. The first task of the job is to determine what the intended audience needs to know to solve a specific problem; the second is to convey that in a format which the intended audience will find communicative. Therefore, keep the intended audience always in mind as you read these reports. Also, look at the report as a whole which leads from the summary, through a statement and discussion of a problem, to recommendations for a solution.

The checklist for evaluating long reports is in Appendix B, pages 495–496.

Technical Recommendations Report: Exercise 15.1

FORK TRUCKS: RECORD-KEEPING
AND PREVENTIVE MAINTENANCE

ABSTRACT

Some of the problems of maintenance for 40 fork trucks at the Columbus Cotton Works plant are discussed in this report. Specifically, the problems of the record-keeping system and a preventive maintenance program are presented. In each area, the problem is described and a solution is recommended. Currently, there is neither a system for keeping maintenance records nor a preventive maintenance program. The recommended solution in each case is to establish such a system. The report includes detailed procedures for implementation of the recommended systems.

by

Georgette P. Burdell

for

English 3023

May 16, 1990

Manuscript specifications for a Title Page with Abstract appear in Appendix A, page 482.

EXECUTIVE SUMMARY

This report examines the fork trucks at Columbus Cotton Works in relation to problems of inadequate record-keeping and the need for a preventive maintenance program. Because both record-keeping and preventive maintenance are valuable to a company, this report recommends that both be established.

The record-keeping system will monitor the types of breakdown, the parts used in repair, and the total cost involved in maintenance. The system will be based on 3" × 5" cards which will be filled out by the mechanic and his foreman after each maintenance operation.

The preventive maintenance program will be a program for maintenance now to prevent future breakdowns. The report recommends three levels of preventive maintenance: 1) The operator will check the fork truck each day before he uses it. 2) Mechanics will check each fork truck after an operating cycle of 100, 500, and 5000 hours. 3) Each fork truck will receive periodic major overhauls of component units.

Both of these programs provide benefits to management. The record-keeping system will provide information by which decision-making will be improved. The preventive maintenance program will save money by decreasing the amount of downtime on trucks.

TABLE OF CONTENTS

1. Introduction ....
2. Discussion—Maintenance ....
   2.1. Record-Keeping ....
      2.1.1. Problems ....
      2.1.2. Proposal—3" × 5" Card System ....
      2.1.3. Benefits ....
   2.2. Preventive Maintenance ....
      2.2.1. Present Conditions ....
      2.2.2. Levels of Preventive Maintenance ....
      2.2.3. Costs ....
3. Comparison of Costs for In-House and Contracted Maintenance ....
4. Conclusions and Recommendations ....
5. Appendices ....
   5.1. Appendix 1: The Maintenance Card ....
   5.2. Appendix 2: The Cumulative Card ....
   5.3. Appendix 3: The Daily Operator Checklist ....
   5.4. Appendix 4: Estimated Costs under Breakdown Maintenance ....
   5.5. Appendix 5: Estimated Frequency of Breakdowns under Breakdown Maintenance ....

LIST OF FIGURES

Figure 1. Comparison of Breakdown and Preventive Maintenance Costs ....
Figure 2. Maintenance Card ....
Figure 3. Cumulative Card ....
Figure 4. Daily Operator Checklist ....

1. INTRODUCTION

The focus of this report is maintenance of the fork-truck fleet at Columbus Cotton Works (CCW), a textile finishing plant. CCW is a large multi-building facility where rolls of cloth are finished and printed. There are 40 fork trucks at the plant which fulfill material-handling needs.

The objective of this report is to recommend improvements in the performance of the present fork-truck operation. This performance will be measured both in terms of cost and in terms of utilization of each fork truck.

The scope of this study includes only two aspects of the fork-truck area: the record-keeping in maintenance, and a preventive maintenance program. Other areas that relate to cost and utilization of fork trucks will not be discussed. Recommendations, with detailed procedures for implementation, will be presented in the record-keeping and preventive maintenance areas.

Research for the study included investigation of the present situation at the CCW plant and discussions with manufacturing representatives for fork-truck companies.

The discussion first describes problems associated with the fork-truck fleet. In two major sections the report proposes a record-keeping system and discusses the benefits of the system. In discussing preventive maintenance the report describes problems associated with the present system, describes levels of preventive maintenance, and gives cost comparisons for in-house and contracted service.

2. DISCUSSION

The maintenance department is responsible for keeping equipment operating at Columbus Cotton Works. An examination of the fork-truck aspect of maintenance revealed two problem areas: (1) the record-keeping system and (2) the preventive maintenance program.

2.1. <u>Record-Keeping</u>: The keeping of records is essential to any organization. It is the historical basis on which decisions for the future are made. Keeping records eliminates the risk of having someone forget necessary data.

2.1.1. <u>Problems</u>: At CCW some records are currently kept in the maintenance department. Labor hours are written up on shop work reports. Purchase orders are written to order needed parts. In the end, however, all this information is sent to accounting for their uses. Maintenance does not keep a record of what it does.

(1) <u>Inadequate Maintenance-Cost Records</u>: It is dangerous to have fork trucks running without data about what it costs to operate them. Unless someone can remember all the maintenance performed on all forty fork trucks, it may cost more to maintain an old truck than to buy a new one. As this study has noted, a record of labor costs and part costs is not kept in the maintenance department. The maintenance department therefore has no way of evaluating potential maintenance costs and consequently has no way of determining whether repairing a truck will be cost-effective.

(2) Inadequate Breakdown Information: Without records of breakdowns, patterns in types of breakdowns cannot be determined. With patterns established, measures could be taken to correct problem areas. Although not directly required for operation, this information lends itself to a more efficient maintenance operation.

(3) Inadequate Economic-Analysis Data: Economic analysis compares the cost of maintaining an old truck to the cost of owning a new one. The lack of cost data prevents any economic analysis. CCW maintenance currently keeps the fork trucks until they fall apart.

2.1.2. Proposal 3"×5" Card System: To deal with these problems, a simple, accurate record-keeping system should be installed. A system using 3" × 5" cards will fill this need. The Maintenance Department will be able to record needed information to use it in its decision-making. The system will use two types of cards.

(1) Maintenance Card: There will be a maintenance card for each maintenance operation. It will contain a detailed record of what happens during each repair. The card will contain the following information (see Appendix 1):

    (a) Date
    (b) Fork Truck Number
    (c) Description of Breakdown
    (d) Parts Used
    (e) Total Labor Hours
    (f) Mechanic's Initials

After the repair is completed, the mechanic will file the card with the foreman maintaining records. The foreman will total the parts costs and convert the labor hours into a labor cost for transmittal to the Accounting Department.

(2) Cumulative Card: One of these cards will be kept for every fork truck. It will contain (see Appendix 2):

    (a) Truck Number

    (b) Assigned Location

    (c) Cumulative Total Cost (Parts and Labor)

    (d) Tally of Breakdown Types

After the foreman completes the maintenance card, he will update the information on the cumulative card.

(3) Filing Procedures: The cards will be filed in a truck-number sequence. The first card in each section will be the cumulative card for a truck. Following it will be the maintenance card for each maintenance operation on that truck.

2.1.3. Benefits: 3"×5" cards are not the only way to keep records. Other options range from a computer-based management information system to another size of file card. Based on capabilities and cost, however, the 3"×5" card system provides the greatest benefit.

(1) Easy To Operate: The 3"×5" card system does not require any complicated mathematical skills. It simply records what has happened and provides information useful in decision-making. The foreman and the mechanic are the recorders. The plant engineer is the person who will analyze the data. The system is essentially a pencil-and-paper operation. No special equipment will be needed except for the file boxes and the preprinted 3"×5" cards. The cost for that equipment would be approximately $200.

(2) Provides Internal Information: A highly desirable feature of the system is that all the needed information is generated

from within the maintenance department. There is no need to ask the accounting department to generate a report of the cost data. This is beneficial in two ways. For one reason, it would prevent the friction of a policy change in the Accounting Department's procedures. Also, records kept in the Maintenance Department would meet their information needs whereas accounting records would provide only costs and not a detailed list of parts.

(3) Provides Complete Records: This system allows the records to be kept in a machine-by-machine sequence. When a certain fork truck needs to be evaluated, it is easy to locate its record. The cumulative card gives up-to-date information on cost of maintenance and types of breakdown. If more detailed information is needed, the maintenance cards for that truck can be checked to determine the parts used in repair, the labor hours, or the date of a certain breakdown.

(4) Monitors Excessive Costs: These records also serve as a control function. With the cost information, it is possible to evaluate a potential maintenance expense. By comparing the expected expense with all the past expenses for that fork truck, maintenance can determine whether the upcoming expense is worthwhile. The exact procedure of comparison is economic analysis but that procedure is beyond the scope of this report.

(5) Monitors Types of Breakdown: The method of record-keeping allows the maintenance department to analyze the types of breakdown which occur. They can plot the trends that are developing and concentrate their efforts in a particular area. For instance, if there are an unusual number of transmissions burning out, the records system indicates that a more frequent changing of transmission fluid is needed. The classification of breakdowns also allows

the computing of unit costs for each type of breakdown. These records give the maintenance department a basis for projecting future maintenance costs.

2.2. Preventive Maintenance: Preventive (or planned) maintenance is exactly what the name implies: performing maintenance now to prevent future breakdowns. There are two savings as a result of preventive maintenance. First, there is 25%–50% less downtime for the trucks. Second, management is able to schedule fork trucks for preventive maintenance, thereby gaining better utilization from the fork trucks.

2.2.1. Present Conditions: Currently there is no preventive maintenance program at CCW. Maintenance is done only after a breakdown occurs. One worker commented that the only time he sees the trucks are when they are towed in for repair. In the past there was a preventive maintenance program. From the comments of the workers, however, it was ineffective. There was no one with the authority to assure that the program was carried out. Therefore, to utilize the advantages of preventive maintenance, it is necessary to install a program similar to the one described below.

2.2.2. Levels of Preventive Maintenance: There are three levels of service in preventive maintenance: (a) daily operator check, (b) regular preventive maintenance, and (c) periodic major overhauls. All three levels are essential to a successful preventive maintenance program.

(1) Daily Operator Check: A high percentage of downtime can be prevented by having the driver check the truck on a daily basis. Although the items that the operator can check may be limited, these few simple things are important to maintain. Because of the

type of driver at CCW, this daily operator check must be simple. Following is a list of the items a driver can check:

  (a) engine oil

  (b) radiator water

  (c) fuel level

  (d) tires

  (e) head and tail lights

  (f) hour meter

After completing the daily check, an operator will fill out a form noting whether each item is satisfactory or needs attention (see Appendix 3). Again, the form has to be simple so that it will encourage an accurate response from the drivers. These forms can be kept at the office of the department to which the fork truck is assigned. Maintenance can be notified of items that need attention and work on the truck can then be scheduled at a convenient time.

  (2) Regular Preventive Maintenance: The operators will not be able to detect all the repairs that should be made. It is necessary to have a mechanic check each machine periodically. Two considerations determine what items are checked, and how often: the operating conditions, and the hours of operation. Operating conditions take into account the dust in the air, the moisture content of the air, the condition on the paths (roughness), and the severity of the loads. Normal operating hours for a fork truck are about 2000–4000 hours per year. According to the literature of the field, under normal operating conditions, maintenance operations should be broken down into three cycles: 100 hours, 500 hours, and 5000 hours. Recommended maintenance operations for each cycle are as follows:

(a) 100 Hours

    differential

    hydraulic lines

    wheel bearings

    hydraulic cylinders

    tires

    chains

    chain rollers

    hydraulic pump and valve

    operating brake

    seat brake

    battery

(b) 500 Hours

    steering-gear unit

    wiring

    hydraulic-system oil

(c) 5000 Hours

    forks

    upright (mast)

    power axle

    engine

    brushes

    armature

    master accelerator

(3) Periodic Major Overhauls: In the different cycles, the emphasis is to keep the trucks in good repair. Periodically, according to the manufacturer's recommendation an item on the truck may

need to be completely overhauled. In the overhaul procedure each piece of the unit is inspected and either cleaned or replaced.

2.2.3. <u>Cost</u>: A preventive maintenance program is not without cost. On the other hand, such a program will save money by preventing breakdowns. Questions which should be answered in this area include: (a) How much preventive maintenance should be done; and (b) should maintenance be done "in-house," or contracted?

<u>Preventive Maintenance vs. Breakdown Maintenance</u>: Each type of maintenance costs money. The task is to find the optimum combination of the two to minimize the total cost. For illustration, see figure below.

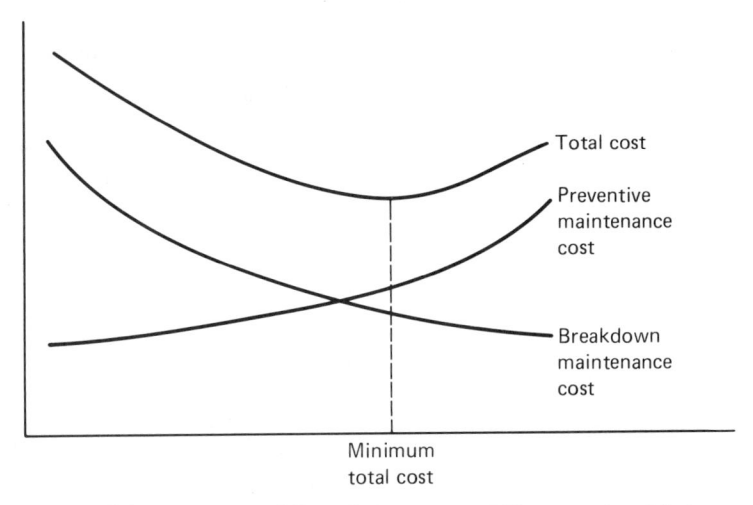

Figure 1. Comparison of Breakdown- and Preventive-Maintenance Costs

4. CONCLUSIONS AND RECOMMENDATIONS

A study of the fleet of 40 fork trucks at Columbus Cotton Works has shown that inadequate records are the major cause for an unnecessarily costly operation. The absence of records on maintenance and breakdowns means, in turn, that CCW has inadequate information about what trucks are available for service, what repairs are likely to be needed, what costs to project for repairs, and when new trucks should be purchased.

Recommendations to address and solve these problems are as follows:

1. to establish a 3" × 5" card file on each fork truck
    a. to enlist the efforts of drivers to note operational flaws, and
    b. to keep a cumulative card index on each truck in the fleet;
2. to use the card file to monitor the condition of the fleet and to plan a preventive maintenance system.

The report further recommends that in-house maintenance can be 9.1% less costly than contracting for an external source of maintenance.

5. APPENDIX

5.1. Appendix 1: The Maintenance Card

Breakdown description		Date	#

Parts	Cost	Remarks:	
	Total parts cost	Total labor hours	Initials

Figure 2. Maintenance Card.

5.2. Appendix 2: The Cumulative Card

#	Where assigned	
Date	Total cost	Record of breakdowns

Figure 3. Cumulative Card

5.3 Appendix 3: The Daily Operator Checklist

Date		Machine #
Operator's name		Supervisor's initials

Items to be checked	OK	*Needs attention*
Engine oil		
Radiator water		
Fuel		
Tires		
Head and tail lights		
Warning light		
Hour meter		
Other instruments		
Horn		
Steering		
Service brakes		
Parking brake		
Hydraulic controls		

Explain items that need repair:

Figure 4. Daily Operator Checklist

5.4. Appendix 4:
Estimated Costs under Breakdown Maintenance

Breakdown	Cost 1st yr.	Cost 2nd yr.
Powerplant	$10549	$12926
Cranking	5281	7651
PM	2859	6678
Brakes	3871	4610
Hydraulics	14582	4152
Ignition	3966	4021
Fuel System	2008	3966

5.4 Appendix 4:
Estimated Costs under Breakdown Maintenance

Breakdown	Cost 1st yr.	Cost 2nd yr.
Steering	5235	3296
Transmission	3650	3111
Cooling	2988	2945
Charging	2214	2845
Roll Cage	928	1512
Clutch	2091	985
Drive Axles	699	746
Cab	707	590
Exhaust	323	483
Instruments	534	428
Air Intake	267	217
Lights	233	150

5.5. Appendix 5:
Estimated Frequency of Breakdowns under Breakdown Maintenance

Breakdown	Frequency 1st yr.	Frequency 2nd yr.
Cranking	168	253
PM	89	240
Ignition	134	186
Powerplant	91	158
Fuel Systems	71	152
Hydraulics	214	149
Brakes	89	133
Cooling	78	132
Charging	68	90
Steering	47	57
Transmission	55	55
Air Intake	27	42
Clutch	25	33
Roll Cage	28	18
Instruments	25	17
Exhaust	12	17
Cab	28	14
Lighting	14	8
Drive Axles	8	5

PRODUCT-WAREHOUSING OPERATIONS OF COLUMBUS CLOTH WORKS:

A COMPARATIVE FEASIBILITY STUDY

Abstract

This report analyzes the economic factors of two alternatives to the current Product-Warehousing operations of Columbus Cloth Works. Based upon comparison of the current system, a new conventional system, and an AS/RS, in terms of after-tax cash flow, present worth, and rate-of-return calculations, the report suggests that neither alternative be adopted. For the future it suggests that the CCW Industrial Engineering Department investigate cost reduction by altering the current system.

George P. Burdell

English 4023
April 26, 1990

Manuscript specifications for a Title Page with Abstract appear in Appendix A, page 482.

TABLE OF CONTENTS

1. Executive Summary ....
2. Introduction ....
3. Problem Definition ....
4. Data Collection ....
    4.1. Costs of the Current System ....
    4.2. Production Values ....
    4.3. Pallet-Load Information ....
    4.4. Seasonal-Demand Information ....
5. Proposed Alternatives ....
    5.1. Current System ....
        5.1.1. Description ....
        5.1.2. Cost Determination ....
    5.2. New Conventional System ....
        5.2.1. Description ....
        5.2.2. Cost Determination ....
    5.3. AS/RS ....
        5.3.1. Description ....
        5.3.2. Evaluation of Design Parameters ....
        5.3.3. Cost Determination ....
6. Economic Analysis of Proposed Alternatives ....
    6.1. Current System ....
        6.1.1. After-Tax Cash-Flow Analysis ....
        6.1.2. Present-Worth and RoR Analysis ....
    6.2. New Conventional System ....
        6.2.1. After-Tax Cash-Flow Analysis ....
        6.2.2. Present Worth and RoR Analysis ....

6.3. AS/RS ....
      6.3.1. After-Tax Cash-Flow Analysis ....
      6.3.2. Present-Worth and RoR Analysis ....
   6.4. Economic Comparison of Proposed Alternatives ....
7. Conclusion and Recommendations ....
8. Appendices ....
   8.1. Calculation of AS/RS Capacity ....
   8.2. Throughput Calculations ....
   8.3. Accelerated Depreciation Factors ....
   8.4. Present-Worth Calculations ....
   8.5. Zollinger Article ....
9. Bibliography ....

1. EXECUTIVE SUMMARY

This report presents a comparative-feasibility study concerning three alternatives for the Product Warehousing operations of Columbus Cloth Works. Limited in scope to economic factors, the study compares the current system, a new conventional system and an AS/RS, on the basis of after-tax cash flow, present worth, and rate-of-return calculations. Because the rate of return on both the new conventional system and the AS/RS is considerably less than CCW's 30% MARR, the study indicates that neither new alternative should be adopted. Instead, it is recommended that the Industrial Engineering Department investigate methods for reducing the annual operating costs of the current system.

2. INTRODUCTION

This report presents an economic analysis of three mutually exclusive alternatives for the warehousing operations of Columbus Cloth Works, currently located in Columbus, Arkansas. The purpose of this report is to describe each of these alternatives and evaluate it on the basis of after-tax cash flow, present worth, and rate of return. Rate of return (RoR) on each incremental investment will then be compared to the MARR (minimum attractive rate of return) required for an investment.

This study is limited to an evaluation of economic factors. Other factors, to which a dollar sign cannot easily be affixed, will be examined in a later study.

This report will describe CCW's warehousing problems, the data collected for analysis, and each of the proposed alternatives in terms of initial investment and annual operating costs. The report then presents an economic analysis of each alternative using three criteria: after-tax cash flow, present worth, and rate of return. Finally the study presents its recommendations and conclusion.

3. PROBLEM DEFINITION

Six interrelated problems characterize the warehousing situation at Columbus Cloth Works. They are as follows:

1. <u>Damaged goods</u>. This problem is caused by multiple handling of loads and by the use of incorrect packaging materials.
2. <u>Inaccurate maintenance of inventory records</u>. For example, the actual number of units of inventory-on-hand may deviate considerably from the recorded count.
3. <u>Poor storage and retrieval system</u>. Many times warehouse workers cannot determine the exact location of an item.
4. <u>Inadequate criteria for determining area allocation</u>. CCW has no predetermined method for arranging items in order of demand frequency or load weight.
5. <u>Inadequate on-site facilities</u>. CCW's on-site facilities house only one-third of the finished product, thus requiring the leasing of two additional warehouses.
6. <u>High annual operating costs</u>. Operating three individual warehousing concerns, CCW has incurred unusually high annual costs: excessive personnel, fork-lift, and utility costs, and unnecessary truck shuttle between warehouses.

CCW might solve these problems by implementing an Automatic Storage and Retrieval System. In its simplest terms an Automatic Storage and Retrieval System (AS/RS) is a warehousing installation which identifies and sorts, dispatches and stores, picks, consolidates orders, and keeps records, all on a completely automated basis. As described in James M. Apple's book entitled <u>Material Handling Systems Design</u>, an on-line computer serves as the control component for the entire system. The purpose of this study is to determine the economic feasibility of such a system and any other alternative systems which would alleviate CCW's warehousing problems.

4. DATA COLLECTION

Pertinent data for the feasibility study include costs requirements for the current system, design considerations (production levels and load information) for the current and the possible alternative systems, and information regarding seasonality.

4.1. Costs of the Current System: The annual warehouse operating costs for the current system are presented below in tabular form.

1978 Annual Warehouse Operating Costs
Current System

Manpower Costs:			$637,000
Direct Labor	$345,000		
30% Fringe Benefits	103,500		
Total Direct Labor		$448,500	
Indirect Labor	145,000		
30% Fringe Benefits	43,500		
Total Indirect Labor		188,500	
Warehouse Costs:			215,000
Leasing	204,000		
Bldg. Maintenance	11,000		
Equipment Costs:			190,000
Lift-Truck Leasing	120,000		
Trailer, Truck Leasing	70,000		
Utilities Cost:			61,000
Electricity	40,000		
Gas	21,000		
Taxes:			20,000
On Truck	20,000		
Insurance:			5,000
On Truck	5,000		
Total			1,128,000

4.2. <u>Production Values</u>: Central Planning released the current production, the anticipated 1983 production, and the peak on-hand inventory levels by product classification. These figures reflect an overall 8% growth rate in the next 5 years. The plant will then be operating at capacity; no further growth in the Columbus Warehouse Division is expected after 1983. These production figures are presented below.

Product Classification	Total Units Produced (in Millions) 1978	1983	Peak On-Hand Inventory
Basket	350	375	49.6
Cluster Weave	290	300	33.4
12 Thread	80	90	9.6
Cluster Thread	5	35	.6
Convenience	200	200	24.7
Total	925	1000	117.9

4.3. <u>Pallet-Load Information:</u> Various sources released the following information regarding the average number of units and weights of pallet loads by product classification. In conjunction with the production data, this information has been useful in determining throughput.

Product Classification	Avg. No. of Units per Pallet	Weight Lbs. per Pallet
Basketweave	7,000	2,300
Clusterweave	17,600	2,750
12 Thread	5,400	1,550
Cluster Thread	21,000	—
Convenience	25,000	—

4.4. **Seasonal-Demand Information:** Seasonal demand and CCW's policy to alleviate the strain it places on production is an important factor in warehousing policy. The following sketch reveals the impact of seasonal demand on production and inventory levels.

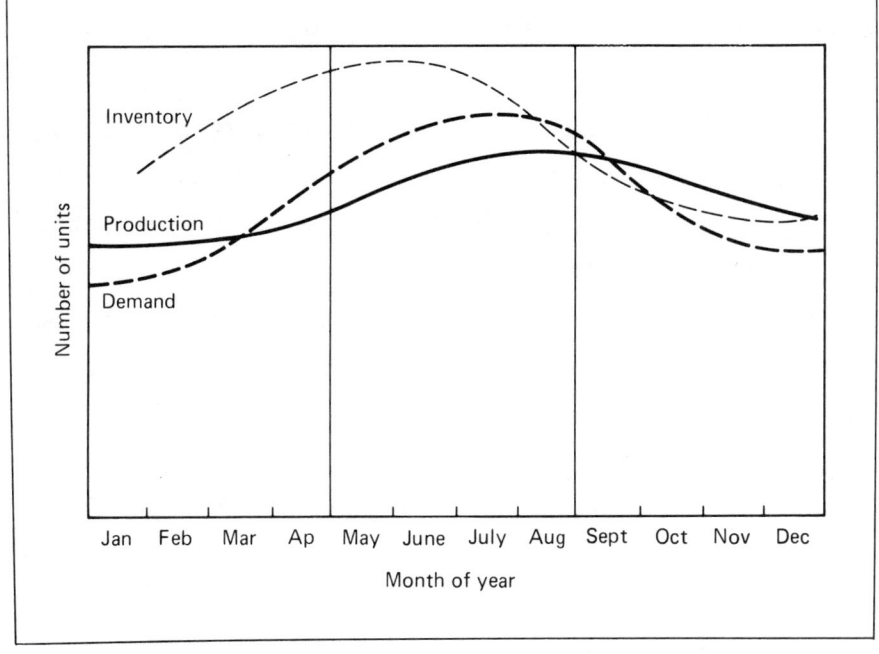

[Graph showing Inventory, Production, and Demand curves across months January through December, with "Number of units" on the y-axis and "Month of year" on the x-axis.]

5. PROPOSED ALTERNATIVES

To solve CCW's warehousing problems this study considers on an economic basis three possible alternatives: the null alternative, that is, no change to the existing system; the construction of a conventional on-site warehouse to store goods currently housed in the leased warehouses; and the construction of a completely automated system.

5.1. Current System

5.1.1. Description: CCW's current warehousing system consists of warehouse #1, conveniently located adjacent to the manufacturing facilities, and two additional warehouses (#2 and #3) located several miles from the manufacturing facilities. Since the two additional warehouses are not located at the manufacturing site, an additional expense is incurred, namely the trailer truck and manpower costs associated with the shuttle of finished goods.

The following data indicates the size of the current warehousing system at CCW. In addition to the 23 employees cited as being involved in physical handling of goods, there are 13 others involved in the clerical and administrative tasks of the warehousing operation.

Warehouse Number	Square Footage (sq.ft.)	Usable Storage Height (ft.)	No. of Employees Involved in Physical Handling
1	75,000	15	8
2	82,000	19	5
3	124,000	18	10

5.1.2. Cost Determination: Because all three warehouses and all equipment are leased, the initial investment cost is $0. The annual operating costs, as described in section 4.1. of this report, are equal to $1,128,000 each year. These annual costs, as are annual costs for every alternative plan, are subject to 7% inflation each year and so do not remain constant in the before- or after-tax cash flow.

5.2. New Conventional System

5.2.1. Description:
As an explanation of warehouse #1, a new, yet conventional fork-lift warehouse will be constructed adjacent to manufacturing facilities. It will be on property owned by CCW's parent company, and so escape the annual leasing costs associated with warehouses # 2 and #3.

5.2.2. Cost Determination:

Table 5.2.2.1.
Initial Costs

Cost Element	Cost Calculation	Total
Land & Site Prep.	149,900 sq. ft. @ $.69/sq. ft.	$103,431
Building	149,900 sq. ft. @ $23./sq. ft.	3,447,700
Investment Tax Credit	10% of ($3,447,700)	(344,770)
Total Initial Investment Cost		$3,206,361

Table 5.22.2.
Annual Costs

Cost Element	Cost Calculation	Total
Fork Trucks	12 trucks @ $7,060/truck	84,720
Direct Labor	25 employees @ $11,500/employee	287,500
Fringe Benefits for Dir. Labor	30% of Direct Labor Cost	86,250
Indirect Labor	8 employees @ $16,000/employee	128,000
Fringe Benefits for Ind. Labor	30% of Indirect Labor Cost	38,400
Electricity and Gas		70,150
Maintenance		12,650
Total Annual Operating Costs		$707,670

5.3. AS/RS

5.3.1. Description: The Automatic Storage and Retrieval System under consideration consists of two stacker cranes, nine rows of racks, pickup and deposit stations, and a method of charging and discharging from the system—a combination of conveyor and fork trucks. The stacker unit consists of a floor-running trolley to which is attached a rigid hoisting column equipped with a carriage. The carriage is fitted with load-carrying, telescoping forks powered for left and right motion. Traveling between closely spaced racks, the stacker unit serves rack openings from top to bottom on both sides of the aisle. Complete computer control is offered by a central console unit.

5.3.2. Evaluation of Design Parameter:
Certain parameters must be evaluated to determine the initial investment of an AS/RS. They are as follows:

1. No. of columns of storage/aisle, n_1. Using the 44" pallet side parallel to the aisle:

$$n_1 = \frac{(369\,\text{ft} - 52\,\text{ft})(12\,\text{in/ft})}{(44\,\text{in} + 8\,\text{in})} = 73.15 \sim 73$$

2. No. of levels/column, n_2.
   Using 85' ceiling level and 7'6" high pallet levels:

$$n_2 = \frac{(85\,\text{ft} - 4\,\text{ft})(12\,\text{in/ft})}{(7.5\,\text{ft})(12\,\text{in/ft}) + 10\,\text{in}} = 9.72 \sim 9$$

3. No. of openings/aisle, n_3.

$$n_3 = (2)(73)(9) = 1314$$

4. No. of aisles for 13,000 openings, n_4.

$$n_4 = \frac{13{,}000\,\text{openings}}{1314\,\text{openings/aisle}} = 9.89 \sim 10$$

5. Bldg. Width, n_5.

$$n_5 = (14.5\,\text{ft/aisle})(10\,\text{aisles}) = 145\,\text{ft}$$

6. Bldg. Length, n_6.

$$n_6 = 369\,\text{ft}$$

7. Bldg. Area, n_7.

$$n_7 = (369\,\text{ft})(145\,\text{ft}) = 53{,}505\,\text{sq.ft.}$$

See Appendix 8.5 for a detailed explanation of these parameters.

Table 5.3.3.1

Calculation of Initial AS/RS Costs

Cost	Variable	Value	Weight	Calculation
Rack cost C_1	load size	119.17 ft^3	3	$C_1 = (3 + 2 + 2) \times$ ($14/opening) \times$ (13,000 openings) $C_1 = $1,274,000$ $C_{1\,1979} = $1,528,800$
Stacker cost C_2	load weight rack weight stor. mach. height load weight logic	3000 lbs 9 loads 85 ft. 3000 lbs. cent. console	2 2 4 2 4	$C_2 = (4 + 2 + 4) \times$ ($13,000/stacker) \times$ (2 stackers) $C_2 = $260,000$ $C_{2\,1979} = $312,000$
Transfer cost C_3	—	—	—	$C_3 = ($40,000/trans.0) \times$ (2 transfers) $C_3 = $80,000$ $C_{3\,1979} = $96,000$
Building cost C_4	clear height	85 ft	2.5	$C_4 = (2.5) \times$ ($17/sq. ft.) \times$ (53,505 sq. ft.) $C_4 = $2,273,963$ $C_{4\,1979} = $2,728,755$

See Appendix 8.5 for a detailed explanation of these cost calculations.

Table 5.3.3.2.

Summary of Initial AS/RS Costs

C_1, Rack Cost	$1,528,800
C_2, Stacker Cost	312,000
C_3, Transfer Cost	96,000
C_4, Building Cost	2,728,755
TC, Total Cost	$4,471,158

Table 5.3.3.3

Initial Costs

Cost Element	Cost	
	Calculation	Total
Land & Site Prep.	53,505 sq. ft. @ $.69/sq. ft.	$ 36,918
Building Cost (C_4)		2,728,755
AS/RS Cost ($C_1 + C_2 + C_3$)		1,936,800
Investment Tax Credit	10% of (4,665,555)	(466,556)
Total Initial Investment Cost		$4,235,917

Table 5.3.3.4.

Annual Costs

Cost Element	Cost Calculation	Total
Fork Trucks	3 trucks @ $7,060/truck	$ 21,180
Direct Labor	14 employees @ $11,500/employee	161,000
Fringe Benefits for Direct Labor	30% of Direct Labor Cost	48,300
Indirect Labor	4 employees @ $16,000/employee	64,000
Fringe Benefits for Ind. Labor	30% of Indirect Labor Cost	19,200
Electricity and Gas		50,000
Maintenance		80,000
Total Annual Operating Costs		$443,680

6. ECONOMIC ANALYSIS OF PROPOSED ALTERNATIVES

Three terms—after-tax cash flow, present worth, and rate of return—must be defined for a proper understanding of the economic analysis. Generally an investment opportunity is described by the actual cash receipts and disbursements that are anticipated if the investment is undertaken. The representation of the amounts and timing of these cash receipts and disbursements is referred to as the investment's cash flow. Theusen, Fabrycky, and Theusen comment at length on cash flow in their book Engineering Economics: "Since taxes constitute a substantial portion of the disbursements that are related to an alternative, it is sound decision making to include the effect of taxes on a cash flow basis." An after-tax cash flow, then, records the cash flow by calculating taxes or tax credits (usually 50% of the taxable income) on the yearly receipts and disbursements and adding these figures to the before-tax cash flow.

As a second method of comparison, present-worth analysis is employed. The present-worth amount is the amount at the present ($t=0$) that is equivalent to an investment's cash flow for a particular interest rate i. Again citing the discussion in *Engineering Economics*, three features make the present-worth amount a suitable basis for comparison. First, it considers the time value of money according to the interest rate selected for the calculation. Second, it concentrates the equivalent value of any cash flow into a single index at a particular point in time ($t=0$). Third, the value of the present-worth amount is always unique no matter what may be the investment's cash flow.

The rate-of-return analysis provides the third and most important method of comparison. In economic terms, the rate of return represents the percentage or rate of interest earned on the unrecovered balance of an investment so that the remaining balance is zero at the end of the investment's life. In relation to present worth, it is the interest rate i for which the present-worth amount is equal to $0. For an investment having the magnitude and risk that an AS/RS does, the rate of return should be approximately 30%.

6.1. Current System:

6.1.1. After-Tax Cash-Flow Analysis

A End of Year	B Before-Tax Cash Flow	C Depreciation Charges	D Taxable Income (B+C)	E Tax Savings (-.5×D)	F After-Tax Cash Flow (B+E)
0	0	0	0	0	0
1	-1,128,000	0	-1,128,000	564,000	-564,000
2	-1,206,960	0	-1,206,960	603,480	-603,480
3	-1,291,447	0	-1,291,447	645,723	-645,723
4	-1,381,848	0	-1,381,848	690,924	-690,924
5	-1,478,578	0	-1,478,578	739,289	-739,289
6	-1,582,078	0	-1,582,078	791,039	-791,039
7	-1,692,824	0	-1,692,824	846,412	-846,412
8	-1,811,322	0	-1,811,322	905,661	-905,661
9	-1,938,114	0	-1,938,114	966,057	-966,057
10	-2,073,782	0	-2,073,782	1,036,891	-1,036,891
11	-2,218,947	0	-2,218,947	1,109,473	-1,109,473
12	-2,374,273	0	-2,374,273	1,187,136	-1,187,136
13	-2,540,472	0	-2,540,472	1,270,236	-1,270,236
14	-2,718,305	0	-2,718,305	1,359,152	-1,359,152
15	-2,908,586	0	-2,908,586	1,454,293	-1,454,293
16	0	0	0	0	0

6.1.2. Present-Worth Analysis

For an interest rate of 15%, the present worth of the current system over a period of 15 years equals −$4,659,636.

6.2. New Conventional System

6.2.1. After-Tax Cash-Flow Analysis

A End of Year	B Before-Tax Cash Flow	C Depreciation Charges	D Taxable Income (B+C)	E Tax Savings (−.5×D)	F After-Tax Cash Flow (B+E)
0	−3,206,361	—	—	—	−3,206,361
1	−707,670	−160,318	−867,988	433,994	−273,676
2	−757,207	−152,302	−909,509	454,754	−302,453
3	−810,211	−144,607	−954,818	477,409	−332,802
4	−866,926	−137,553	−1,004,479	502,239	−364,687
5	−927,611	−130,499	−1,058,110	529,055	−398,556
6	−992,544	−124,086	−1,116,630	558,315	−434,229
7	−1,062,022	−117,994	−1,180,016	590,008	−472,014
8	−1,136,363	−111,902	−1,248,265	624,132	−512,231
9	−1,215,909	−106,451	−1,322,360	661,180	−554,729
10	−1,301,022	−101,000	−1,402,022	701,011	−600,011
11	−1,392,094	−95,870	−1,487,964	743,982	−648,112
12	−1,489,540	−95,870	−1,585,410	792,705	−696,835
13	−1,593,808	−95,870	−1,689,678	844,839	−748,969
14	−1,705,375	−95,870	−1,801,245	900,622	−804,752
15	−1,824,751	−95,870	−1,920,310	960,310	−864,440
16	+1,440,299				+1,440,299

6.2.2. Present-Worth and RoR Analysis

For an interest rate of 15%, the present worth of the new conventional system over a period of 15 years equals $5,925,454. The rate of return (RoR) on the incremental investment over a 15-year period is about 10%.

305

6.3. AS/RS:

6.3.1. After-Tax Cash-Flow Analysis

A End of Year	B Before-Tax Cash Flow	C Depreciation Charges	D Taxable Income (B+C)	E Tax Savings (−.5×D)	F After-Tax Cash Flow (B+E)
0	−4,235,917				−4,235,917
1	−443,680	−560,210	−1,003,890	501,945	•58,265
2	−474,738	−466,813	−941,551	470,775	+3,963
3	−507,969	−408,357	−916,326	458,163	−49,806
4	−543,527	−350,448	−893,975	446,987	−96,540
5	−581,574	−292,732	−874,306	437,152	−144,421
6	−622,284	−235,368	−857,652	428,826	−193,458
7	−665,844	−178,278	−844,122	422,061	−243,783
8	−712,453	−120,993	−833,446	416,723	−295,730
9	−762,325	−90,595	−852,920	426,460	−335,865
10	−815,688	−85,956	−901,644	450,822	−364,866
11	−872,786	−81,590	−954,376	477,188	−395,598
12	−933,881	−81,590	−1,015,471	507,735	−426,146
13	−999,252	−81,590	−1,080,842	540,421	−458,831
14	−1,069,200	−81,590	−1,150,790	575,395	−493,805
15	−1,144,044	−81,590	−1,225,634	612,817	−531,227
16	+1,038,217				+1,038,217

6.3.2. Present-Worth and RoR Analysis

For an interest rate of 15%, the present worth of the AS/RS over a period of 15 years equals − $5,051,786. The rate of return on the incremental investment over a 15-year period is between 13% and 14%.

6.4. Economic Comparison of Proposed Alternatives

The following table summarizes the economic analysis of the proposed alternatives.

	Current System	New Conventional System	AS/RS
Present Worth at i = 15% over 15 years	− $4,659,636	− $5,925,454	− $5,051,786
Rate of Return over 15 years	———	10%	13%–14%

7. CONCLUSIONS AND RECOMMENDATIONS

Neither the new conventional system nor the AS/RS yields the MARR (minimum attractive rate of return). If Columbus Cloth Works commits capital to one of these alternatives it will have to forego investment in a more lucrative project, one which yields the MARR or a rate of return greater than the MARR. For this reason this report recommends adopting the null alternative, that is, maintaining the current system. As a future project for the CCW Industrial Engineering Department, the report further suggests a study of means to improve the cost-effectiveness of the current system.

8. APPENDICES

8.1. Calculation of AS/RS Capacity

Product Class	Peak Inventory (Millions) A	Avg. No. of Units/Pallet B	No. of Pallet Loads A/B
Basket	49.6	7,000	7085.71
Cluster Pack	33.4	17,600	1897.73
12 Pack	9.6	5,400	1777.78
Cluster Clip	.6	21,000	28.57
Convenience	24.7	25,000	988.00

Total No. of Pallet Loads 11,777.79

(11,777.79 Pallet Loads)(1.08 growth) = 12,732.75
$\sim$ 13,000 Pallet Loads

8.2. Calculation of AS/RS Throughput Capacity

Product Type Units	1978 Production Units/ (Millions)	Avg. No. of Units/Pallet	Pallet Loads per Year
Basket	350	7,000	50,000
Cluster Pack	290	17,600	16,477
12 Pack	80	5,400	14,815
Cluster Clip	5	21,000	238
Convenience	200	25,000	8,000
Total	925		89,530

$$\text{No. of Pallets per Working Day} = \frac{89,530 \text{ pallets/year}}{246 \text{ working days/year}}$$

$$= 364$$

$$\text{No of Pallets IN per hour} = \frac{364 \text{ pallets/working day}}{24 \text{ hours/working day}}$$

$$= 16$$

Reading and Evaluating Technical Reports 309

$$\text{No. of Pallets OUT per Hour} = \frac{364 \text{ pallets/working day}}{16 \text{ hours/working day}}$$

$$= 23$$

Multiplying the number of pallets IN and OUT by a peak factor of 1.33 we have

Peak No. of Pallets IN per Hour =   (16)(1.33) = 22
Peak No. of Pallets OUT per Hour =   (23)(1.33) = 31

Two stacker units, dual command, could accommodate the throughput.

8.3. Accelerated Depreciation Factors

Year	Equipment Factor	Building Factor
Start-up	.1250	0
1	.2188	.0500
2	.1741	.0475
3	.1473	.0451
4	.1205	.0429
5	.0938	.0407
6	.0670	.0387
7	.0402	.0368
8	.0133	.0349
9		.0331
10		.0315
11		.0299
12		.0299
13		.0299
14		.0299
15		.0299

8.4 Present-Worth Calculations

End of Year	15% PW Factor	Current System After-Tax Cash Flow	Current System Present Worth	New Conventional System After-Tax Cash Flow	New Conventional System Present Worth	AS/RS After-Tax Cash Flow	AS/RS Present Worth
0	1.0000	—	—	−$3,206,361	−3,206,361	−$4,235,917	−$4,235,917
1	.8696	−$1,128,000	−$980,908	− 273,676	− 237,989	+ 58,265	+ 50,667
2	.7562	− 1,206,960	− 912,703	− 302,453	− 228,715	3,963	2,007
3	.6575	− 1,291,447	− 849,126	− 332,802	− 218,817	49,806	32,747
4	.5718	− 1,381,848	− 790,140	− 364,687	− 208,694	96,540	55,202
5	.4972	− 1,478,578	− 735,149	− 398,556	− 198,162	144,421	71,806
6	.4323	− 1,582,078	− 683,932	− 434,229	− 187,717	193,458	83,632
7	.3759	− 1,692,824	− 636,333	− 472,014	− 177,430	243,783	91,638
8	.3269	− 1,811,322	− 592,121	− 512,231	− 167,448	295,730	96,674
9	.2843	− 1,938,114	− 551,006	− 554,729	− 157,709	335,865	95,486
10	.2472	− 2,073,782	− 512,639	− 600,011	− 148,323	364,866	90,195
11	.2150	− 2,218,947	− 447,074	− 648,112	− 139,344	395,598	85,054
12	.1869	− 2,374,273	− 443,752	− 696,835	− 130,238	426,146	79,647
13	.1625	− 2,540,472	− 412,827	− 748,969	− 121,707	458,831	74,560
14	.1413	− 2,718,305	− 384,096	− 804,752	− 113,711	531,227	69,775
15	.1229	− 2,908,586	− 357,465	− 864,440	− 106,240	531,227	65,288
16				+1,440,299	+ 177,013	+1,038,217	127,597
Total			−$4,659,636		−$5,925,454		−$5,051,786

9. BIBLIOGRAPHY

Ackerman, Kenneth B.; Gardner, R.W.; and Thomas, Lee P. Understanding Today's Distribution Center. Washington, D.C.: Traffic Service Corporation, 1972.

Allred, James K. "Step by Step through an AS/RS Justification." Material Handling Engineering. October, 1976.

Apple, James M. Material Handling Systems Design. New York: The Ronald Press Company, 1972.

Thuesen, H.G.; Fabrycky, W.J.; and Thuesen, G.J.:Engineering Economy. Englewood Cliffs, New Jersey: Prentice-Hall, Inc., 1977.

White, John A. "Justifying Material Handling Expenditures." Proceedings—AIIE 1977 Spring Annual Conference.

Zollinger, H.A. "Planning, Evaluating and Estimating Storage Systems." Proceedings—Advanced Material Handling Seminar, April, 1975.

Manuscript specifications for Bibliography and End Notes appear in Appendix A, pages 477–479. Forms for End Notes and Bibliography appear on pages 455–457 in Chapter 24.

chapter 16

Technical Presentations

The introduction to this text gives an overview of the various forms which engineering and technological projects take: the creative idea for solving a problem, brainstorming sessions, telephone conversations, letters, memorandums, reports. In the course of a project, the technical content will assume many configurations: written and oral, short and lengthy, exhaustively planned and situational.

A technical presentation, like a major proposal or a long technical report, marks a major phase in a project. The basic thing to remember about the technical presentation as an oral form is this: the technical content is the same as it is in any technical document. General considerations of organization, length, selection of detail, and use of graphics are very much the same whether you are writing or speaking. The crucial difference is that the oral presentation is a performance; yet the technical matter should be thoroughly familiar to you.

The chief task, then, becomes learning how to present technical content in a different form. As with any written form, the most important work is outlining, a process which lies outside the actual process of writing. Planning and good organization are even more necessary for the oral presentation.

FACTORS TO BE MANAGED

Before you open your mouth to speak, you should have already managed the following factors.

Acknowledging the Fear of Speaking

So much has been written about the fear of speaking as one of the common afflictions of humankind, that we will deal with that first. Let us do so by breaking that fear into its components.

Unless you know that you fear speaking, do not assume that you will be afraid; do not allow the attention given to stage fright to become a self-fulfilling prophecy. To give an example, the saddest movie of my childhood was *Lassie*; everyone knew that everyone cried during that movie. One child in the neighborhood went to the movie carrying more than one handkerchief; when the title and credits began to appear on the screen, he began to cry. He wept copiously at the first frames and continued to weep throughout the movie. In my experience, many speakers react the same way: "Everybody is afraid of public speaking; therefore, I also must surely be afraid to stand up before a group and talk." Don't be like the child who went prepared to cry in a movie; don't assume that the experience must be and will be frightful.

Before you begin to do anything about any fear you may have of speaking, first give yourself permission to feel those feelings. Consider also that you may have good reason to fear speaking. Three reasons apply in particular to engineers or technologists:

a) If you are sufficiently expert on any subject to deliver a presentation before an audience, you have to do highly concentrated study to develop that expertise. Such study, especially given the exactitude which most technological projects require, is like a centripetal force drawing you inward to a specific point.

   The necessity to talk about what you know to an audience is a strong counter-force, a centrifugal force which flings you outward, however unwillingly, to communicate the results of concentrated study. To write you must shift gears from one mode of communication to another; to speak smoothly and well, you must shift into an overdrive gear. So account for that necessary outward shift when you think ahead to a technical presentation.

b) Consider also the shift from drawings, calculations, lab experiments, and field observations to words as your primary mode of expression. To shift from the more familiar technical modes of description and expression to words immediately adds a variable—one, moreover, which many engineers either distrust or find more imprecise. Neither the written word nor the spoken word can be crafted to the fine tolerances you so love.

c) As if it were not enough to shift from the concentrated work of an E/T and to shift into words as your medium of expression, speaking adds the greatest variable of all: the human variable of the listening audience immediately present.

Remember that what you know is a constant which underlies whatever communications mode you may use and whatever communications situation you may confront. Writing and speaking call upon you to shift into modes which you may find less to your liking. If in these paragraphs I have been able to give your fear a name, own your right to be a bit on edge, or even terrified. Those feelings are normal for you. Do not expect that you will ever entirely overcome some degree of nervousness about presentations. Most people feel it; in fact, if they don't, they may lack an edge that makes for a more effective presentation.

Managing the Fear of Speaking

If you do approach the necessity to speak in public with fear, then accept that as a normal and widely experienced reaction. Allow for your fear as a given factor in the overall management of your presentation, and manage it in these ways:

a) Fear is physical reaction, so deal with the physical symptoms of fear: breathe deeply, relax your arms and shoulders. As a student of mine said, "You are telling me to do the same things I did before races when I ran track in high school: breathe, relax, limber up." Take a walk, run up a flight of stairs, do something physical to burn off some of the physical tension.

b) An athlete or a singer prepares for a performance; you too should give attention to rest and diet. To illustrate the importance of physical energy: a student of mine completely failed in his first speech—he was disoriented, poorly organized, and finally, silly at the lectern. A later conference revealed that he had eaten breakfast early in the morning on the day of the speech, but by 2:30 p.m. he was tired and so hungry that he lacked the energy levels required to make a speech. No wonder he performed poorly. Part of his assignment for the next speech was to take care of himself physically; his grade improved from F to A, and he has discovered that he actually enjoys speaking.

Perhaps you cannot banish a feeling of physical nervousness entirely, but you can do much to manage physical factors which contribute to it.

Managing the Physical Setting

You can also do much to assure your physical comfort during a presentation.

a) Physical factors which lie very easily within your control relate to the physical setting of your speech. If you are going to speak in a classroom, return to it after classes to stand in the space and acclimate yourself and your voice to the space. Manage the factor of standing up and hearing the sound of your own voice in that space, for one of the unfamiliar and uncomfortable parts of speaking can be simply the sound of your own voice at greater volume within the attentive silence of an audience. If you are speaking in another setting, arrive early and familiarize yourself with the physical space.

b) Another manageable factor of a physical setting is the lectern. If you are tall, learn to stand back a step from the lectern so that the audience does not get a view of the top of your head as you bend to check your notes. If you are short, stand closer to the lectern and consider moving to either side of the lectern from time to time during the speech so that you are not hidden from the audience; even better, consider placing your notes on a table so that you are not blocked from an audience.

If you are nervous, the lectern itself will hide your shaking knees; so, whenever possible, use a lectern. To conceal the shaking of your hands, rest them on the sides of the lectern. Press your fingers together tightly so that they will not have the "play" to tremble.

c) If you will be dealing with a microphone, make certain that you are comfortable with it. With some microphones, the audience will lose the sound of your voice if you turn your head while speaking; so make certain that you can avoid such spotty transmission. If, by some chance, you have to hold a hand microphone like a singer and if you are nervous, hold your elbow closely to your side and hold your thumb parallel to the shaft of the microphone to diminish the shaking of your hand. If possible, use a lapel microphone so that mobility will be one thing less for you to worry about.

The most important factor to remember is becoming informed beforehand about what the physical conditions of your speech will be. If you plan ahead so that such a factor does not surprise you, you can better diminish or handle any nervousness you may have.

In considering the general physical factors and equipment used in a speech, draw an analogy with sports. Would you ever consider playing a round of tournament golf unless you had first surveyed the course? Would you play a tennis match with a racquet you had simply picked up at random without knowing something about the size of the grip, the weight, and the tension of the strings? Would you pick up a shotgun for the first time just before the first clay bird was released from the trap and expect to score a hit? No, on all counts—if you are at all interested in making a good showing. The very same principles apply to the use of any equipment in a speaking situation. You will perform better if you have thoroughly mastered the equipment you will use.

Managing Speech Notes as a Physical Factor

Making notes for a speech is a physical factor, not just a matter of writing. It too can be managed. The most important thing to remember about notes is this: you are not writing prose which will be read as you are reading this book. Write on only one side of the page, and design the page for ease both of reading and of physical handling. Underline key words, indent a series of key points or phrases, write key words in the margin of a prepared text so that you can use that as an outline. At the bottom right-hand corner of the page write a key word of what comes at the top of the next page. In sum, make the notes a working document for yourself so that you can use them comfortably and unobtrusively.

Some texts advise that you outline each major point on a 4 × 6 note card. If you do that, do not shuffle and tap the cards on the lectern. Others advise the use of 8½ × 11" pages—and most speakers you see in public will use this method; they will carry the notes in an attractive folder or loose-leaf binder. On the lectern, slide the page you've finished to the left rather than turning or flipping the page; in this and other ways, avoid reminding the audience that you have notes. Whatever you do, prepare a set of workable notes for yourself, and prepare to use them in a way that is not distracting to the audience.

Managing A/V Equipment and Graphics as a Physical Factor

If you are going to use a chalk board, a flip chart, or any kind of audio-visual equipment to illustrate your speech, make sure of two things: that you are thoroughly comfortable with their use in the physical setting, and that you have smoothly coordinated the text of your speech with the illustrations. You will lose your audience, and your presentation will lose momentum, if you make long pauses to shift between your speech text and the graphic illustration. A speech, like a song, depends somewhat upon the momentum of an uninterrupted flow.

Up to this point we have said nothing about what you are going to say and how you are going to say it. How well you do both can be influenced by how well you manage many other factors leading up to the actual writing and delivery of the presentation. Writing and delivery are also matters of management: managing the content of your presentation, managing time, and managing the attention of the audience.

Managing the Material for an Oral Presentation

In discussing all written forms, we have discussed the division into parts and the management of the parts of any written document. The same rules apply to an oral presentation—and even more so, for the listeners' ability to understand what you say depends largely upon how you arrange and manage the data. Just as you make the design of data on a page clearly visible by means of lists, underlining, and other management techniques, in a speech you must let the listeners know at all times where you are and where they are in the presentation. Following are some ways to manage the material in a speech.

a) As you would for the introduction in a 500-word theme, or for the summary which begins a lab report, or for the Executive Summary which begins the technical report, first give the listeners an overview of what you will be covering. Be specific in how you do this; address the audience directly, "I want to discuss with you four major points which are essential in understanding (subject) ." or "This presentation discusses (subject) . To give you a detailed description/understanding of all the factors which go to make up (subject) , I'm going to divide this presentation into three major parts, and illustrate each part with (graphic techniques) ."

By making clear to the listeners where you will be taking them, you prepare them, and you also establish a basis for managing both your material and your listeners' attention. Don't feel that you are being too obvious or that you will be too repetitious; instead, remember this essential factor about an oral presentation: speaking establishes a very fragile bond with your audience; they can tune you out at any instant unless you carefully reduce the parts of your presentation to manageable increments.

b) As with the body of any written work, create strong logical transitions between the parts of an oral presentation. If you break the first major part of the body into three subheadings, pause at the end of the first part to review briefly what you have said and then introduce the second part. Use a model similar to this: "In the first part of our discussion of (subject) , we have established (points a, b, c) . The next section will carry us into a discussion of _____ and _____." "Now that we have established (points a, b, c) , another major factor/consideration/point to consider is (subject of part 2.)"

Two factors make such an approach especially necessary when you are dealing with technical content. For one: the material is difficult to comprehend and requires sustained and directed attention, even for experts in the field. For the other: in any oral presentation, you must manage the fragile bond of attention. The person who fell peacefully to sleep at point I-A may rouse himself to re-enter the stream of discourse; or the person whose eyes glazed over while she worried momentarily about her cash flow may get hold of herself and try to renew her attention. You know yourself that you do not and cannot always follow a lecture with complete and uninterrupted concentration. Don't think that an audience will be any more willing and able to listen to your oral presentation than you yourself usually are, and allow for that potential inattention in the way you present material.

Managing Audience Attention

Another especially important factor about any technical presentation is that very frequently you are selling an idea to an audience who may be largely non-technical. This audience factor greatly increases the necessity to be clear and well organized. Moreover, you do not have the sales techniques available to you that you would have if you were selling an automobile or a vacuum cleaner. Sales in engineering and technology must rest upon the persuasive force of the clear, thorough, accurate, and compelling technical explanation of what the problem is and how you are best suited as an expert to solve it. This characteristic of persuasive technique places an enormous weight upon your ability to manage material and manage the attention of a listening audience.

For all these reasons, carefully plan and manage the body of any kind of oral presentation.

a) Rely upon A/V devices and other graphic aids in any technical presentation. This is necessary not only because the material lends itself to such treatment, but also because it is appropriate for the audience; technically oriented people are usually better able to perceive and process visually than verbally so using visuals can help you manage the listeners' attention. Be especially aware of the following techniques.

   (1) If you have a printed handout to illustrate your speech, plan very carefully when and how you will distribute it. Clearly explain to the audience what it is and what purpose it serves. Also, use and refer to the handout as a guide to what you are saying. Do not, repeat do NOT, distribute a handout which is not directly useful to you; if you do, you have just given the audience a way to tune you out, read the handout while you are speaking on another part of the topic, or even write a grocery list while you are hard at work at the lectern wondering why this audience is so inattentive.

   As a rule, a transparency on an overhead projector or even work on a chalk board are more effective than a handout. By using those, you maintain better control of the listeners' attention and the flow of information to them.

   (2) Introduce a slide or a chart verbally before you show it. That is, let the audience know what to expect. Then show the slide, or introduce it in a context you have already prepared. "The first chart/slide will show _____." "The next chart leads us another step/introduces a second/third point about (subject) ."

Factors to Be Managed

(3) Introduce charts in sequence, rather than display all of them at once. Do this as a way of centering the listeners' attention upon the one chart you are discussing. Do the same with points you may list on a chalk board.

(4) Be certain that you stand near any visual aid that you use. Do not stand at the lectern with your notes and point several feet to item 3 on an outline on chart or chalk board. If you are using an overhead projector with a transparency, point to the area you want the audience to attend to; do not rely on verbal directions alone to direct the listeners' attention. Again, there are many visual distractions within the visual aids; any listener is perfectly free to tune you out and muse upon what chart 4 means while you are busily explaining chart 2. What this comes down to once more is the necessity to manage the listeners' attention.

b) The conclusion of an oral presentation must also contain careful management devices—like the conclusions of a 500-word theme, a professional article, or a technical report. Make the various points of your speech add up to a series of conclusions or recommendations. You have led your listeners point-by-point through the presentation; now you must establish the destination you were heading toward. Most of the techniques of other forms of discourse apply here, especially the necessity to be clear and specific.

Managing Time

From first to last, time is an important factor in speaking. We have discussed brevity and a well-proportioned page design in the memorandum and the letter, but nowhere are the elements of brevity and proportioned design more important than in speaking. For that reason attend closely to two factors: how long your prospective audience has asked you to speak, and how long you will be able to hold the concentrated attention of the audience.

If the appointment to make a presentation includes a time limit, adhere strictly to it. Rehearse and time your presentation, give careful attention to selection of important details, allow time for questions at the end or digressions to illustrate points in the body of the speech. But do not, repeat DO NOT, impose upon an audience past the allotted time. If you have ever sat through a lecture or a sermon you thought would never end, or if you have been stuck with a garrulous drunk or a monologuing dinner partner, you know the feeling, of being trapped while someone speaks on past the end of your attention span; don't create that feeling in an audience to whom you speak.

Just as the young engineer in the introduction had to limit what he wrote about generators specifically to what the audience needed to know, limit your presentation in terms of the listeners' needs. More so for a spoken presentation than for a written one, those needs may be very limited. The work of writing and planning can then be a very severe discipline indeed; recall poet May

Swenson's comment about selection of detail, "We all have to murder our darlings." Therefore, since you cannot say *every*thing, select the elements which will give you the most thorough coverage with the best overall effect.

As a purely practical note, wear a watch, have a clock in view—however you manage that, know at all times how much time has elapsed and how much time you have remaining. Unless you are very good at using a digital readout on a watch, the traditional clock face configuration is easier to refer to during a speech.

GROUP PRESENTATIONS

Very often a technical presentation which sells either a project or a company's services will require a team of presenters. If you are in that situation, pay particular attention to these factors above and beyond all the other considerations that we have listed and discussed.

Design Conference

Meet together as a group to decide how the work is to be assigned. Decide as a group the major points you must cover and how the work of each individual will form a component of an overall design. Do not, repeat DO NOT, work individually on your presentations until you have first met to determine what the overall work will entail. If it is a project on which you have already had assigned tasks, then each of you has an area of expertise, but even so the physical setting, the audience, and the time allotted may necessitate many shifts in responsibility for presenting portions of the subject.

Presentation

If a team is making the presentation, appoint one person to introduce the individual presentations and orient the audience to the connections among them. This person can:

(1) begin by introducing each member of the group with a sentence about how each one's contribution will make up the presentation as a whole.

(2) say something like "I will begin by" (then describe what you will do just as you would introduce any speech).

(3) briefly recapitulate each individual presentation, just as one would conclude any speech, and then introduce the next speaker; make certain that a strong transition prepares the way for the next speaker.

(4) plan carefully who will give the summation of the presentation. This may be the coordinator who introduced the topic and the speakers; it may be the senior member of the firm who steps in then to give an overview and final persuasive conclusion. Whoever this person is, this speaker should be the most adept of the group; for, especially if this is a presentation with sales as its purpose, the summation is very much like the closing argument of an attorney to a jury. While different in form, the decision of the prospective client(s) is nevertheless a kind of verdict.

To conclude this section on how to present technical subjects to an audience, it may be well to rely upon the root definition of the term *manage*. The word dervies from *manus* which means *hand*. To manage is to "have well in hand." The very ancient meaning of that term was to manage, or have well in hand, a spirited horse or team of horses. A coachman driving six horses had always to be aware of the action of each horse in the team in order to manage them or keep them well in hand. For a coachman, or for a single rider, the hands on the reins are crucial because it is largely through the reins that the manager sends and receives signals. Obviously, a speaker is not driving an audience like a team of horses, but the management task is somewhat the same. You are in charge of a group of people whose attention can stray at any instant; you have planned a route; it is your primary responsibility to see to it that the group remains concentrated in its attention and direction, and that you all arrive at the desired destination united in understanding and opinion. This is a sizeable management task, just as managing a team of spirited horses would be. Do it once, and you'll understand why many people become nervous when they stand up to deliver an oral presentation: you're in the driver's seat and you take upon yourself the responsibility for whether all the listeners arrive at the same point of understanding at the same time. Only by careful planning and management can you assure that they will.

chapter 17

Professional Articles

The first sustained writing an engineering, technology, or science student is likely to do is the lab report. From this common experience, demands for sustained writing tend to divide into two tracks: the E/T follows along a track of lab reports and moves on to the short technical report and the formal technical recommendations report; science majors tend to move along a track toward the thesis, the dissertation, and the professional article which publishes the results of research. However, both technological professionals and scientific professionals may elect to write professional articles for publication in specialty journals.

The professional article differs from the lab report and the technical report in this regard: one *elects* to write such a paper and submit it either for publication or for reading at the meeting of a professional organization. The lab report is the end result of an experiment; the technical report is the chief means of conveying to a client the substance and significance of a technical project. The professional article, in contrast, is not the final logical end of professional activity even though a large body of knowledge in any profession is communicated by that means. The writer of a professional article often intends it to aid her professional advancement. The outcome is that this professional activity communicates the results of research, reports the lessons of professional experience, and shares insights with other specialists in a field.

In form, the lab report and the technical report have certain prescribed features which appear in a prescribed order. The professional article is less formal in its construction and, in fact, varies widely in subject, in style, in forms of documentation, and in whether the article is documented at all. The form a professional article takes is determined almost entirely by where the writer plans to submit the article: a magazine, a journal of research, transactions, the proceedings of a professional meeting, or some other publication. The selection of subject and the style are also determined by the prospective placement of the manuscript.

This range of variables means that the would-be writer must do a very thorough analysis of the prospective audience beforehand. And while the writing process is not as formal as that involved in producing a lab report or a technical report, the professional article exists within its own system; therefore, it can and should be carefully planned and managed.

The subject matter for a lab report or a technical report is prescribed by an experiment or project. Professional articles, however, may be on almost any professionally related subject. Generally they fall into three categories: articles to report the results of research, articles to give general information to professionals, and articles to persuade.

The field of Allied Health—nursing, physical therapy, dietetics, medical technology—offers an example of how published articles vary in subject for the various sub-groups within that large profession. An article in the journal of the American Physical Therapy Association might be a highly technical and objective article which reports the use of new techniques in treating athletes with injured knees. The reading audience would be made up almost entirely of physical therapists who might then directly apply those techniques to their work with patients. Therefore, such an article would be very similar in tone, format, and level of technical information to a lab report or to the body of a technical report. An article in a publication of the American Dietetic Association might also be a formal, highly technical report of the results of research which could be applied directly to the work of dietitians interested in sports and nutrition. In that sense, basic research in any profession can become applied research when the results are published for colleagues within a narrowly defined specialty.

Articles of more general information also appear in professional journals. Health-care professionals must keep thoroughly informed of government regulations and of the changing conditions under which any profession operates. Therefore, an article to inform a group of readers of the impact of new regulations governing prescription drugs in sports medicine would be of interest to readers of a publication like the *Journal of Allied Health*. With increased interest in sports medicine, an article by a teacher or administrator might show how a professional curriculum could adapt to such a trend. In either instance, the experience and insights of a professional in dealing with common tasks and common concerns become the subject matter of an article written and submitted for publication.

It is important here to emphasize that a discipline in engineering, technology, or science does not exist under ideal laboratory conditions. Developing knowledge within a given profession calls for technical reports; such advances also call for information about and discussion of how the profession works with society at large or with the specific groups which it serves. These concerns create a need: workers who are involved with such concerns must become writers who share information and insight with the larger professional community.

There are other, less significant articles which inform as well. I can cite in particular an article in the *Journal of Allied Health* which described how someone in that field might get a job overseas. The writer had served in several foreign assignments and was thus able to give specific information on how to inquire about such jobs, what salaries and term of service to expect, and some of the major issues involved in practicing a profession in a different culture with a different language. Thus the most individual experience in a profession can be translated into an article of general interest for a reading audience of colleagues.

Political issues and discussion of professional issues arise in most professional groups. In this case, an article to persuade can be the outgrowth of a professional's direct involvement with current issues. Some groups within the profession of Allied Health may question the negative effects of sports competition on boys under the age of 10. Another professional might do a survey of court cases involving younger athletes. Any profession interfaces with individuals, with large organizations, with government, and with society as a whole, so issues for discussion and debate always arise. Professional articles thus operate in a variety of ways beyond the specifically technical reporting of information.

As this extended example based on the Allied Health field shows, the range of subject and purpose for a professional article is almost endless for a well-informed and innovative member of any profession. Consider that electrical and electronic engineers have nearly three dozen journals and transactions, each representing a rather narrowly specialized band within that entire discipline, and that there are publications which deal with specific applications of engineering and planning, such as in the fields of corrections and of retail and commercial construction. A prospective writer in any professional specialty has a wide range of opportunities. The range of opportunities can become almost a disadvantage, for a professional article is in no way as formally prescribed in its construction or as narrowly specified in its content as the lab report or the technical report. In addition, all the concerns of organization, format devices, where and whether to use illustrations, and how and whether to give end notes and bibliography are likely to vary from publication to publication, even within a well-defined profession like Allied Health. For electrical engineers, the IEEE has a standard style sheet to prescribe form in publications in that area. For the most part, however, the writer must either follow the specifications published in a journal or infer from reading a wide selection of articles in a journal what form a submitted article should take.

Therefore the nature of any professional article is such that we cannot break it down into component parts as we would the other forms included in this text. The discussion which follows, however, explains how a writer can design and manage a system for publishing an article.

PUBLISHING A GENERAL-INTEREST ARTICLE

Most professionals develop, as a result of experience, general insights and expertise which can be communicated to other professionals. Indeed, professional publications and professional meetings exist to make such communication possible. Because of that connection with a profession as a whole, the professional article is more audience-oriented in broad general terms than the report of the results of research.

REPORTING THE RESULTS OF RESEARCH

The best way to assure publication of the results of a research project is to build the writing of the article into the research work itself. In fact, the preliminary questions about beginning a research project are the same ones a writer asks before planning an article.

Subject: WHAT?

1. What is the idea? (or Statement of the Problem)
2. Has this been done before?
   To answer that, both researcher and writer consult bibliographies and avail themselves of computer searches for preliminary bibliographies. A further step a writer takes is to inquire of experts in the field or to query an editor about whether an article is needed on a particular subject. These basic preliminary searches underlie both the writing process for a professional article and the undertaking of a research project.
3. How does this idea relate to and build upon previous research? Exploring such connections is another common basis for research for the thesis writer, the researcher, or the professional in search of an article topic.

4. What will be required to do the project? What resources are available?
   - methods
   - materials
   - personnel
   - investigative instruments
   - lab procedures
   - time
   - money

All these basic considerations contribute to making the business of planning a research project and that of planning a professional article the same kind of task initially.

Audience: WHO?

1. Who is the potential audience?
   It is here that the writer branches off from the researcher, for a researcher more nearly sets out to address a problem in and of itself. For any creative professional, projects suggest themselves as a natural outgrowth of study and investigation. A researcher who wants to proceed further toward an article must ask further questions about the audience. Will they be

   colleagues

   practicing professionals

   members of government

   organizations in industry

   an organization or special interest group within a profession?

2. What publications for the target audience are most likely to be receptive to the results of this study? Professional journals and magazines may have a rather specialized readership; thus the answer to such questions can establish a well-defined target audience and also the publications which serve such a readership.

   To make an effective audience analysis, the prospective writer must apply these questions:

1. Who is the audience?
   - members of a profession at large?
   - an organization or special interest group within a profession?
2. What is the audience's level or nature of expertise? How does a readership use its expertise:
   - in becoming generally informed?

- in the practice of a profession?
- in teaching?
- in administration?
- in production and marketing?
- in policy making?
3. What is the level of the audience's need for the substance of the article:
    - to generate discussion—insight into common problems which inhere in a profession?
    - to suggest application—new ways of viewing established ideas and procedures?
    - to present theory—advances in the state of the art in a profession?

Purpose: HOW?

What will be the focus of the article? Subject and audience determine the answer, in part, but the specific publication can also yield answers to this question. A formal journal of research will hardly publish a "how to" paper for teachers of basic courses, nor will a slick publication aimed not only at professionals but also at consumers publish a lengthy and detailed paper on an obscure research topic. The prospective writer should therefore determine whether the substance of the article will be to report, to inform, or to persuade. Will the article be:
- detailed and directive for the reader who wants to learn *how*?
- general and descriptive for the reader who wants to learn *what*?
- theoretical for the reader who wants to learn *why*?

What will be the tone of the article? Tone, or the writer's attitude toward the subject, is an elusive quality. However, most publications will have a somewhat uniform tone and approach. *Time* magazine is probably most strongly identified with a characteristic tone, style, and overall point of view, but even *Scientific American* and *Science* have characteristic tones. To identify the tone of magazines which may be possible publishers for a professional article, survey:
- whether the articles are written in first person: *I, me, my;*
- whether the publication uses the jargon or the formal vocabulary of a profession;
- whether the publication uses illustrations, and what type: drawings, photographs, technical illustrations;
- whether you receive the feeling through reading the articles that the writer is addressing *you* as reader or is writing *about* a subject;

- whether the article seems to be devoted only to informing the reader, or is also interested in entertaining the reader;
- whether it attempts to shape the reader's opinion.

All of these factors help a writer to determine the general tone to set in writing an article for submission.

Publication: HOW?

Once a writer has selected a specific publication, a close reading of the publication determines how the article should be prepared. The discussion of purpose gives a partial answer, but most magazines have elements of format which can be analyzed for an answer:

- What is the average length of the article?
- What is the layout: number of columns to the page? illustrated? subheadings in bold type? These are particularly important considerations, for a writer can prepare an article with paragraphs which will set up well in that format, add subheadings, include illustrations, and tailor the article to the publication. To determine the average paragraph length, count the words in one inch of copy down a column. Are the paragraphs relatively short, are they long and dense, are they broken with lists and numbered items? Showing an awareness of how the copy will actually look when it is set in type is absolutely essential, and an experienced editor will acknowledge the author's awareness as she reads copy.
- Most important of all, does the magazine have a page of suggestions and/or specifications for writers who submit copy? If so, the article submitted should follow these *to the letter*. Many professional publications are edited by other professionals (who volunteer that work as a professional activity, but who are not editors by profession). Therefore, such publications are simply not set up to do extensive copy-editing of submitted manuscripts. In fact, some print by a photocopy process which transfers the submitted manuscript directly to the publication page. All these considerations place heavy responsibilities upon the writer to prepare "camera-ready copy" targeted for the specific publication.

Publication: WHERE?

Aside from the researching, writing, and submission of an article, there are many other external considerations which can influence its acceptance:

- publication schedule: whether a publication is issued weekly, monthly, quarterly, or as funds allow, will greatly influence the demand for manuscripts.

Acceptance of an article in a quarterly may mean a publication date more than a year after acceptance, a real consideration in the sciences and technology.
- submission rate: more prestigious publications may have large numbers of submissions which influence both the time required for an acceptance or rejection response and the time required for publication.
- age of publication: some newer publications which are not yet well established may be more likely to accept an article than the older, better established journals with a large submission rate.

The writing itself is only a very small part of a total planning and management picture; every sustained writing project presents a sizeable management task. The writer of a technical report also writes memorandums and letters in a total correspondence system; similarly, the author of a professional article who wants to see it published also manages a project, which includes:

1. analysis of possible publications;
2. first query to the editor;
3. submission of the manuscript;
4. waiting for a response—and waiting, and often waiting still more;
5. correspondence about the accepted manuscript and needed revisions;
6. further correspondence about the manuscript;
7. directions and supervision for a typist;
8. postage, the right sized envelopes, trips to the post office.

To manage that process well, keep a correspondence log to track the activity on a manuscript. For, in essence, all of the work surrounding the publication of a manuscript can be called systems design and management. A professional comes to see the task of writing publishable papers as a skill which must be developed and maintained over a lifetime as an essential component in professional success.

EVALUATING PROFESSIONAL ARTICLES

The three professional articles which follow are included to illustrate the importance of audience analysis in all professional writing. The first two articles are my treatment of the same subject for two different publications—the *Southern Humanities Review*, which rejected the article, and the *Georgia Journal of Science*, which published the paper in 1977. The third article, by Georgia Tech student Alice Miller, is an example of a scientific article delivered to others in a close circle of student engineers.

Taken together, the three articles illustrate the range of possible styles: from literary style, with Modern Language Association style in footnote references and bibliography; to scientific documentation; to a report of original research, using only one reference source to provide a general definition and background to the study.

In reading each article, note particularly the following points of style:

1. Introductory techniques: Which article shows the greatest awareness of an identified audience? Would such a link with an audience enhance one's chances of having an article accepted for publication or reading at a professional meeting?
2. Body techniques: In the body of the first Lorimer article, many names are cited which do not appear in the second version of the article. How would the audience determine whether an author created a broad historical context or concentrated only on a narrowly defined topic?
3. Style and Tone: In 1, 2, 3 order, which of these articles is most objective and scientific in tone? Which is least objective, or more affective?
4. Documentation: Note the techniques of documentation in the first two articles on Lorimer. Can you think of practical reasons, related to an audience, for why different forms are used for the humanities and for science?
5. Concluding techniques: Is there a different appeal to audience or a different degree of objectivity in each of the articles? Why in terms of audience?

Manuscript specifications for preparing a Professional Article appear in Appendix A, pages 474, 475–481. End Note and Bibliography Forms are cited and discussed in Chapter 24, pages 455–457. Manuscript specifications for End Notes and Bibliography appear in Appendix A, pages 477–479.

Example 17.1

A FURTHER NOTE ON DR. JOHN LORIMER'S INTEREST IN SCIENCE

In 1970 Robert R. Rea and Jack D. L. Holmes published an article in the <u>Southern Humanities Review</u> entitled "Dr. John Lorimer and the Natural Sciences in British West Florida." They described Lorimer as typical of the amateur scientist of the eighteenth century. His studies of the flora of West Florida earned him election to the American Philosophical Society in 1769. The naturalist William Bartram was his guest in Pensacola. In addition to pursuing a professional interest in botanical pharmacognosy, Lorimer interested himself in charting sections of the Gulf Coast and lands in Mississippi. In 1766 he determined the longitude of Pensacola, using his observations of the eclipses of the satellites of Jupiter in making his calculations. For several years he maintained a complete record of observations of the weather in West Florida. He was a householder, owner of a large tract of western land, and a member and president <u>pro tempore</u> of the West Florida Assembly—in short, just the sort of man of affairs and amateur scientist whom we see so often in the eighteenth century making extensive observations of the world around him.[1]

Rea and Holmes, in dealing with Lorimer's interest in things terrestrial, did not note his observation of the 1769 transit of Venus. A letter to a British correspondent forwarded to the Royal Society of London reveals him as one of the amateur scientists so captivated by that phenomenon. His observation and the subsequent report serve to relate him to a larger scientific community. He was one of the men who made up the learned world which looked toward the transits of Venus in 1761 and 1769 as momentously significant phenomena in the solar system.

It is difficult to overemphasize the intensity of interest which these transits aroused in the eighteenth-century scientific community. Learned men were just then beginning to grasp the immense breadth of Newtonian physics and were thoroughly awed by their newly-found power to understand what they perceived to be the innermost secrets of the universe.

Since 1716, the year of the Royal Society's publication of Edmond Halley's paper explaining the significance of the coming transits, scientists had planned for them as rare opportunities to gather data leading to the accurate calculation of distances in the solar system. Observing the transit of Venus across the sun from widely separated points on the globe would provide a known base length and known angles between the equal sides of a triangle; finding the distance between the earth and the sun then became a matter of finding the altitude of an isosceles triangle. Once they could determine the astronomical unit by this means, astronomers could in turn determine the distance of other planets from the sun and thus the size of the solar system.[2] This was considered the solution to the "final problem" of astronomy in the generation after Newton.[3]

Adding urgency to the desire to observe the transits was the infrequency of their occurence. A transit of Venus had occurred in 1639, before the learned world had become aware of the significance of the phenomenon; hence, no data had been recorded then. It was known that Venus would traverse the sun's disk in 1761 and in 1769, but not again for 105 years, and thereafter in pairs eight years apart at alternating 121½ and 105½ year intervals.[4]

By 1761, scientists showed considerable interest in viewing the transit, as evidenced by the 120 detailed observations reported to various scientific societies from 62 stations around the globe.[5] Because the transit was not visible south of Newfoundland, the colonial government of Massachusetts financed an expendition to St. John's which was led by John Winthrop. There he made a detailed and relatively accurate table of observations later published in the Philosophical Transactions of the Royal Society, the only report of the phenomenon from America.[6]

The 1761 transit had been so widely reported that the 1769 transit was greatly anticipated, especially by American colonial scientists who hoped that their contribution of data would distinguish their community in the scientific world. As scientists began to collect instruments and prepare observation stations, newspapers and almanacs carried accounts of how amateurs might view the transit. People all over the colonies prepared smoked glass and all manner of small telescopes in anticipation of June 3, 1769, the date of the transit. In Cambridge, John Winthrop prepared to make his second observation of a transit, having failed to gain a second appropriation from the Massachusetts government—this time for an expendition westward to Lake Superior where that position would afford a better view.[7]

Winthrop's Harvard group was without lenses for one of its telescopes, for parts had been returned to Philadelphia, after repairs in London, too late to be forwarded to Cambridge. However, David Rittenhouse fitted them into a telescope which he readied for use at a station sponsored by the American Philosophical Society on the Rittenhouse farm at Norriton, twenty miles from Philadephia. In the months before the transit, Rittenhouse began construction of an observatory, worked to adjust his clocks for maximum accuracy, and made careful calculations to determine the longitude and latitude at Norriton—even running chains from a point in Philadelphia where the longitude and latitude were known exactly.[8] This kind of preparation was being repeated in both settled and isolated places all over the world; the 1769 transit produced 138 observations from 63 positions.[9] With the next transit of Venus expected in 1874, the 1769 transit was for the eighteenth-century scientist a last chance.

This fact added to the care which men like Winthrop and Rittenhouse took with their preparations. Their activities stand in real contrast to John Lorimer's situation. Less than a month before the June third transit, Lorimer was by what he termed "an absolutely military order" obliged to transfer from a relatively comfortable station in Pensacola to a dangerously unhealthy garrison at Mobile. It was dirty, it was swampy, it was fever-ridden. It was a place known then as a "graveyard for Britons."[10] At a time when the learned world was united in a common interest, Lorimer keenly felt his isolation. As he wrote to a friend, "You may be assured that an epistle from an ingenious friend tho' ever so short is like a Cordial in this lonely spot." He added plaintively, "... be so good as to acquaint me of what is new."[11]

Lorimer not only lacked sufficient time to make preparations to view the transit from Mobile, but the weather was so cloudy in late May that he was unable to make the necessary observations for calculating the longitude and latitude and for checking the accuracy of his clock. Without making these preparations he could not collect useful data.

To add to Lorimer's difficulties, the day of the transit was stormy. He reported, "In the forenoon it rained. It was impossible til after 3 p.m. to discover whereabouts the sun was, and then it was only through a cloud that I could see the body of Venus which was considerably advanced on the Sun's Disk.... The sun soon disappeared and was not to be seen until about 5. I had then a pretty distinct view, and could see it through some flying clouds for a full half hour, when it entirely disappeared."

Despite such difficulties, Lorimer made an attempt at a formal report, sending an account of his observations in a letter to a friend in England. The letter makes it evident that had Lorimer been in Pensacola, whose longitude and latitude he knew, he would have been able to collect more usable data on the transit. Sightings were best made by groups, and in Pensacola Lorimer not only could have worked with his friend, cartographer George Gauld, but also would have had the use of Gauld's instruments: a reflecting quadrant, which was used to make sightings essential in setting clocks, and a large surveyor's theodolite, which was used in measuring altitude. His letter also noted the advantage of Gauld's station on Pensacola Bay, which afforded a distinct view of the horizon. Lorimer included in his letter a partial account of Gauld's observation which had been sent to him in a letter.

Although Lorimer's English correspondent forwarded his observations to the Royal Society, the letter evidently was not of sufficient value to be read to the Society or to be printed in the Philosophical Transactions. In his book on the eighteenth-century transits of Venus, Harry Woolf makes no mention of Lorimer's report in his compilation of twenty-two sightings in colonial America. Thus until now Lorimer's report had remained among informal accounts of the transit which appear in journals and collections of letters.

John Lorimer viewed the transit of Venus as an amateur scientist who had already received a measure of recognition from the American Philosophical Society. Later in his life, the Royal Society would recognize his study of earth magnetism and the compass. Professors Rea and Holmes have helped to piece together a picture of Lorimer's life as a surgeon serving in a remote outpost of the British colonies.

Another item in the catalogue of hardships suffered in British West Florida is Lorimer's frustration at being in a remote spot where he could not participate fully in one of the most significant events to interest men of science in the eighteenth century. Until now his letter has been filed away in the archives of the Royal Society, and it emerges as no more than a human-interest footnote to a major scientific event.

FOOTNOTES

[1] Robert R. Rea and Jack D. L. Holmes published a paper in Southern Humanities Review in 1970 (vol. 4, number 4, pp. 363–372) entitled "Dr. John Lorimer and the Natural Sciences in British West Florida." While they include much information on Lorimer's interest in science as cited on this page, they did not mention his account of the transit of Venus.

[2] Harry Woolf, The Transits of Venus (Princeton: Princeton University Press, 1959), p. 17.

[3] Ibid., p. vii.

[4] Raymond Phineas Stearns, Science in the British Colonies of America (Urbana: University of Illinois Press, 1970), p. 654.

[5] Woolf, p. 21.

[6] Philosophical Transactions 52 (1761–1762), 272–279.

[7] Brooke Hindle, The Pursuit of Science in Revolutionary America 1735–1789 (Chapel Hill: The University of North Carolina Press, 1956), p. 147.

[8]Frederick E. Brasch, "The Royal Society of London and its Influence upon Scientific Thought in the American Colonies," The Scientific Monthly 33 (1931), p. 464.

[9]Woolf, p. 21.

[10]Robert R. Rea, "'Graveyard for Britons,' West Florida, 1763–1781", Florida Historical Quarterly 47 (1969), p. 358. Rea details the events and difficulties of Lorimer's medical practice.

[11]This and subsequent quotations from Lorimer are taken from a letter designated "Letter on the Transit of Venus, dated Mobile June 24, 1769," contained in the American Philosophical Society collection of Royal Society materials. Royal Society designation: Letters and Papers V 49 275. American Philosophical Society designation: History of Science Film Number One, Reel Eight, Frames 4344–4357.

[12]Rea and Holmes, p. 370.

BIBLIOGRAPHY

Brasch, Frederick E. "The Newtonian Epoch in the American Colonies (1680–1783)," American Antiquarian Society, Proceedings, New Series 49 (1939), 314–332.

———"The Royal Society of London and its Influence Upon Scientific Thought in the American Colonies", The Scientific Monthly 33 (1931), 336–355, 448–469.

Cohen, I. Bernard. Some Early Tools of American Science. Cambridge: Harvard University Press, 1950

Hindle, Brooke. The Pursuit of Science in Revolutionary America 1735–1789, Chapel Hill: University of North Carolina Press, 1956.

Rea, Robert R., and Jack D. L. Holmes. "Dr. John Lorimer and the Natural Sciences in British West Florida," Southern Humanities Review 4 (1970), 363–372.

Rea, Robert R., "'Graveyard for Britons,' West Florida, 1763–1781", Florida Historical Quarterly 47(1969), 345–364.

Stearns, Raymond Phineas, Science in the British Colonies of America. Urbana: University of Illinois Press, 1970.

Winthrop, John. "Observation on the Transit of Venus, at St. John's in Newfoundland," <u>Philosophical Transactions of the Royal Society of London</u> 52(1761–1762), 272-279.

Woolf, Harry. <u>The Transits of Venus</u>. Princeton: Princeton University Press, 1959.

Example 17.2

Dr. John Lorimer's Observation of the Transit
of Venus at Mobile in 1769

Maxine Turner

English Department

Georgia Institute of Technology, Atlanta 30332

ABSTRACT

Dr. John Lorimer, a British Army surgeon serving in West Florida, reported a sighting of the transit of Venus of June 3, 1769. He wrote from his station at Mobile to an unnamed correspondent to forward his letter to the Royal Society. Hampered by bad weather and insufficient instruments, Lorimer collected no useful data. Instead, his letter is among the annals of the many amateur sightings of the eighteenth-century transits of Venus.

A familiar figure in accounts of the eighteenth century is the man of affairs and amateur scientist who was alive to the phenomena of the world around him. One such man was British Army surgeon John Lorimer, who made his assignment to British West Florida an occasion for extensive scientific observation.

Lorimer's studies of the flora of West Florida earned him election to the American Philosophical Society in 1769. The naturalist William Bartram was his guest in Pensacola. In addition to pursuing a professional interest in botanical pharmacognosy, Lorimer interested himself in charting sections of the Gulf Coast and lands in Mississippi. In 1766 he determined the longitude of Pensacola, using his observations of the eclipses of the satellites of Jupiter in making his calculations. For several years he maintained a complete record of observations of the weather in West Florida. He was a member and president pro tempore of the West Florida Assembly (Rea and Holmes, 1970).

His observations of the transit of Venus, and the subsequent report, reveal him as one of the men who made up the learned world which looked toward the transits of Venus in 1761 and 1769 as momentously significant phenomena in the solar system.

It is difficult to overemphasize the intensity of interest which these transits aroused in the eighteenth-century scientific community. Learned men who were just beginning to grasp the immense breadth of Newtonian physics were thoroughly awed by their newly-found power to understand what they perceived to be the innermost secrets of the universe.

Since 1716, the year of the Royal Society's publication of Edmond Halley's paper explaining the significance of the coming transits, scientists had planned for them as rare opportunities to gather data leading to the accurate calculation of distances in the solar system. Observing the transit of Venus across the sun from widely separated points on the globe would provide a known base length and known angles between the equal sides of a triangle; finding the distance between the earth and the sun then became a matter of finding the altitude of an isosceles triangle. Once they could determine the astronomical unit by this means, astronomers could in turn determine the distance of other planets from the sun and thus the size of the solar system (Woolf, 1959). This was considered the solution of the final problem of astronomy in the generation after Newton.

Adding urgency to the desire to observe the transits was the infrequency of their occurrences. A transit of Venus had occurred in 1639, before the learned world had become aware of the significance of the phenomenon; hence, no data had been recorded. It was known that Venus would traverse the sun's disk in 1769, but not again for 105 years, and thereafter in pairs eight years apart at alternating 121½ and 105½ years intervals (Stearns, 1970).

By 1761, scientists showed considerable interest in viewing the transit, as evidenced by the 120 detailed observations reported to various scientific societies from 62 stations around the globe (Woolf, 1959). Those reports created great anticipation as the 1769 transit neared, especially among American colonial scientists who hoped that their contribution of data would distinguish their community in the scientific world. People all over the colonies prepared smoked glass and all manner of small telescopes in anticipation of June 3, 1769, the date of the transit (Winthrop, 1761–1762). This kind of preparation was being repeated in both settled and isolated places all over the world; the 1769 transit produced 138 observations from 63 positions (Woolf, 1959). With the next transit of Venus predicted for 1874, the 1769 transit was a last chance for eighteenth-century scientists.

While full-scale expeditions and entire communities made preparations to view the transit elsewhere in the world, John Lorimer was isolated in West Florida. Less than a month before the June third transit, Lorimer was transferred from a relatively comfortable station in Pensacola to a dangerously unhealthy, dirty, swampy, fever-ridden garrison at Mobile (Rea, 1969).

Lorimer not only lacked sufficient time to make preparations, but the weather was so cloudy in late May that he was unable to make the necessary observations for calculating the longitude and latitude and for checking the accuracy of his clock. Without making these preparations he could not collect useful data.

To add to his difficulties, the day of the transit was stormy. He reported, "In the forenoon it rained. It was impossible til after 3 p.m. to discover whereabouts the sun was, and then it was only through a cloud that I could see the body of Venus which was considerably advanced on the Sun's Disk ... I had then a pretty distinct view, and could see it through some flying clouds for a full half hour, when it entirely disappeared" (Lorimer, 1769).

Despite such difficulties, Lorimer sent an account of his observations in a letter to a friend in England. The letter makes it evident that had Lorimer been in Pensacola, whose longitude and latitude he knew, he would have been able to collect more useful data on the transit. Sightings were best made by groups, and in Pensacola Lorimer not only could have worked with his friend, cartographer George Gauld, but also would have had the use of Gauld's instruments: a reflecting quadrant, which was used to make sightings essential in setting clocks, and a large surveyor's theodolite, which was used in measuring altitude. His letter also noted the advantage of Gauld's station on Pensacola Bay which afforded a distinct view of the horizon.

As it happened, neither Lorimer nor Gauld was able to make accurate observations of the transit or to record useful data. Surely Lorimer was among the better informed of those who viewed the transit, but he lacked the instruments, the conditions, and sufficient aid to make a full report. Although his English correspondent forwarded his account of the phenomenon to the Royal Society, it evidently was not of sufficient value to be read to the Society or to be printed in the Philosophical Transactions. In his book on the eighteenth-century transits of Venus, Harry Woolf makes no mention of Lorimer's report in his compilation of 22 sightings in America. Thus until now Lorimer's report has remained among informal accounts of the transit which appear in journals and collections of letters.

John Lorimer viewed the transit of Venus as an amateur scientist who had already received a measure of recognition from the American Philosophical Society. Later in his life, the Royal Society would recognize his study of earth magnetism and the compass.

Ironically, one of the most significant events to interest men of science in the eighteenth century found Lorimer in a remote spot where he could not participate fully. His report, instead of being a distinguished contribution to science in the Age of Newton, has been until now filed away in the archives of the Royal Society. It emerges as a human-interest footnote to a major scientific event. It serves also as a reminder of what every scientist knows only too well—the experiment which fails.

Georgia Journal of Science 35: 136–139, June, 1977

LITERATURE CITED

Lorimer, John. 1769. Letter on the transit of Venus. Amer. Phil. Soc. Royal Society Letters and Papers. 49:275.

Rea, Robert R. 1969. Graveyard for Britons, West Florida 1763–1781. Fla. Hist. Quat. 47:345–364

——— and Jack D. L. Holmes. 1970. Dr. John Lorimer and the natural sciences in British West Florida. Souther Humanities Rev. 4: 345–372.

Stearns, Raymond Phineas. 1970. Science in the British colonies of America. University of Illinois Press, Urbana. 654.

Winthrop, John. 1761–1762. Observation on the transit of Venus, at St. John's Newfoundland. Royal Society Phil. Trans. 52: 272–279.

Woolf, Harry, 1959. The transits of Venus. Princeton University Press, Princeton, 258 p.

Following is an example of a short professional article produced for a national meeting of the Student Section of the Society of Women Engineers. Georgia Tech student Alice Miller is the author.

Example 17.3

ELECTROPHORETIC CONSOLIDATION OF SLIMES

Introduction

Phosphorus is an essential plant nutrient and hence necessary for food production. Currently, the United States produces approximately 40% of the world's supply of phosphate rock, which is the primary source of phosphorus for fertilizers. Florida accounts for 80% of the production in the United States, or approximately 32% of the world total of phosphate rock. Most of this production is in the Central Florida area around Bartow.

The phosphate occurs in a matrix of phosphate rock nodules, sand, and clay, along with minor amounts of other minerals. The matrix is mined with large electric draglines and transported to the benificiation process, where the phosphate rock is separated by washing, screening, and flotation.

The disaggregated clays and water form a waste slurry termed slimes, which, at 3.4% solids, is placed in above-ground impoundment areas, where the clay is allowed to settle. The recovered water is recycled to the process.

The clays are micron- and sub-micron-sized particles with large and active surfaces which hydrate, and hence prevent settling to over 18-20% solids under natural conditions. This material is fluid, and the volume is extremely large because of the contained water, so dams must be built for its retention. This condition limits reuse of the land and is an environmental hazard.

Slime consolidation has been studied by numerous investigators over the last 40 years. Today, the efforts are more intense than ever to solve the problem of waste disposal.

Theory and Purpose

As defined by the McGraw-Hill Encyclopedia of Science and Technology, "Electrophoresis is an electrochemical process in which colloidal particles are made to migrate under the influence of an electric current. Charges are borne on the particle surface, and can result either from absorbed ions that have been taken from the surrounding water or from charged atoms or groups of atoms that are an integral part of the chemical structure of the particle. By virtue of these surface charges, a colloidal particle will move toward an electrode of opposite charge." (1971, IV, 599)

Work done on phosphate slimes from Tennessee deposits have demonstrated that electrophoresis is effective on some types of clays. On the basis of these earlier results, the same kind of procedure was applied in tests on Florida slimes.

The purpose of these experiments was first to determine if direct current flow through diluted slimes would result in improved settling rates and compaction of the clay over the conventional methods, and then, provided the technique was successful, to conduct tests to optimize its physical factors, such as electrode configuration, current densities, and voltages. The tests were also set up to observe changes in pH and conductivity.

Test Description

For the initial tests, slimes at 3-6% solids were placed in a plexiglas cell (see Figure 1) containing a grid anode and a grid cathode which were placed at various heights in the cell. Every 15 minutes, over a period of 3-4 hours, voltage, amperage, and percent-solids readings were recorded. Every hour pH and conductivity readings were taken. A reference cell, identical to the electrostatic cell except that voltage was not applied to the electrodes, was used as a standard to allow normal gravity settling of the slimes.

After the initial runs, slimes at 4.5% solids were placed in the test cell with an electrode separation of approximately 8 cm. Direct current tests were run with the voltage ranging from 3 to 12. The ten tests ran from 3-4 hours each, with voltage, amperage, percent-solids, and conductivity readings taken every 15 minutes. Every 30 minutes, pH readings were taken.

A test using slimes previously thickened to 7.3% solids was conducted for 3.75 hours with voltage, amperage, percent-solids, and pH readings taken.

Experiments were also conducted to identify an anode material that would neither deteriorate nor generate a large amount of electrolysis. For each test in which a different anode material was used, the anode and clay-water mixture were weighed separately, and a current at 12 volts was passed through the cell for a number of hours. When the test was over, the anode and clay-water mixture were again weighed to determine the weight loss in each due to electrolysis.

Results

It was observed that when a current at 12 volts passed through a cell containing slimes at 4.5–4.7% solids for a period of 3.5 hours, the slimes consolidated to approximately 6.37% solids (see Figure 2), a concentration considerably higher than the 5.17% solids obtained in the reference cell after the same amount of time. This is a 23% improvement in compaction. Twelve volts was found to be the optimum voltage for the slimes to reach the highest percent solids. It was also noted, for the plate separation tests, that the best results occurred with a plate separation of approximately 8 centimeters.

In the experiment using the previously thickened slimes, it was observed that the slimes continued to settle at 12 volts, but that the percent-solids increase over the same amount of time was smaller than for the slimes tested at a lower percent solids (see Figure 3).

The conductivity readings, taken at the top, center, and bottom of the cell, were usually highest at the bottom of the cell and lowest at the top. It was also observed that the pH increased as a function of both time and voltage (see Figure 4). The conductivity and pH data were complex, varied, and difficult to interpret.

In all tests, significant evidence of ongoing electrolysis was visible in the form of H_2 and O_2 bubbles. Due to the electrolysis, the anode was substantially consumed, and water was lost by evaporation. Of the materials tested for the anode, carbon rods lost the least amount of weight during a specific test period and also minimized the amount of water lost as a result of electrolysis (see Figure 5).

Conclusions and Recommendations

As a result of the tests, it was determined that electrophoretic consolidation of slimes is successful in attaining a higher percent solids than normal gravity settling of slimes. The rate of settling of the slimes is affected by the voltage being applied to the cell, the higher voltages being the optimum in reaching the highest percent solids over a given amount of time. However, the cost per dry ton to consolidate slimes increases as the voltage increases (see Figure 6).

After a length of time (usually 2-3 hours) the rate of consolidation of the slime tends to decrease, but the rate is still faster than that of normal gravity settling. This could be a result of the clay particles packing tightly around the anode, reducing the current flow, and causing a decrease in the strength of the electric field.

Carbon, of all the materials tested, appeared to be the best material for the anode rods, because it reduced both electrolysis and anode decomposition. Reducing the amount of electrolysis benefits the economy of the operation.

The following recommendations were made:
1. conduct additional tests at higher voltages;
2. study the effects of electrode size and current density;
3. run tests using an alternating current;
4. test the carbon-rod anode to determine if it is as effective as metal in consolidation of the slimes;
5. study the pH and conductivity measurements further, to determine their significance; and
6. determine a method for continuous bench-scale tests.

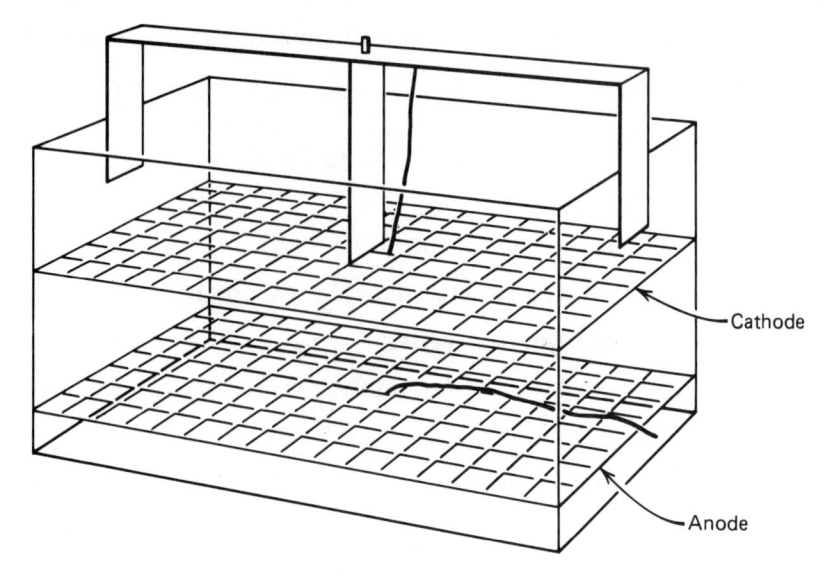

Dimensions (W × L × H) : 20.3 × 38.2 × 24.5 cm

FIGURE 17.1 Bench-Scale Electrophoresis Cell

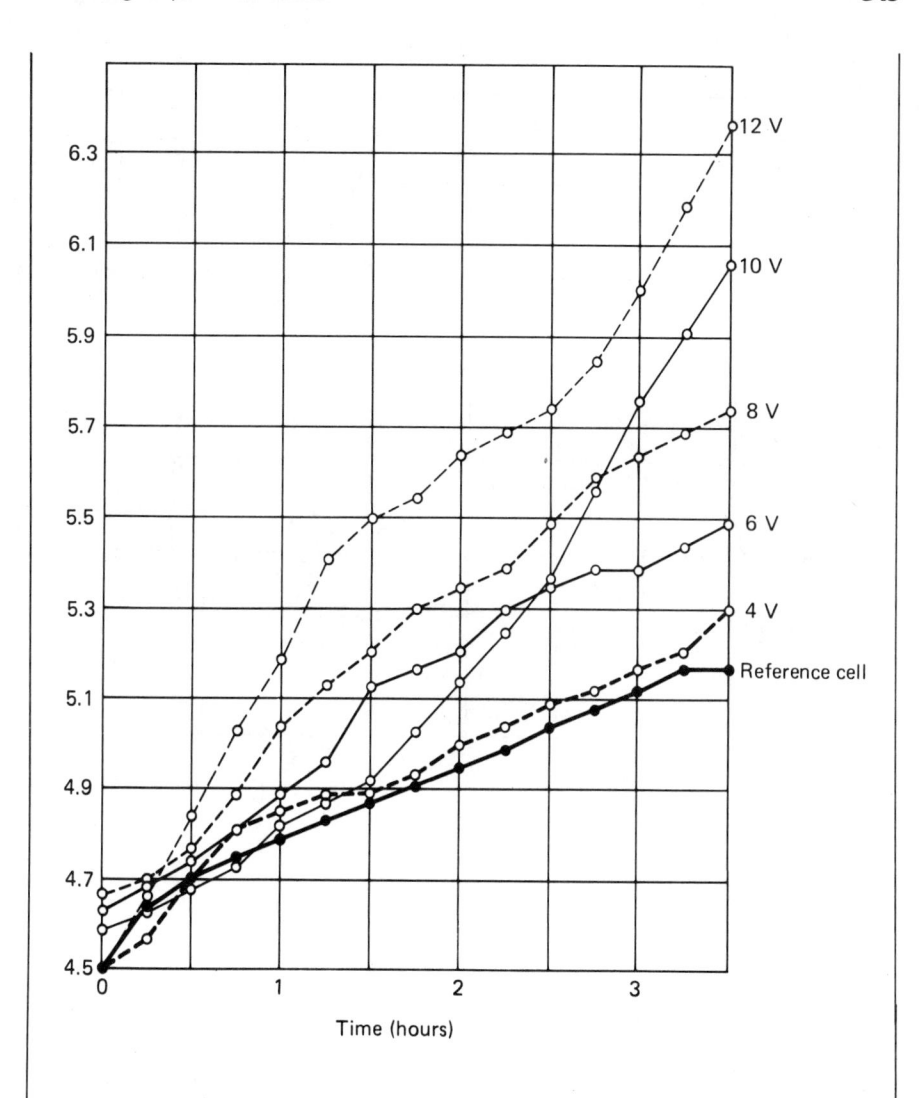

FIGURE 17.2 % Solids vs. Time for slimes settled at varying voltages

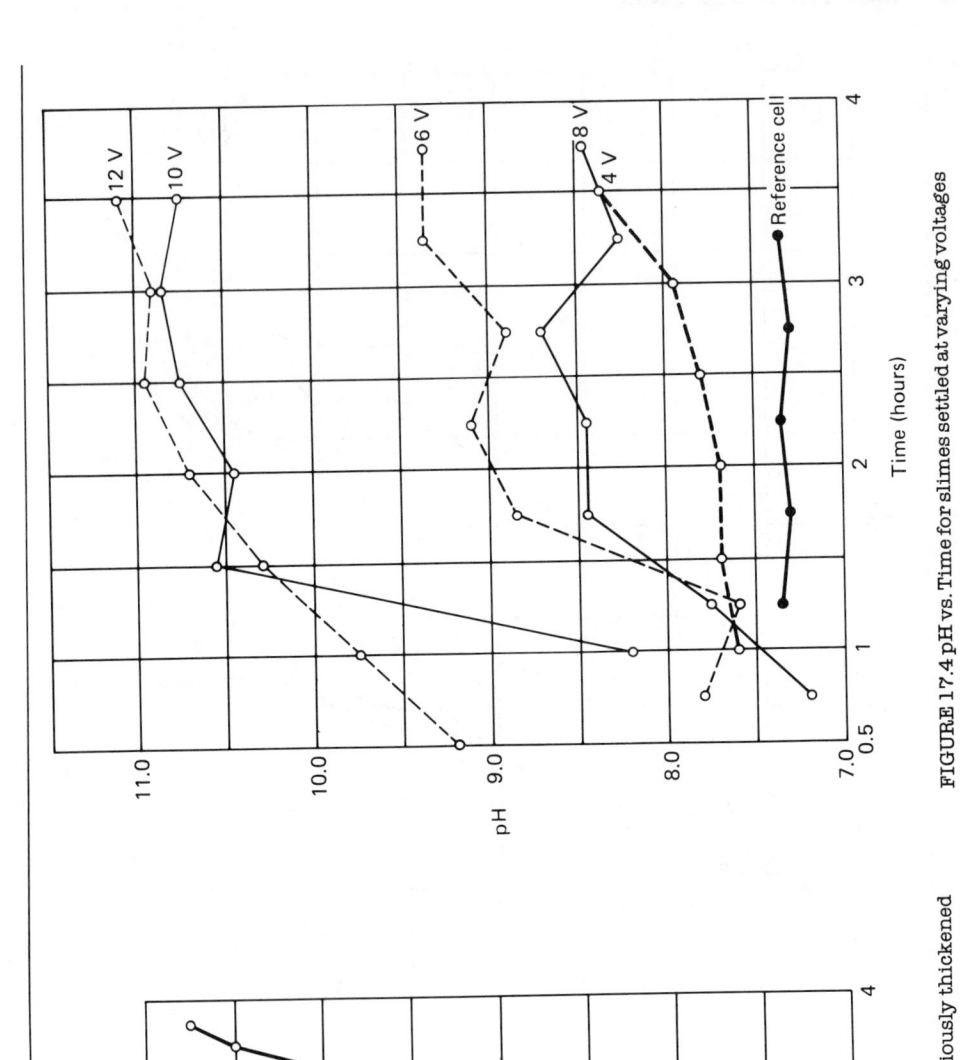

FIGURE 17.3 % Solids vs. Time for previously thickened slimes at 12 Volts

FIGURE 17.4 pH vs. Time for slimes settled at varying voltages

Date: 8-10 to 8-11-81   Electrode Material: 2 carbon rods
Conditions: Electrophoresis Cell —Clay Slurry @ 4.5 initial % solids.
Voltage: 11.8 V dc   Current: 350 mA
Ambient Temp: 23.1 deg. C
Initial Slurry Temp: 23.5 deg. C   Final Temp: 22.9 deg. C
Beginning Time: 8:30 AM
Ending Time: 8:00 AM   Total Time: 23.5   Hrs.

Electrode	Clay-Water
Beginning	Beginning
Weight = 12.6 grams	Weight = 5647.2 grams
Ending Weight = 12.4 grams	Ending Weight = 5601.8 grams
Weight Loss = 0.2 grams	Weight Loss = 45.4 grams

Clay-Water

$$100 \times \frac{45.4 \text{ Wt. Loss (g)}}{5647.2 \text{ Beginning Wt. (g)}} = 0.80 \text{ \% Weight Loss}$$

$$\frac{0.80\%}{23.5} \frac{\text{Weight Loss}}{\text{Total Time (Hrs)}} = .034 \text{ \% Weight Loss/Hr.}$$

Electrode

$$100 \times \frac{0.2 \text{ Wt. Loss (g)}}{12.6 \text{ Beginning Wt. (g)}} = 1.59 \text{ \% Weight Loss}$$

$$\frac{1.59}{23.5} \frac{\text{\% Wt. Loss}}{\text{Total Time (Hrs)}} = .068 \text{ \% Weight Loss/Hr.}$$

Figure 17.5.   Electrode Evaluation

$ per Dry Ton to Settle Slimes

Voltage	Cost	Length of Test	Change in % Solids
3 volts	$ .09	3.5 hours	+0.67%
4 volts	.17	4.0	0.81
5 volts	.27	3.75	0.61
6 volts	.57	4.0	1.09
7 volts	.53	3.75	0.81
8 volts	.68	3.75	1.12
9 volts	.65	3.25	1.19
10 volts	.90	3.75	1.55
11 volts	1.00	3.75	1.80
12 volts	1.83	3.5	1.87

FIGURE 17.6

part four

A Technical Writer's Handbook

18. Words
19. Sentences
20. Paragraphs
21. Graphics
22. Outlines
23. Technical Persuasion
24. Researching and Writing a Library Paper
25. Endnotes and Bibliography

Technical writing is the study of how to write specific forms for an audience in specific situations, not a course in general or basic writing skills. While such a definition may be accurate, it does not account for a technical writer's need for a quick reference source or a general guide to developing writing skills. Therefore, this last section addresses the specific concerns of technical writers in basic writing skills.

In arrangement and content, this section includes most of the features of a standard writing handbook: vocabulary and usage; sentences and paragraphs; outlines; and even those traditional components of most writing texts, an introduction to elementary logic and a guide to writing a term paper.

This handbook differs from other such guides, however, in that it is specifically related to the demands of technical writing. Since letters most often require a sensitivity to appropriate diction, the section on words gives an appropriate emphasis for the technical writer who must write letters. Since technical writers must deal more than other writers do with numbers, time, and motion, the section on words also gives greater emphasis to such words. Since all technical writers must make decisions about when and how to use graphics in the logical development of a prose work, the chapter on graphics common to most technical writing texts is included with the paragraphs as a component in the process of writing. Since all technical writers must sell their work in one way or another, the discussion of logic is limited to the concerns of the technical writer who must translate technical expertise into effective marketing techniques.

Assuming that writing is an individual lifetime skill like golf, tennis, or playing a musical instrument, the handbook contains directions for the motivated individual who wants to build communications skills over the long term.

chapter 18

Words

This writing lab on words works in some ways like the glossary of usage in a standard composition textbook. More important, however, the writing lab identifies areas of special concern for the technical writer, and the lab also outlines a lifetime program of independent study to keep the technical writer current on usage. Unlike many formulas which technically educated people use and can always depend upon, language is constantly shifting and changing. Like an ecological system, the language is subject to many alterations of time and fashion; new words will crop up and enjoy currency for a time only to be swept away and replaced by others. Language is a living system always in dynamic process. Any user of the language, therefore, must be like a natural scientist, always observing and noting the current state of words and how they are used.

The primary areas of concern for technical writers are 1) Standard English, 2) Current Business Usage, and 3) Precise Technical Usage.

STANDARD ENGLISH

Spoken English varies from place to place and from group to group, but written English must be standard. This is particularly true of technical and business English:

1. because words are standard in meaning in much the same way a wrench is made to a specific English or metric measure;
2. because computer applications of language call for more precise standards of usage;
3. because the international communications community requires standard usage;
4. because many long technical documents are team efforts and as such cannot accommodate individual eccentricities of style.

Technical writers, whose natural medium may not be the written word, often stray from standard English into colloquialisms and unidiomatic usage.

Colloquial Usage

Written language is more formal and must be more standard than spoken language; for speech does not always transcribe as precise, standard usage.

Note the typical colloquial expressions translated into standard usage below:

Colloquial	*Standard*
a lot	many
back up	enforce
couple of	two
doesn't even	does not
ends up as	results in
just a few of	some
looked at again	reconsidered
maybe	perhaps
pick	select
show up	arrive

Writers who use dictation must be especially careful to avoid a colloquial tone in letters. A letter of three or four short paragraphs may be no more than 300 words long, yet we speak at about 100 words per minute. It is all too easy for the person dictating a letter to lead a reader through long dense paragraphs, as in the example below.

> "Joe, you were most kind to be so cordial in your reception to discuss the Hayes contract and I think we really got down to some basic issues about cost which will save us a lot of time and could save a ton of money as the project progresses. As you said, though, we have to keep a close eye on timing and scheduling, especially as the winter rain comes on with the possibility of delays and days off for the construction crews and the fact that mud slides are hard to handle on a job with a soil type of this type."

Standard English

People talk that way, but talk does not transcribe as effective style in letters. The person dictating can also be unaware of how the spoken passages will transcribe as a design on a page, a factor we have already discussed as crucial.

The essay and much imaginative writing have the flavor of good talk—that is a style, in fact, that I have aimed for in much of the exposition of this text for college users—but one who is learning to write must develop a sharp awareness of what works in writing as different from what works in speaking.

Unidiomatic Usage

An idiom in any language may be entirely illogical in terms of grammatical explanation, but certain words or word combinations simply work a given way in a given language. Idioms cannot really be taught; they are learned by exposure to language as we read and listen to practiced speakers. The list below illustrates some of the most common ways an inexperienced writer reveals a general unfamiliarity with precise, tasteful usage. The following sentences show some of the most common idiom faults:

- I am *anxious* to begin. (*eager* is the more idiomatic usage)
- The *consensus* of opinion is. (*consensus* alone implies "of opinion")
- This chair is *different than* the other. (different *from*)
- The *reason* is *because* it rained. (*reason* is *that*)
- I was *very impressed*. (*much impressed,* or *impressed*)
- In regards to. (*With regard to*)

This short list merely illustrates how unidiomatic expressions appear in language use; no text can cover every possible instance. Mistakes with idioms will most often occur with prepositions *in, on, after, before, against, toward, up, down, across,* etc. Be certain to watch fine shades of meaning in such words.

Technical writers who are not frequent readers are most likely to betray a general unfamiliarity with how words go together on a page. A student who does not make a conscious effort to read and write except for a writing class is like a tennis player who participates only in tournament play.

Some writers betray their unfamiliarity with language when they write sentences like these:

- This letter records my *past encounters* with technical writing.
- My lack of skill in typing has *tampered with* the accuracy of my report.
- The company is *endowed with* competent personnel trained in techncial writing.
- Our company *desires to partake* in this work.
- All other aspects of particle beam technology *are* outside the scope of this report.

- Replacing the part will be *high just for the labor.*
- If the engine *had a little more* power, it would be sufficient.
- Switching technology must be developed that is *in an order of magnitude better in performance than that* currently available.
- The *insurvivability* of people is probable because the *lethality* of the weapon is strong.

These unidiomatic expressions are not mistakes which can be corrected as one would correct subject-verb disagreement. Only consistent attention to language will ensure that your language will become generally more idiomatic.

Correct Usage

Some matters of usage are gramatically right or wrong. Since English has few inflections, writers tend to falter at inflected pronouns.

a) *myself/yourself:* these are *only*
   1. reflexive: I embarrassed myself. or

   intensive: I myself will do it. You yourself are to blame. *Never* use them where you would use "I" or "me."

b) *I/me:*

   David and *I* will look forward to seeing you.

   Thank your for seeing [David and] *me.*

c) *who/whom:*

   The contractor *whom* we hired: we (subj.) hired (vb.) *whom* (obj.)

   The contractor *who* worked for us: *who* (subj.) worked (vb.)

Try to be as natural and unaffected as you can, for efforts to be "correct" can lead to mistakes like "*whom* do you think you are, *anyways*?" A good rule is: "when in doubt, use *who.*" It has been said that it is the mark of a half-educated fool to mis-use *whom.* It is the mark of a fool, period, to say "Hopefully, David and *myself* will go, how about a person such as *yourself*?"

What to do about sentences which combine *anyone* ... *their* is a vexing question among English teachers. In strict grammatical terms, such words as *anyone, every, none, anybody* are singular: "*None* of us *is* entirely right." "*Everybody* must make his own decision". Some teachers haven't troubled themselves with such distinctions for years; others have duly marked such agreement mistakes as wrong. More recently, the emotions aroused by the feminist movement have caused the National Council of Teachers of English (NCTE) to adopt the usage: "*Everybody* must make *their* own decision," as a way to avoid having to decide between a possibly sexist single pronoun and an awkward "his or her."

Judging from my reading of business and technical prose, I suggest a compromise that lies between the very liberal stand of the NCTE and the very conservative stand others take:

When using words that imply more than one person, use *their*: "Everybody must do their part" "None of us are entirely correct."

When using words that contain the suffix *one*, use *he* or *she:* "*One* is responsible for *his* actions." "Every*one* must solve *her* problems." Alternate *she* and *he* in paragraphs rather than to use the awkward "he or she" "him or her."

None of us has the final and absolute answer. What is suggested here is a common-sense solution which makes some allowance for the convenience of using *their* at times while preserving the logic of using *he* or *she* when *one* appears before it in a sentence. Doing this in your writing requires you, of course, to exercise your memory and your judgment—and that is what any writing always requires. Be guided also by your professor's more immediate observations of trends in usage.

That which is "correct," as distinct from that which "communicates effectively," represents a division between schools of thought in English usage. Therefore, what is "correct," in the sense of proper and appropriate for a given audience in a given situation, very often satisfies the immediate need of technical communication.

CURRENT BUSINESS USAGE

The business letter is a form of writing particularly subject to the use of trite expressions. That factor makes learning fresh and effective usage particularly important. The chapter on letters discusses how to begin a letter. The list which follows illustrates:

How Not *to Begin a Letter*

- Enclosed please find ...
- Enclosed herewith ..
- Please find enclosed ...
- Pursuant to our recent conversation ...
- We are in receipt of your letter dated ...
- Per your request ...

The next list illustrates:

How Not *to End a Letter*

- Thanking you in advance . . .
- Your consideration will be appreciated.
- The favor of a reply is requested.
- If I can be of further assistance, please do not hesitate to call.
- Thank you for your attention to this matter.
- Your attention to this matter will be appreciated.

Very frequently a beginning writer's attempt to sound businesslike will result in an almost ludicrous style. What follows is an example of a student's answer to a letter asking that he describe his academic and work experience so that I could determine his content mastery for a technical writing course:

> "I have received your request for a treatsie of my academic undertaking. With reference to that request, I will expand upon the information provided by the registrar's office. I am a senior in Civil Engineering. Within this framework I have studied all aspects of Civil Engineering, with a heavy emphasis upon structural design. This emphasis has not caused an exclusion of other sub-field knowledge. I have, within the past twelve months and including the coming quarter, taken or am in the process of taking every senior elective normally offered by the School of Civil Engineering. This assimilation of a broad-based knowledge of C.E. should:
>
> 1. Allow me to converse with professionals in other Civil specialties from a position of less than total ignorance; and
> 2. Increase my personal marketability within the entry-level job market.
>
> "While I expected to expound primarily upon the art and methods of structural design, it should not surprise you to receive reports from me ranging from groundwater utilization to foundation design to sanitation engineering."

The greatest danger for a writer who sets out in that tone and style is that he has set a pace which he may not be able to maintain. In the first line, he falters with the spelling of *treatise,* and *undertaking* really should include an *s*. Had he pitched his tone and style at a somewhat lower level, such errors, while obvious, would be received with more sympathy. When he assumed a somewhat grandiose tone, however, he left himself almost no margin for error; such a style, if attempted, must be done to perfection. Therefore, aim your style at a level and a tone which you can comfortably and confidently maintain. Otherwise, you become like the musician who sets the tempo so fast that he greatly increases the likelihood of making mistakes, or like the miler who begins at a fast pace only to stumble at the three-quarter mark.

In a world where bad letters abound, you will distinguish yourself more easily and favorably by writing simply and directly in an unadorned style than by striving for heights of eloquence.

PRECISE TECHNICAL USAGE

Technical writing so frequently deals with numbers, quantities, and time and motion that a writer must make very careful distinctions in usage.

Numbers

Rather than to adhere to more traditional spelling and hyphenation rules, base your decisions about numbers upon the reader's need to perceive a numeral effortlessly in a technical document.

Traditional: When he was twenty, he spent two years in Europe before returning to the family home on 42nd Street.

Technical: They spent 40% of their budget buying 20 planes which had been obsolete for two years.

Reduce the margin of error; do not ask your reader to translate a word into a number in a technical document. At the same time, spell out numbers used in a non-technical sense. Do *not* begin a sentence with a number: 400 people were hired. Do *not* use a locution such as "five (5)" except where contracts require both words and numerals. The rule does not tell you what to do in every instance; it gives you the basis for making a judgment. Above all, BE CONSISTENT.

Words Related to Quantities

In general usage, even careful speakers have grown lax about making distinctions between *number/amount* and *few/less*. Dealing in quantities, however, makes necessary a greater degree of care among technical writers.

Number and *few* are used for items which are counted.

> A number of spectators gathered around the wreckage.
> Numbers of problems increase with each new design.
> Fewer problems resulted from less complex designs.
> Few parts are required in this design.

Amount and *less* are used for substances which are measured out.

> There is less stress upon this design, and fewer stresses generally on this type of aircract.
> The amount of stress depends upon the number of angles.

Over/more than: Over is almost colloquial. It is better to say "more than three years," "more than $50 million."

Plus/in addition to: as with "over/more than," this preference produces a better sound. Not "a major in electrical engineering, plus a communications certificate," but "in addition to a major in electrical engineering, a communications certificate," or "a major in electrical engineering and, in addition, a communications certificate."

Words Related to Cause

In general usage, *because of, as,* and *due to* are sometimes used interchangeably. Distinctions should be more refined in technical writing.

Because introduces an adverbial phrase which tells why and how:

> The wing failed because of poor design.

Due to introduces an adjectival phrase which describes a cause:

> Icing due to high altitudes caused the wing to separate.
> It was not an accident due to pilot error.
> (These examples are only to illustrate the usage; "caused by" would be preferable. "Due to" is too much used, and usually misused.)

As introduces a clause pertaining to time and motion, not to cause:

> As the plane made its final approach, the flaps acted as a brake as the plane lost air speed.

Verbs

Technical writing most often relates to action. Technical writers should therefore be especially sensitive to shades of meaning in verbs which can make a great difference in the action implied:

Will implies that the writer confidently expects the occurrence;

Can implies that an action is possible, but not inevitable;

Could and *should* are more conditional and less positive than "can" and "will";

Shall is almost always reserved to specify an order or a contractual obligation.

Examples

    The concrete *will* be poured and dry by March 7.

    The concrete *can* be poured and dry by March 7.

    The contractor *should* delay 10 days.

    The contractor *can* delay 10 days.

    The concrete *shall* be poured by March 7.

    The pump *will* operate at 10 psi.

    The pump *can* operate at 10 psi.

    The pump *should* operate at 10 psi.

    The pump *shall* have a maximum operating pressure of 10 psi.

Passive Voice

Traditionally, formal style and technical writing have required the use of passive voice. A major reason for such usage is the ban upon personal pronouns (*I, me, my*) and upon a *you* point of view. First a student in freshman English writes essays in the first person point of view; then writing in technical courses forbids the first person, so the natural tendency is to shift into passive voice: "results have been tabulated," "calculus is not capable of being learned by students;" the logical third step in the progression produces business letters which begin, "The letter written by you has been received by us."

The key to avoiding passive voice is to substitute a concrete noun in the subject (or actor) position in the sentence:

    This *report tabulates* the *results* of a study which ...

    The *technician* first *removes* the *cowling* and then uses a socket wrench to loosen the bolts on . . .

Tests show that . . .

Applying Hooke's Law *proves* that . . .

Benoulit's Equation demonstrates . . .

The following *chart classifies*—on the basis of . . .

Especially in technical writing, people identified as *I, me, my* are not the only actors. Workers—identified as engineers, technicians, experimenters, researchers, operators—perform actions. Tests, procedures, investigations, documents, reports: each of these also can occupy the subject position as actor in a sentence.

Colloquial Verbs

An inexperienced writer will frequently use long verb phrases instead of strong verbs, as the following sentences illustrate:

There is *much more needed to be done.*

The coating *keeps* the fabric *from getting* torn and dirty.

The experiments *are nearly done.* They *have been divided up into* stages.

What further investigations *are expected to be conducted?*

The data base *needs to contain* a description of test conditions.

While your technical education may supply you with technical nouns and adjectives, make sure that your use and understanding of strong verbs keeps pace with the remainder of your vocabulary.

Perhaps the most important point to remember is this discussion of words relates to how one learns these distinctions in usage: not in a chapter like this, but through consistent explosure to language. Increasing your awareness of language in the world around you, even by a very small percentage, will help you in turn to improve your communications skills. Such effort will also help you to develop and maintain a facility with language.

The assigned exercises were constructed in recognition of the need to work daily in small increments to build a skill. The exercises also recognize that writing is a lifetime skill which we never cease to develop and maintain.

EXERCISES

Listening

The *purpose* of this assignment is to increase your awareness of language; the *plan* to follow appears in the steps listed below.

Exercises

1. Tune your radio to a news and information station if you receive such broadcasts in your area. Use driving time and leisure time to listen and program your mind to the patterns and usage of expository prose. Many such stations have short features on gardening, home repairs, medicine, psychology, and economics as well as news, weather, and sports coverage. Each kind of program will expose you to a specialized use of language.

2. If you cannot receive news and information, turn your radio to a variety of stations. Rock, country and western, or easy-listening station: each will have a particular pace, tone, and selection of vocabulary for its major audience. Learning to be aware of variations in usage will sharpen your ability to make the judgments necessary to write well.

3. This assignment is not designed to keep your ear glued to the radio. Instead, re-direct at least some of your normal listening times to programs which will enrich your exposure to language.

    a. Begin with local newscasts each evening. For the local news, divide a page into three columns headed News, Sports, Weather. During the news segment, list any special words or any transitional devices characteristic of news broadcasting. During the weather report, list special technical words, phrases, and transitional devices characteristic of weather reports. During the sports report, repeat the exercise, carefully noting the level of formality and the specially coined words characteristic of sports broadcasts. Doing this exercise for a week will help you to sharpen your awareness of differences in tone, of usage levels, and of special vocabularies characteristic of a given subject.

    b. Extend your listening to the national news after a week of careful attention to the local news. As each story is aired, classify it according to subject: politics, defense, economics, people in the news, sports, etc. Under politics, list special vocabulary, repeated phrases, and transitional devices. Do the same kind of list for the other topics to determine how tone and usage differ from one topic to another.

    If you have difficulty writing a summary and conclusions to an essay, a letter, or one part of a report, pay very close attention to how a correspondent concludes his report aptly with a single, often clever, sentence or two.

    On both local and national news, pay close attention also to how people speak in interviews. Within half an hour, you may hear an Iowa farmer discuss hog prices, a welfare mother take her text upon budget cuts, a government official discuss a new law, a physician describe a medical discovery, and a woman cadet describe the rigors of education at a service academy. These will reveal not only varied accents, but varied levels of usage, varied tones, and a variety of idioms and special vocabularies. As you become aware of a fuller range of language usage, you can better make judgments about language within your own range of writing and speaking. Such awareness is also invaluable in the audience analysis necessary for all technical writing.

c. Major networks and educational channels all schedule public affairs programs. These are especially helpful to your study of language because you can hear experts in sustained discussion of complex issues. A person's grasp of detail, the tone and techniques of persuasion, the interaction of the speaker—all of these offer an opportunity to see adult professionals communicate their ideas. Not only in the use of language, but as models of dress, posture, personal interaction, and general demeanor, government officials and acknowledged experts offer models for communicating as an adult professional in the working world.

Many public affairs programs offer transcripts by mail at a nominal cost. To order a transcript of a program which you have recorded on a tape or video cassette offers an opportunity to give intensive study to how people communicate their ideas. A transcript can also illustrate how the cadences of spoken language, however expert and formal, differ from those of most written language.

To repeat the note following the radio assignment, this assignment of attending more closely to television is not designed to have you spend endless hours before a television set. Listen intensively to *no more than* one hour a day of newscasts and to *no more than* three additional half-hour public affairs segments during a week. Once you feel comfortable with one exercise, phase in the next. Like isometric exercises or weight lifting, these exercises are designed for consistent use in small increments.

Reading

The *purpose* of the reading assignments is to help you program your mind with how words work on paper. Good writing begins with reading. The *plan* for being a more careful reader is laid out below.

1. Spend some time each day reading a daily newspaper. If you do not normally spend time reading, 15 minutes every morning and afternoon will soon make a difference in your sensitivity to language. As with listening closely to radio and television, reading various segments of a newspaper will make you more aware of levels of usage, specialized vocabulary, and the like. In addition, these reading assignments will illustrate particular skills you may need to develop.

    a. If you have difficulty writing the first sentence of a letter and compactly including all the information it must contain, read the first paragraph of every story on page one of a newspaper. The "lead paragraph" of a news story is much like the first paragraph of a letter; it tells who, what, where, why, and when, without wasting space.

b. Letters from readers are usually chosen for their brevity. Keep up with the letters column and learn to make judgments about how well a person writes and why a particular letter may have been chosen for publication. In a large city daily, to have a letter selected is to win a point in a highly competitive game, so you may be reading the work of serious amateur writers.

c. A daily newspaper also includes a variety of articles on specialized topics, such as medicine, science, and economics. How a reporter writes about a complex subject for a very large popular audience gives the E/T writer good examples of how to translate technical expertise into readable—and salable—reports. Such stories often make very skillful use of example, analogy, and figurative language in order to clarify complex subjects for the general reader.

d. Classified ads offer good examples of concise language. A technical or business writer should avoid a "telegraphic" style which makes it seem that the writer is paying for each word; nevertheless, the selection of detail and the ordering of detail within a real estate ad or an automobile ad offer concise examples of what details to include and how to arrange details to the best effect.

e. For slow readers, the width of the newspaper column is very helpful for increasing speed of scanning from left to right in the reading process.

Limit these exercises to short segments during the day, at least one in the morning and another in the evening if possible. Do not limit yourself to your local paper, but explore the *New York Times*, the *Wall Street Journal*, and the *Christian Science Monitor*.

2. For more general reading, read a weekly news magazine. *Time* and *Newsweek* have general coverage of world and national news, and specialties like religion, sports, movies, and books. As with radio, television, or a newspaper, such a magazine can give you the opportunity to explore how different subjects call for different styles and vocabulary. Be aware of an elaborate style in *Time* and, to a lesser degree, in *Newsweek*; be aware also that such a style is inappropriate for most writing that you will do in your technical career.

U.S. News and World Report is probably the most useful model for the would-be technical and business writer because its style is understated. *Business Week* is similar. These magazines are especially helpful because of two additional characteristics: the use of lists, bold-faced subheadings, and other typographical devices is very similar to what technical writing calls for; and the graphics used to illustrate stories are excellent examples of the interplay between the verbal text and the chart or graph. Through such reading, even if you are not highly aware of these devices of style and typography, you may develop on some level a better sense of how to place material on a page.

In addition to the specific techniques to be learned from reading a lead paragraph in a news story, or from observing the relationship between print and graphics, reading can offer general opportunities to study how writers write. Hidden away in a news story there may be an especially apt definition of a technical term; an editorial or a long article may contain a logical classification and division of a topic; a story on two conflicting points of view may make use of good technique in comparison and contrast. The writer may not have said to himself, "Here I'm going to whip out my notes on rhetoric," but people who have mastered the craft of logical thinking and of translating logical thought into words will employ traditional rhetorical patterns in their writing.

These exercises in listening and reading should occupy you for not much more than an hour a day in half-hour or fifteen-minute segments and for an additional two or three hours per week in half-hour to one-hour segments.

CONCLUSION

For a further study of words and for reference, be informed about the standard reference works. *The Chicago Manual of Style* answers most questions about usage and manuscript design and preparation. For spelling, a good dictionary of recent date will provide most answers. For very quick reference, the Gregg Company *20,000 Words* and *35,000 Words* list only spelling with syllables—and the word processor, which arbitrarily ends a line with a machine's disregard for when to hyphenate, makes knowing when and how to divide a word important. For technical words, the McGraw-Hill dictionaries for individual engineering disciplines—electrical, mechanical, etc.—are a valuable source, as are technological handbooks and code books.

The general approach of this text, however, is to encourage the mature professional practice of keeping current through reading literature in your technical discipline and remaining aware of how words and terms—and all conventions related to communicating in your field—are used. The final arbiters for a word which does not appear in standard sources are, first, a recent refereed journal in the discipline or, second, a practicing professional who regularly uses the word in its technical sense.

chapter 19

Sentences

The sentence is the basic unit of communication in English. Therefore, good writing must always rest on a foundation of good sentence structure. Technical writing places two significant demands upon sentence structure:

1. The frequent use of numbered lists in memorandums, longer letters, and reports makes the sentence, and not the paragraph, the basic component in many technical documents.
2. Technical content must include so many terms and details for thoroughness and accuracy that some sentences are necessarily quite long. Good sentence structure is required, then, to bear the weight of masses of complex technical detail.

The sentence exercises in this Writing Lab are designed to help technical writers do three things:

1. to understand the sentence as a structure with component parts which may be inserted or removed to create an effective structure;
2. to develop greater competence in sentence writing by imitating sentence patterns; and
3. to learn to evaluate the structure of sentences and make effective revisions.

BASIC SENTENCE STRUCTURE

The	engineers	built	a	bridge.
	(subject)	(verb)		(object)
	ACTOR	ACTION		RECEPTOR

That is the fundamental structure from which all other configurations of the sentence develop. Writing sentences is a matter of adding other components to that basic unit or of rearranging the word order within that unit.

ADDING COMPONENTS TO A SENTENCE

1. At the Beginning

In the spring, the engineers built a bridge.
After the flood receded, the engineers built a bridge.
Using surplus materials, the engineers built a bridge.
Racing against a deadline, the engineers built a bridge.
Under the auspices of the Corps of Engineers, the engineers built a bridge.
When the bids had been awarded, the engineers built a bridge.
Although the weather was rainy, the engineers built a bridge.
Because the contractors were on strike, the engineers built a bridge.

2. In the Middle

The engineers who were hired by the Army built a bridge.
The engineers, under the supervision of a Major, built a bridge.
The engineers, racing against a deadline, built a bridge.
The engineers—overworked and underpaid as they were—built a bridge.
The engineers, under the auspices of the Corps of Engineers, built a bridge.

The engineers who cleared the road built a bridge.

The engineers—civil, mechanical, and electrical—built a bridge.

3. At the End

The engineers built a bridge in the spring.

The engineers built a bridge after the flood receded.

The engineers built a bridge, using surplus materials.

The engineers built a bridge because the old one was swept away in the flood.

The engineers built a bridge which spans the river.

The engineers built a bridge despite repeated delays.

The engineers built a bridge as soon as the bids were awarded.

Viewed as structure, each of those sentences is made up of a basic component and of other components added to change the meaning and function of the sentence in some way.

MULTIPLE COMPONENTS IN SENTENCE STRUCTURE

1. Coordinating Conjunctions

The engineers built a bridge, AND the contractors built a road.

The engineers built a bridge, BUT the flood swept it away.

The engineers built a bridge, FOR the old one had been demolished.

The engineers built a bridge, SO the Army awarded them another contract.

The engineers built a bridge, YET the villagers still feared the river.

Coordinating conjunctions help to connect components in a multiple structure. They tend to express simple, equal relationships, however, rather than more complex relationships between statements. One of the great weaknesses in student writing is the use of too many coordinating conjunctions. Think of "and" as analogous to "+", and "but" to "−". Just as you have learned other symbols to express more complex relationships between numerical values, you must go beyond the plus and minus of coordinating conjunctions as a way of expressing logical relationships in sentences.

2. Correlative Conjunctions

The engineers built BOTH a road AND a bridge.
The engineers NOT ONLY built a bridge, BUT they ALSO built a road.
EITHER the engineers will build a bridge OR they will build a road.
NEITHER the engineers NOR the contractors could build the bridge.

Correlative conjunctions give a kind of double bonding to relationships in a sentence which coordinating conjunctions alone cannot do.

3. Conjunctive Adverbs with Semicolons

The engineers built a bridge; HOWEVER, they had a cost overrun.
The engineers built a bridge; THEREFORE, their work had ended.
The engineers built a bridge; THEN they built a road.
The engineers built a bridge; OTHERWISE, the city would have been isolated.
The engineers built a bridge; STILL the river was a threat.
The engineers built a bridge; the contractors did a construction review.
The engineers built a bridge; the construction review, however, revealed serious design flaws.
The engineers built a bridge; their success, therefore, speaks for itself.

LEARNING TO WRITE SENTENCES

Exercise

The term which describes this exercise is "imitation." We learn to write and speak by imitating others. In the years before educators theorized that dull, repetitive labor warped young psyches, imitation was a commonly used method

Learning to Write Sentences 373

for teaching writing. A more common term for imitation is *copying*; that is, simply copy a series of sentences or a passage, and, in that way, program your mind with the patterns and structures you need to master.

1. Copy each sentence in the section on sentence structure. By doing so, program your mind with the ways in which the basic component of a sentence can be fitted with other components at the beginning, in the middle, or at the end. Pay careful attention to commas and semicolons as you copy. (If you have special difficulty with sentences, do this every day for a week.)

2. Select a technical subject familiar to you and write 3 sentences in each of the patterns in the section on sentence structure. Here your task is to imitate the structure of the sentence rather than to copy it word for word. Be sure that you use a technical subject—the same one throughout—so that you can better see how structure affects the interpretation of substance.

3. Below are additional sentence patterns which vary the basic subject/verb/object or actor/action/receptor pattern. Copy each of these; once you have copied them, then write 3 sentences on a technical subject in each of the patterns.

    a. *That the deficit in rainfall has lowered the water table* is evident to the engineers. (noun clause as subject)

    b. *What to do in case of another prolonged drought* is not clear. (noun clause as subject)

    c. *Whom the engineers will consult* has not yet been decided. (noun clause as subject)

    d. *Seeding the clouds with dry ice* may be effective. (gerund phrase as subject)

    e. *Recording the annual rainfall* has occupied scientists since Colonial times. (gerund phrase as subject)

    f. *To ration water* may be the next necessary step. (infinitive phrase as subject)

    g. *To police the use of water* will be the duty of the Sheriff's Patrol. (infinitive phrase as subject)

    h. Many government agencies—*Civil Defense, the Fire Department, the State Patrol, and others*—have cooperated in this effort. (apposition sentence)

    i. Churches, schools, civic groups—*indeed, all segments of the community*—have cooperated in this effort. (insertion sentence)

    j. *Compression, ignition, exhaustion*—these are the effects of the internal combustion engine upon gasoline. (list followed by a dash)

    k. *More money, more personnel, greater government support*—all these are requisite factors for success. (list followed by a dash)

l. The major factors to consider are these: *time, money, and space.* (colon to introduce a list)
m. The primary consideration is saving lives: *reducing the number of auto accidents and improving after-crash treatment.* (colon to expand an idea)

Even a very accomplished writer may not use all of these patterns in a lengthy manuscript. In fact, it is sometimes suggested that only one sentence in four in technical prose should vary the subject/verb/object pattern. The varied patterns are good to have in one's arsenal, however, when prose has to carry added impact.

PUNCTUATION GUIDE

Commas

1. After a long introductory phrase, the careful writer uses a comma.

    On the west side of the property, there is a stream.
2. When the writer uses a long introductory clause, a comma is necessary.

    After we have considered all the bids, we will make an estimate.
3. A writer places a comma between two independent clauses joined by a coordinating conjunction (*and, but, or, nor, so*), and she is careful not to separate the second half of a compound predicate from the rest of a sentence. To do that separates the subject from the verb and can be confusing to the reader.

    We will finishing collecting the data on Monday, and Tuesday morning we will begin writing the report. We will have to set all other work aside and concentrate until this is finished.
4. Using an introductory participial phrase, the writer also uses a comma.

    Pending completion of the project, we will keep a staff on duty.
5. Non-restrictive phrases and clauses, called non-essential by some writers, are set off by commas.

    Proof-reading, which is the last step in the process, cannot be omitted.
6. Subordinate elements which are essential should not be set off.

    Proof-reading is a step which cannot be omitted from the process.
7. To avoid misreading, the writer uses a comma.

Semicolons

1. Conjunctive adverbs (*therefore, however, instead, rather, then, still*) can make the relationship between statements clearer; however, their use requires a semicolon.

    Ordinarily we can complete such a job in a week; however, a shortage of supplies makes that impossible in this instance.

2. A conjunctive adverb may not always be necessary; two independent clauses separated by a semicolon can be effective.

    Bill Johnson will deal with budget; John Jones will handle scheduling.

3. In many cases the conjunctive adverb follows the semicolon; in other cases, however, it appears later in the second clause.

    We usually meet our schedule; in this instance, however, we are delayed.

4. Sometimes independent clauses in a compound sentence are internally punctuated because of month, day, year; city, state, and country; or such non-restrictive elements as introductory phrases and clauses; so the writer uses a semicolon before the coordinating conjunction because it is a stronger dividing mark than the comma.

    The meetings will be held in St. Louis, Missouri; Austin, Texas; and Sante Fe, New Mexico. The dates will be April 26, 1990; June 5, 1991; and July 14, 1992.

Colons

A colon indicates this: something is to follow. It can introduce two things: a list or a statement which expands the previous statement.

   Our course in this matter is clear: litigation is the only alternative.

   We have three primary considerations: time, money, and energy.

Dashes

To indicate a sharp break in the sentence, to serve as informal parentheses, to introduce a re-phrasing or summary—these are the uses of the dash. (The dash is like its name—it should be used sparingly.)

The third phase—in fact, all phases of the project—will be completed within the specified time.

Budget, scheduling, personnel—these are the three areas of conflict.

A partial refund is all that we can make—and all that is due you.

INFORMAL PROCEDURE FOR TESTING SENTENCE STRUCTURE

A writer who does not know formal grammar may nevertheless sense that a sentence contains a flaw in design. This happens when a sentence seems to run too long or when a reader must reread a sentence to get the sense of the statement. The following tests may indicate that a sentence needs revision:

1. A typed sentence which runs more than 2½ lines may need revision, either into two sentences or into two clauses joined by a semicolon and a conjunctive adverb.
2. A combination of *and* and *but*, of two or more *and*s, or of *but* and *so*, or any such multiple use of coordinating conjunctions, may signal a run-on sentence.
3. A combination of *and* and *which*, of *and* and *because*, of *but* and *which*, or of *but* and *because* may also signal a run-on sentence.
4. A comma followed by *which* or *because* near the end of a sentence often signals a poorly constructed sentence.
5. Two or more words ending in *ing* in a sentence can be a flaw. (This is termed the *boing boing* effect.)
6. An uneven number of commas in a sentence may also signal a design flaw. (For this test I am indebted to my Georgia Tech colleague of the Department of Physics, Dr. O'Shea; it is therefore termed the O'Shea Effect.)

While these tests are not formal grammatical standards, they can serve as a quick and simple way to identify sentences possibly in need of revision.

COMMON DESIGN FLAWS IN SENTENCES

Any inexperienced writer should especially note the following flaws in sentence design.

Faulty Use of Coordinating Conjunctions

The data is not readily available, *but is* necessary.

Companies try to notify banks to stop payment on checks, *but are* not always successful.

The process is inconvenient for use on foundations, *but is* often employed for curing concrete blocks.

The cue had a high degree of personnel safety, *and was* based on flashbulb technology.

The Loose Adjective Clause or "Which Afterthought"

The software method adds additional lines to the program, *which is* important when the availability of memory is a concern.

Faulty Parallelism

Fibers from undyed carpet specimens are dissolved in orthochlorophenol *then mixed* and used to prepare a calibration curve.

The experiment required breaking down the problem *then offering* a practical solution.

Dangling Modifier Followed by Passive Voice

After examining all the available data, it is evident that the weapon is not only feasible but necessary.

By following the procedures, a fabric can be returned to good condition.

Run-on Sentence

When using Portland cement in the grout, there can be problems with the grout shrinking slightly upon curing, therefore, leaving the mass of the machine resting on the leveling wedges.

Awkward Structure

All functions are presented, beginning with the concrete and finishing with the grouting operation.

The procedures followed when taking a data count were reported to the supervisor.

Once the foundation has been designed and its components formed and cured, it is time to position the machine and grout it into place.

Unnecessary Comma

The data base is designed to be efficient in its structure, and easily modified.

Misuse of Semicolon

The primary aim was to gain greater efficiency; although, the data base was designed to do other tasks.

Absence of Semicolon

The wing shape was standard however the weight exceeded specifications.

chapter 20

Paragraphs

The paragraph is a logical component within technical documents; the frequent use of paragraphs identified by sub-headings makes the paragraph an important functional unit as well. While paragraphs in technical writing require the traditional features of unity, logical transitions, and appropriate length, there are further concerns which make the paragraph an important component in longer technical documents.

1. The paragraph is a strong component in technical documents because a report or a memorandum may not be read serially and laid aside; instead, the document may be randomly accessed, referred to, or read only in part. The well constructed paragraph identified by a sub-heading thus serves an essential function because any technical document is a *working* document.

2. For standard technical discussions, many companies now compile "paragraph banks" for storage on tapes or disks. A writer who uses a mixture of standardized paragraphs with his own writing must be able to produce a paragraph as a matching component. Since word processors do not have an infinite capacity for changes, the paragraph can become the chief component in revising copy.

3. Technical content also calls for very logical patterns of thinking to convey material which has its own physical order in the technical work itself. For

379

most technical activities there are rhetorical patterns which supply written analogies for technical tasks:
- what is this a part of? classification
- what are the parts? division
- what is it? definition
- what does it look like? description
- how does it work? process
- why does it work this way? cause and effect

The answers to those standard questions about technical activities are both rhetorical patterns of thinking and specific techniques for writing, as the following collection of paragraphs demonstrates.

CLASSIFICATION AND DIVISION

Classification is an upward movement which answers the question "What is this a part of?" Being able to relate a single object or idea to the larger group of which it is a part lends perspective. Division is a downward movement which answers the question "What are the parts?" Being able to identify the parts of an object or of an abstract subject is a first step toward understanding the subject and writing about it.

Basis

An orderly process of classifying and dividing is impossible without first determining the basis. A basis is like a common denominator which serves as a basis for comparing a group of fractions. In the same way, selecting cost as a basis, or time, or efficiency of operation, provides a basis for organizing research data. That pattern of organizing data becomes, in turn, a way of defining your task and a way of ruling out extraneous material. In contrast to taking a "shot-gun" approach, selecting a basis is like sighting through the cross hairs and concentric circles in a sighting mechanism.

Division of Objects: Example 20.1

For the purposes of this energy-reduction study, the building was divided into the following areas: floor, roof, and wall, with glass areas treated as a percentage of gross wall area. The heating and cooling apparatus was divided into two systems for each program, and both systems were input as air-to-air pump types. In Program 1 the perimeter areas of each floor were assigned to System 1; the interior areas were assigned to System 2. Each system was input both as multi-zone and as single-zone systems. As an exception, the Courtroom and the second-floor conference rooms were input only as single-zone systems. As an addition, energy-consuming equipment was not directly affecting the heating and cooling loads was input as base utilities for each program so that the program could account for such equipment as a part of total energy consumption.

How

1. Getting organized is the first step in any job. Here the space to be studied for energy reduction is first divided into logical parts.
2. The space is divided on the *basis* of those areas and that equipment which will be subjected to study.
3. Note the exception stated toward the end of the paragraph. Stating exceptions to the parts noted in a division is a standard feature of this kind of writing; it allows for the fact that every part of a subject will not fall absolutely into one division or another. Noting an addition allows for incomplete division also.

Why

1. As an aid to thinking, the dividing process saves time because the writer knows exactly what direction to take.
2. As an aid in the writing process, the writing itself should become almost a matter of filling in the blanks once the pattern of division is established. Just as a certain kind of equation works out according to the same pattern every time, technical writing becomes almost a way of simply and unobtrusively demonstrating facts with words.
3. As an aid to reading, division has that same clockwork pattern for the one who reads technical material. Because technical matter is difficult to read, set up for the reader a pattern which you repeat in various sections: space, time, cost, personnel, or other basic factors. Retain that pattern; for the reader probably needs that order of parts, however boring it may seem to you as writer.

Division of Groups: Example 20.2

> The medical profession classifies the various types of health services into primary, secondary, and tertiary medical care. These categories are based on the levels of medical treatment, qualifications of the health-care professionals, and facilities equipped for the delivery of each level of care. Primary medical care is defined as the initial contact with the health care provided; as such it is synonymous with care and treatment delivered in a physician's office. Secondary health care is defined as that care which is delivered primarily to medical and surgical patients in a community hospital. Tertiary medical care goes beyond these levels as treatment delivered by specialists in a sophisticated medical center.

How

1. Laying the groundwork for studying an organization also requires an initial division of the subject into its logical parts.
2. The basis for division in the sample paragraph is the kind of health care. Each division is also defined.
3. In the total report on health care, this paragraph—although it is written as division—serves the purpose of classification: the upward movement which answers the question "What is this a part of?" As the report progresses, the writer branches off to discuss primary care in a process of division that breaks that part of the topic into more and more detailed sub-sections. This paragraph, then, is only one stage in the entire process of division. It first classifies to lend perspective to the details which are to follow.
4. Through classification and division, create a broad initial perspective for the reader before branching out into the more detailed parts of your document. This perspective is especially useful when you are dealing with groups within an organization or functions which an organization performs.

Why

This kind of thinking and writing is perhaps most useful at either end of a long writing project. In the study which leads to a sustained writing project, you as writer gain some overall grasp and understanding of both the general topic and the specific area you will address. Good classification and division writing communicates this overall grasp of a subject. At the other extreme, the reader who reads the final draft will be impressed by the comprehensive background of a narrow subject.

Division of Ideas: Example 20.3

This study of health care for Idaho students shows that the Department of Education has the following alternatives:

1. providing all health and medical services in-house;
2. contracting with Department of Human Resources facilities;
3. contracting with community resources and health care providers;
4. participating in a group health-insurance program;
5. participating in federal funding programs;
6. combinations of 1, 2, and 3.

How

1. These alternatives are divided on the basis of actions which the organization may take.
2. All the study and data-gathering and thinking has the object of breaking the initial problem into a series of ideas.
3. Note the parallelism: *providing, contracting, participating.*

Why

1. The difference between a good technical document and a bad one is that the bad document is a one-dimensional recitation of facts, whereas good writing takes the second step in technical writing: it *interprets* data.
2. After years of sitting in meetings, my perception of the difference between a leader and a follower is that a follower sees only the parts; the leader sees the parts in relation to the whole: "This problem is a sub-division of a larger problem" (classification); "to solve this, we must take three actions" (division). That kind of thinking ability also makes the good writer.

DEFINITION

Communication cannot exist unless writer and reader agree upon what each word means. Even when writing to a technical audience, a technical writer must make sure that the reader will interpret a term in the intended sense. To avoid any possibility of misunderstanding, it is important that the specific connotations of key non-technical terms be defined.

Technical Definition: Example 20.4

> The compressed air dryers are the heatless-desiccant moisture-absorption type. This type of dryer removes moisture from the compressed air by using desiccant to absorb the moisture in a pressure chamber. The collected moisture is then removed from the desiccant by pulling a vacuum on the chamber during regeneration.

How

1. Begin with the term itself and any descriptive words which qualify the terms. First, the reader must know "What is it?"
2. The example above defines primarily by explaining how the terms in the definition operate in the process.
3. At first reference, spell out letters and abbreviations: BTU's (British Thermal Units), Department of Offender Rehabilitation (DOOR). Thereafter, use the shorter form.
4. Place an unfamiliar or relatively new technical term in quotation marks at first reference. For example, retro rockets were referred to by the use of "retro rockets" in the early days of the space program. Use your judgment about when increased familiarity with a term makes quotation marks no longer necessary. How journals in your discipline do this is your best guide.
5. Underline a word referred to as a word: "the word _safety_ as it is used here means ..." Such an underlined word does not perform its usual function in a sentence; instead, it is thereby distinguished as a term whose function is to be discussed.
6. Your own familiarity with a term reads into it the needed punctuation: "heatless desiccant moisture absorption compressed air dryers." A reader to whom the term is unfamiliar needs commas and hyphens. For someone unfamiliar with the term, punctuation helps: "heatless-desiccant moisture-absorption, compressed air dryer."

Definition

a. The comma rule is that descriptive items in a series are punctuated: "a restricted, fortified, correctional facility." Place those commas, however, according to the *rhythm* of the series of terms, *not* strictly by the rule. Otherwise, the rule may produce "grand, old party".

b. The hyphen rule is that two nouns used as a single descriptive term should be hyphenated: "a moisture-absorption compressed air dryer". Some dual terms, however, are hyphenated and others are not: if "moisture-absorption" why not "compressed-air"? The answer is twofold: how familiar the term is in general use and whether the reader needs the hyphen to aid his reading of those two words as a single term. Again, your professional journals show how you should do it.

NOTE: Numbers 3, 4, 5, 6 in this section are not to be memorized and dumbly followed, but to be recalled and wisely applied according to your best judgment in a given situation related to a given audience.

Why

1. In contrast to an object (like an air dryer sitting there busily drying air), a word is usually treated as only an abstract symbol which refers to an object. If you and I agree that henceforth we will call that air dryer "banana," the word *banana* would call to mind that same air dryer sitting in the same place busily drying air. Because a word is not an object, you must make sure that the word you use calls up in the reader's mind the object *you* have in mind. Otherwise, there is no communication.

2. Technical definition must be precise because each term must evoke as unambiguously as possible the actual object it refers to. In technical writing, as in law, linguistic tolerances are very precise.

Non-Technical Definition (Be Sensitive to the Psychological Effects of Words Upon Your Audience): Example 20.5

It should be noted that the phrase "security control" in no way implies the exclusion of the public from the facility. The dam project was designed in a way which encourages public use. Security and control measures are needed, however, to minimize possibilities of injury to visitors and to reduce the facility's vulnerability to damage by vandals and theft.

How

1. To avoid the possibility of misunderstanding, distinguish the meaning you intend from other meanings often borne by that term. Some texts call this "stipulative definition"; that is, you stipulate the sense in which you are using the term here. Other texts called this "defining by negation"; that is, you predict that the reader may interpret the word in a way that you do not intend, and for that reason, you begin by saying that the word means *not* this, and *not* this, but *this*.
2. A further note on punctuation rules: whereas you underline a single term, it is easier to place a phrase like "security control" within quotation marks. Whether a word processor will underline is a consideration which may dictate quotations marks for all terms and phrases.

Why

Audience analysis includes being able to predict how a given word will affect a given audience. In the example above, the Department of Offender Rehabilitation would read "security control" with an entirely different attitude and an entirely different set of needs and expectations. Careful definition prevents (literally: "goes before") misunderstanding.

DESCRIPTION

A verbal description of a building or a site requires the same proportioned relationship of the parts and the same selection of important details which a drawing requires. Logically dividing the space and carefully selecting details help the reader of a verbal description to visualize the space, just as they help the viewer of a drawing to determine what it represents.

Building: Example 20.6

> The building construction is masonry. Walls at the ground floor level are block-and-brick construction, and the second- and third-floor walls are composed of precast concrete panels. The structural system is poured-in-place reinforced concrete. The first floor is a slab on grade and the upper floor and roof slabs are waffle type.
>
> The interior walls are of gypsum board over fiberglass batt insulation. Interior partitioning consists, for the most part, of floor-to-ceiling, movable panels spaced on multiples of the basic 5-foot building module.
>
> Glazing is single-pane, fixed, tinted, float plate glass. The ground floor has large areas of store-front glass at the north and south entrances, the Post Office lobby, and the west exterior wall. The upper-floor windows are nominally 6 feet high by 2 feet wide, regularly spaced on 5-foot centers around the entire building perimeter.

How

1. Note that the building is first classified as masonry and then divided area-by-area, using construction material as the basis for the division.
2. This description of a building involved in an Energy-Reduction Study moves logically from bottom to top and from exterior to interior. The logical development is analogous to the movement of a camera slowly panning an area; it sweeps some areas to provide a context and pauses at other areas which are more important.
3. Select details related to your purpose in describing the building; in the example above, the selection of construction details relates to the needs of an energy-consumption study.
4. *Style:* Note the pattern of short, rather simple sentences. If the entire process is like a camera panning, each sentence is like a frame of film. The reader has enough to do as he notes the parts, the relationship of the parts, and the construction of the parts. Unless you are an excellent writer, the reader cannot also deal with an elaborate style.

Why

1. Describing a space where work is to be done provides the reader a clear visual picture of the context where the work is to be done.

2. The purpose of a written document determines the selection of detail. A company which had been asked to make a bid to sand-blast the masonry or wash the windows would give the same relationship of parts (top to bottom or exterior to interior), but the selection of detail would vary because of the different purpose.

Site: Example 20.7

Adequate security precautions are imperative to protect the intake area of the dam from intruders. This area of the dam is highly susceptible to trespass because of its location next to the public parking lot and the look-out point.

The only barrier between the parking area and the intake area is one chain-link fence, an obstacle so easily overcome that it permits access to the five control houses in the intake area. These structures, which house large quantities of equipment, are vulnerable to vandalism because the door hinges located on the outside of the doors may be easily removed to allow rapid access to the equipment inside. The water at the intake area also poses a danger to the life of an intruder and creates an imperative need for appropriate security measures here as well.

How

1. The selection of details relates to the purpose of the report: recommending better security control measures for a dam and recreational area.
2. This site description is less passive than the description of a building. Here the details are visual, but they also exist in a causal relationship in describing the problem. Such careful description is an important part of all types of writing.

Why

Selection of detail and tone relate to the purpose of the report. A brochure written for campers who might visit the same recreational area would be much different from this report on security control to protect both them and the site.

Space and the Reader

In describing a building or a site, do not assume too much knowledge on the reader's part. The reader may frequently see a building he himself owns, but your task as a report writer is to have him see that structure through your eyes as you define and describe a technical problem related to that building.

Readers in a large organization may never have seen the building or the site you describe. So don't assume that their owning it and their planning to spend a million dollars on it mean that they are technically informed about the structure.

Space and the Writer

Remember that you may well have far greater abilities to visualize than many readers. You no doubt call up a precise image when you read "6 sq. ft.," but that means nothing to some readers. Thus approach writing about space with an eye toward the probable difference between your ability to visualize and your reader's limitations.

PROCESS

Your reader must know *how* a physical or mechanical process operates before you can go on to explain how to solve a problem related to a piece of equipment or to an entire system. Explaining how a process works is a matter of dividing the process into its logical steps. Process writing is most effective if you also explain the principle involved in *why* each steps works as it does.

Process: Example 20.8

> The existing air dryers use compressed air to regenerate the desiccant chamber. The air is first passed through the desiccant to carry away moisture during the regeneration process; then it is exhausted to the atmosphere. The 16,500 cubic feet per minute of exhausted air is 3% of the total capacity of the air dryers, or approximately 500 scfm. This air must be supplied by the air compressors as a part of the total cost of providing compressed air for the entire process.

How

1. Write three levels of the process to fulfill three purposes.
   a. Describe the *steps* in the process; these actual steps in the mechanical process provide the organizational pattern for process writing.
   b. Explain the *principle* of operation in the steps, or *why* each step operates as it does: to create a vacuum, to speed a mechanical operation, to dry and thus regenerate the desiccant.
   c. Demonstrate the *implications* of the process for your purpose in writing the report: increased efficiency, reduced costs, or the like.
2. *Style:* enumerate the steps in a process by using words like *first, second, then, when, after.* Do not use *firstly, lastly,* or any *ly* word except *finally.* Beware of *ing* words in process writing also, for they create a construction and a rhythm that can be hard to read. For *turning* substitute *to turn;* for *using* substitute *which uses.* If a passage reads awkwardly, test it for the "boing boing" effect which *ing* words can create.

Why

1. The technical description of the *steps* themselves is necessary for your own reference and important also for the technical reader.
2. An intelligent non-technical audience who must read the report—and who may not know a *psig* from a *scfm*—can understand the *principle* you explain without necessarily understanding the technical steps you describe.
3. The natural goal of the descriptions from the two viewpoints above is your explanation of the *implications* of this process in terms of time, money, and the like.
4. In sum, these three levels allow for different parts of your work: defining and solving an engineering problem, explaining it to technical and to non-technical audiences, and selling your work by the explanation you give the reader of the wider implications of the process.
5. As with other forms, include enough detail to give a sufficient idea of the entire process, but highlight those details related to your work. Avoid a lengthy, one-dimensional account which records data without giving an interpretation of them.

CAUSE AND EFFECT

The physical operations of cause and effect can be so complex that it is important to simplify an explanation of cause and effect by using traditional rhetorical patterns:
- single cause—single effect
- single cause—multiple effects
- multiple causes—single effect
- multiple causes—multiple effects

Technical: Example 20.9

> The present waste flow is below design by 60%, or approximately 476,000 gallons per day. This creates retention times within the clarifiers and oxidation ditches that are approximately twice the design retention times. These long retention times within the clarifiers allow more time for the suspended colloidal solids to settle to the bottom of the unit before they are removed. The long retention times within the oxidation ditches allow more time for the microorganisms to oxidize and remove the organic matter. This results in high solids and organic matter removal which, in turn, produces an effluent that meets the NPDES permit standards.

How

1. The difference between cause-and-effect writing and writing about a physical or mechanical process is one of emphasis. Good cause-and-effect writing requires that you describe the process and explain the principles which make the process work as it does.
2. The purpose of cause-and-effect writing is to explain *why*, whereas the purpose in process writing is to explain *how*.
3. The basis for division for process leads toward division into a series of steps (which in a mechanical or physical process necessarily includes cause and effect). As a process of division, the basis for cause-and-effect writing carries process a step further by dividing to discover lists of causes and lists of effects.

4. *Style:* Use causation terms like *because, since, thus, therefore, for that reason, with the result that, when ... then.* Do not use *as* for *because* or *due to* for *because* in cause-and-effect writing. The misuse of *as* can be particularly confusing to a non-technical reader who reads *as* correctly as a word which relates to time and motion, not to causation.

Why

To explain why is one of the most difficult and most necessary activities in any profession. For that reason, any cause-and-effect writing must be carefully and clearly done.

a. A clear statement of cause and effect on the mechanical or physical level becomes a beginning point in a problem-solving process which may have cost savings as its ultimate effect.

b. An explanation of cause and effect creates a common ground of understanding with a technical reader.

c. An explanation of cause and effect can win the confidence of a non-technical reader. A clear explanation of what has caused a problem and what will be required to solve it can enable this reader to make an informed decision on whether to sign a contract and a check.

Persuasive: Example 20.10

> The proposed industrial-effluent utilization action can be accomplished by forming a closed water loop between the industrial-waste treatment plant and the using points. It is possible for this closed-loop system to produce two effects. First, this will eliminate the discharge to the Red River. This, in turn, will mean that Apex Co. will be free from water quality regulations as administered and enforced by the State's Environmental Protection Division and the Federal Environmental Protection Agency. Second, the formation of this closed loop will also reduce the amount of water Apex Co. must now purchase from the city of Plattsburg. At present usage rates, this reduction will produce an annual cost savings of approximately $60,000.

How

1. A basic statement of cause and effect is this: "if *a* happens, then *b* will follow." In explaining the technical causes and effects, this statement can be made with a high degree of certainty. Explanations to a client about the ultimate results of technical recommendations often cannot contain the same degree

of certainty. Therefore divide what can be proved technically from what can only be recommended and predicted persuasively. The technical is the present object; the recommendation is only the future possibility—however well grounded in technical data that possibility may be.
2. *Style:* temper your verbs: *can* instead of *will*, "should be able to," "can be shown that," "under existing conditions ought." Such words are not evasions, but realistic in terms of what statements of cause and effect can and cannot do. (Refer to the more complete discussion of verbs in the Writing Lab on Words.)

Why

1. Physicists and engineers who make statements about physical objects can do so with a great deal more certainty than philosophers, theologians, or even economists. Historically, however, even physicists have disagreed about physical laws of causation; and engineers understand that even engineering is, to a degree, an inexact science. That is why it is so important in cause-and-effect writing to divide what can be clearly proven and established from what cannot be.
2. Rhetoric, as a process of dividing and organizing thoughts, can lend clarity, but that is not the same as absolute technical accuracy. You should be able to see by now that rhetorical patterns are not infallible. As illustrated by the relationship of process writing and cause-and-effect writing, the patterns overlap.

EXERCISES

Reading Exercise

Refer to the assignment memorandum in the Memorandum Chapter, read the student examples of memorandums, and repeat that exercise. In the reading you do in your own academic major, develop a greater awareness of how the traditional rhetorical patterns are components in what you read.

Writing Exercise

Select a technical topic and write a paragraph in each of the rhetorical patterns. Once you have written a definition, a division, a process description, and so on, you may be surprised at how well constructing these individual components produces a complete discussion and description of any given topic.

chapter 21

Graphics in Technical Writing

The extensive use of graphics in technical writing requires a special discussion of graphics as a skill basic to all technical communication.

Engineering and architectural drawing and a wide range of other graphic forms are a familiar part of any technical work. Graphics within a written document, however, are not the same as a set of plans for a building or a shop drawing for manufacturing a machine. In contrast to the uses of graphics most familiar to E/Ts, technical illustration is subordinate to verbal expression in technical writing. Both in its physical dimensions and in its logical function, an illustration works within a text as much as a paragraph does.

To communicate to a wider audience, especially to clients or to management, requires words as the primary medium. The person who writes a technical report may first have visualized the technical problem, rough sketches may have been the first basis of a preliminary conversation about the project, and certainly drawings will direct each step of the work on the project. But words, not graphics, are paramount in the writing which communicates about the project.

Within a written technical document, graphics have rhetorical functions: they define and clarify, exemplify, classify and divide, compare and contrast,

Graphics in Technical Writing 395

describe process, illustrate the parts, and clarify the relationships among parts, in any technical subject.

The actual drawing of illustrations is a technical skill related only tangentially to technical writing. The production and placement of graphics within a finished document is also a skill related to technical writing and not a part of the writing skill itself. Hence the emphasis in this text is upon the rhetorical functions, rather than the mechanics, of illustrations: how they work to advance the development of a technical explanation, and how they enhance a reader's understanding of a written text.

Partition

To show a logical division of parts is basic to all technical communication. The drawing which shows the relationship of parts in a testing device thus serves to give an overview of the relationship of parts:

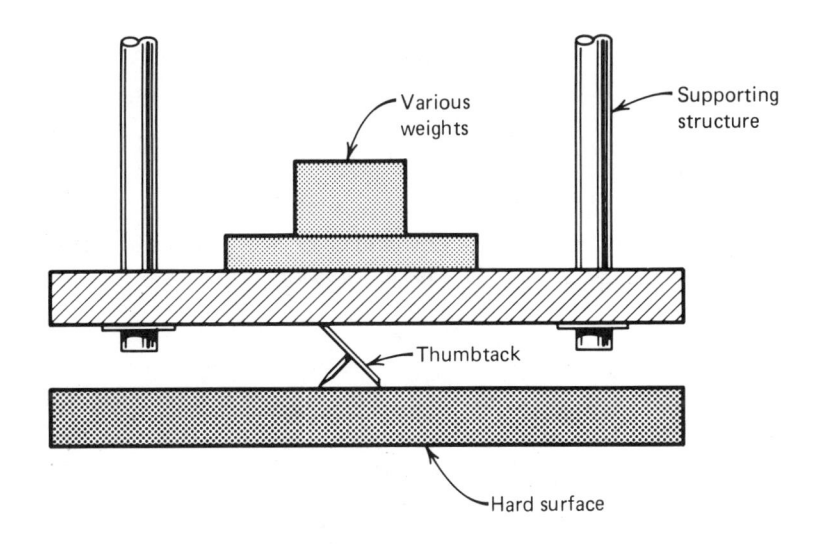

An exploded drawing shows the parts of a device at a point just before assembly; this arrangement communicates the interrelationship of parts before the writer goes on to discuss the process whereby the device operates:

Strain gauge balance

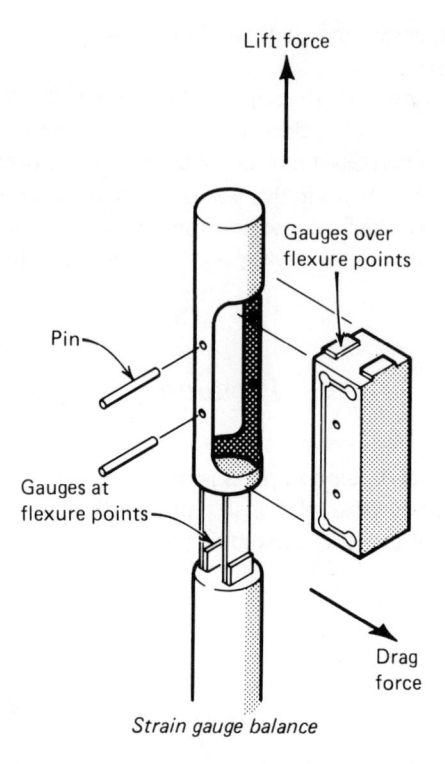

Exploded drawing.

To further divide the description of any kind of device requires, rhetorically, an example or, technically, a detail of a larger drawing:

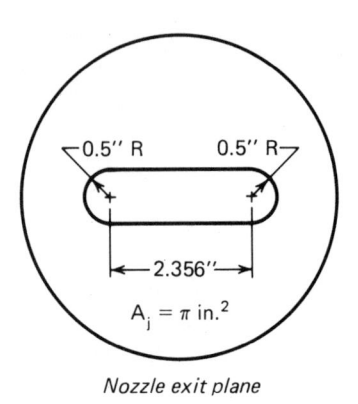

Nozzle exit plane

$A_j = \pi \text{ in.}^2$

Detail.

Graphics in Technical Writing

Specifically technical communication about the interrelationship and operation of parts in a design calls for more complex graphics. The drawings which follow are from a technical report on the aerodynamics of a wing design. The graphic illustrations function technically in a way that neither words nor calculations, working together or separately, can. Thus in this series of three examples which follows, a writer can fulfill the basic rhetorical purposes of **defining** and of giving **illustrations** and **examples**.

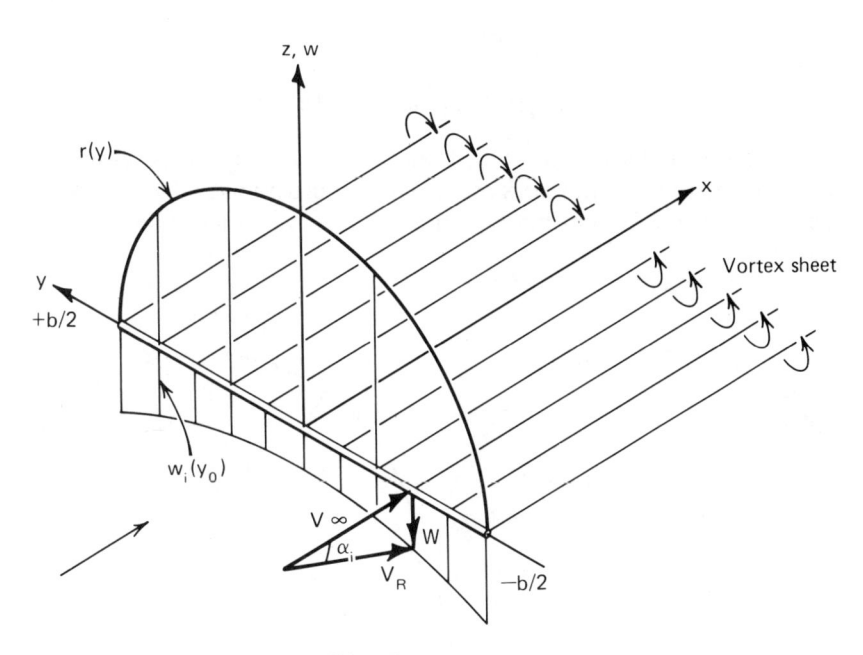

Wing Representation.

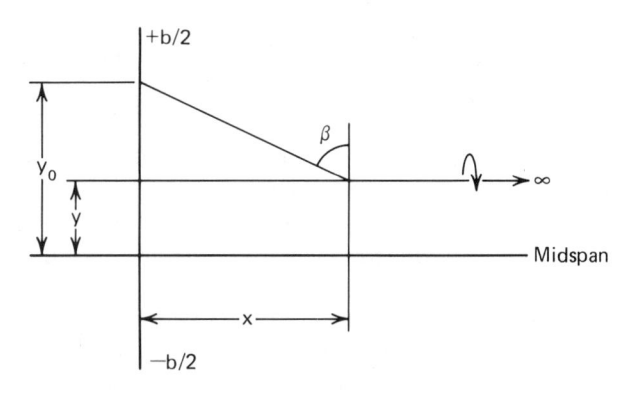

Downwash.

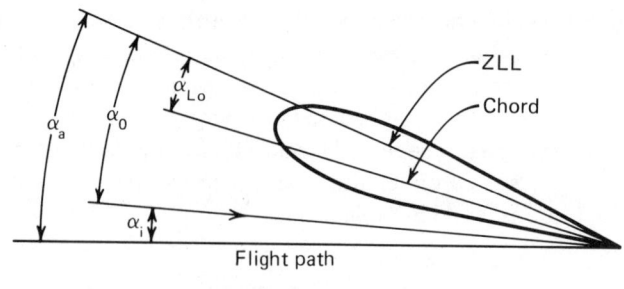

Angles of Attack.

Classification and Division

Graphics are useful also in the basic work of **classifying** and **dividing** large bodies of data to make the work of explaining more manageable. The **pie chart** is especially useful in showing how groups are divided into parts, each on a different basis:

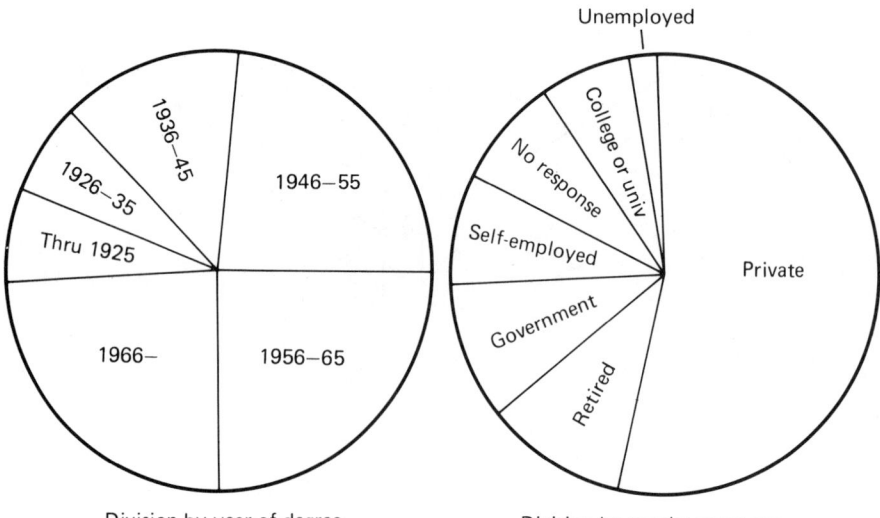

Division by year of degree   Division by employment status

Such use of graphics becomes a rapid and effective means of conveying different configurations of the same body of data. Technical accuracy is, of course, always important, but the essential purpose is to arrange the data for rhetorical clarity. The first concern, then, is the nature of the audience and not the nature of the technical subject itself.

The **bar chart** is also very useful in giving a rapid and effective division into parts. To arrange data on a different basis results in a clear reconfiguration of the data. In the bar charts below, two versions of the results of an alumni survey published in 1975 in the University of Michigan *Technicum* show how regrouping can carry a reader very rapidly and effectively through a given subject.

Graphics in Technical Writing   399

Division on the basis of continuing education courses.
- Education — 82
- Lib arts — 199
- Business — 812
- Medicine — 38
- Military — 57
- Law — 86
- Other — 732

Division on the basis of leisure activities.
- Religious — 1157
- Local civic — 1027
- Youth — 590
- Athletics — 246
- Politics — 231
- Social — 227
- Ecological — 107
- Business — 104
- Charity — 103

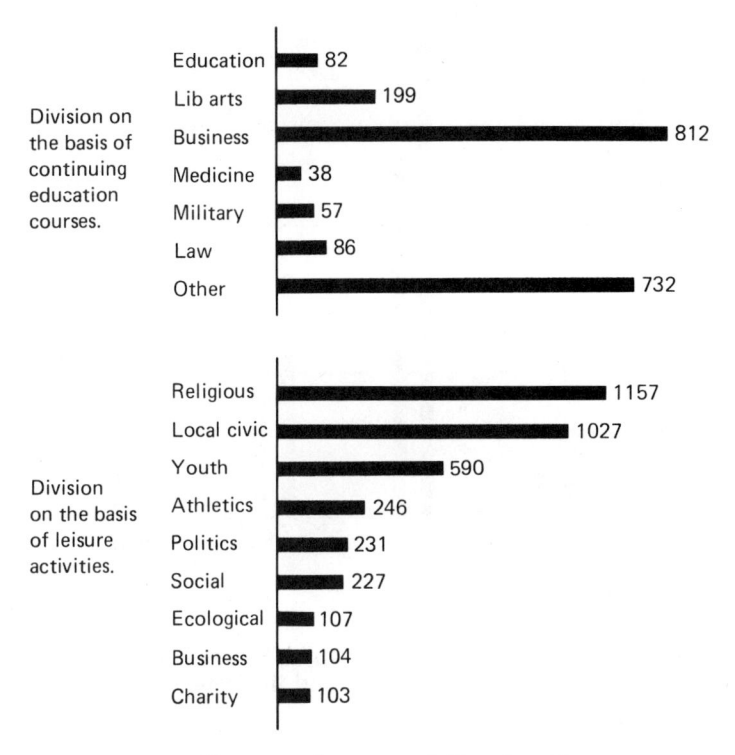

Comparison and Contrast

Comparison and **contrast** can also be displayed graphically. A reference to the alternating and opposing methods of comparison and contrast outlining in the section on outlines will show by comparison how efficient the use of bar charts can be to accomplish the same rhetorical arrangement of data. The chief difference is that the graphic mode does not allow for discussion; the thought pattern, however, is exactly the same.

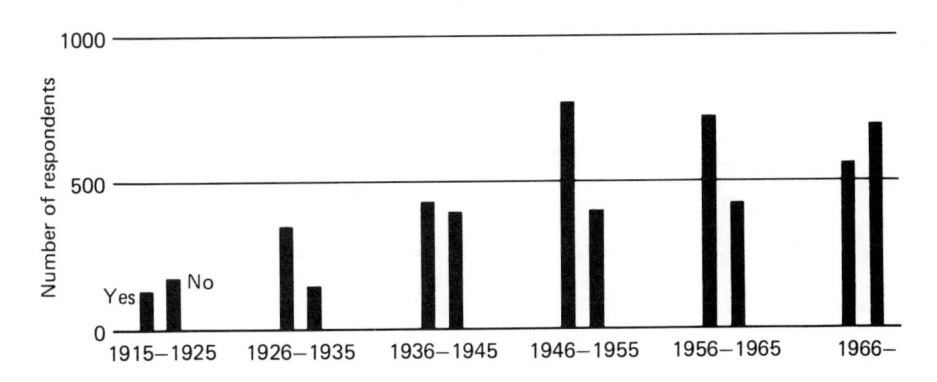

Number of respondents

Yes / No by year ranges: 1915–1925, 1926–1935, 1936–1945, 1946–1955, 1956–1965, 1966–

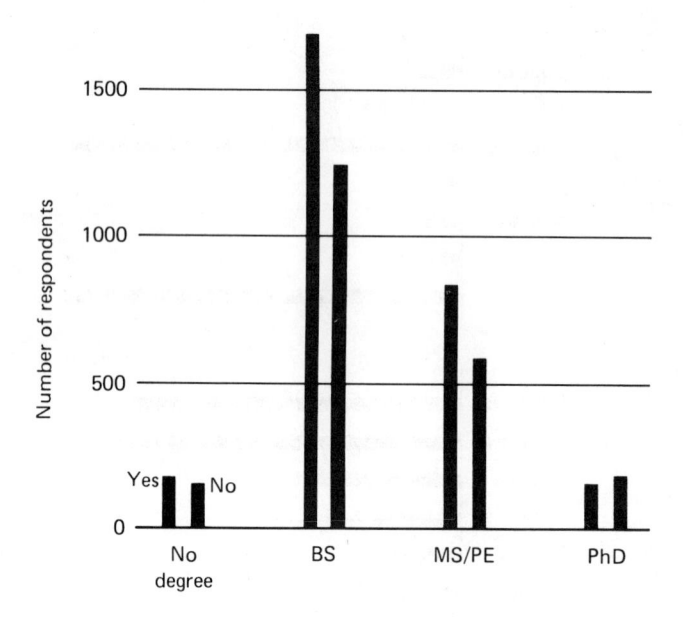

Comparison and contrast in a more precise technical sense shows a comparison and contrast of experimental and theoretical observations about wing performance:

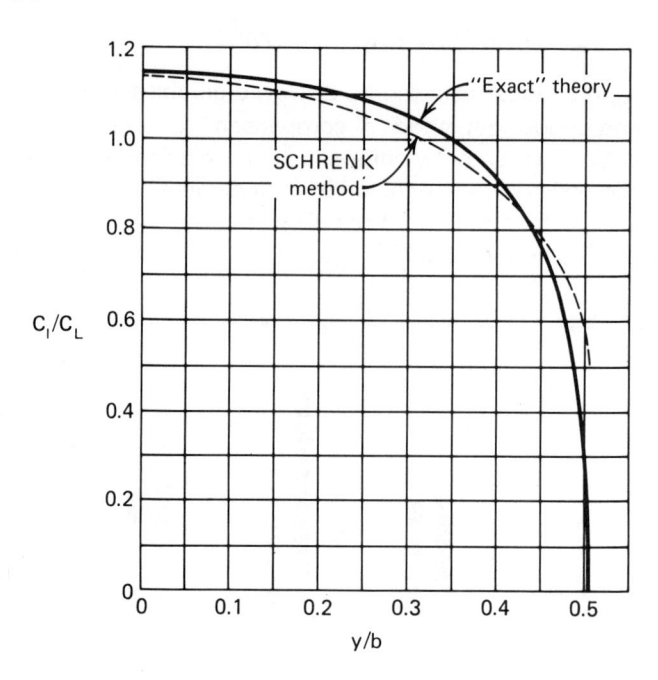

Graphics in Technical Writing

A graph can **compare** and **compare** lift and drag in combination:

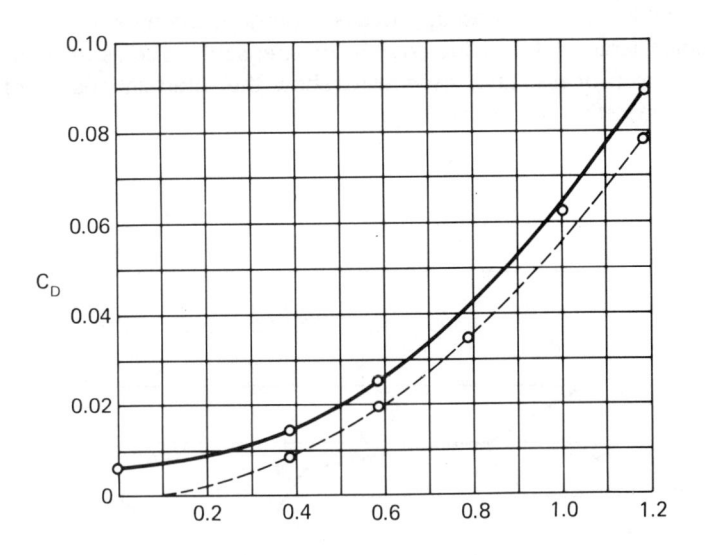

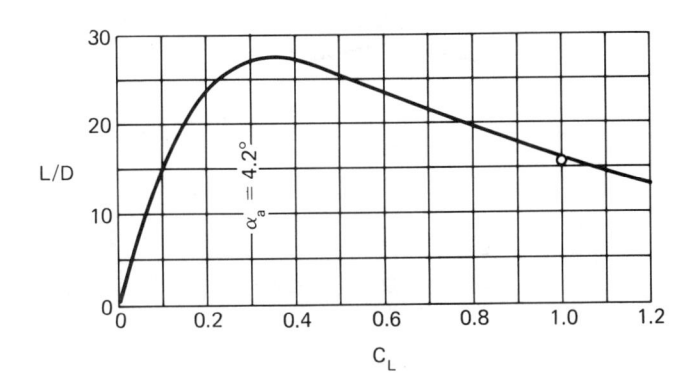

Process

Describing and discussing process is perhaps the most common task of the technical writer. A trend line chart is indispensable; it can summarize a long discussion of process, and, especially, show the interrelationship of several processes operating simultaneously.

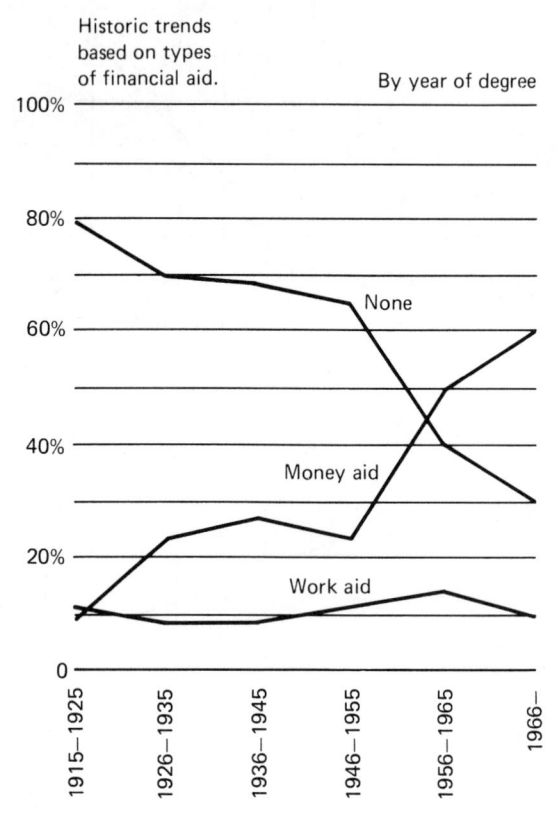

Historic trends based on types of financial aid.

As the illustration shows, such a chart is especially effective in demonstrating how trends change over time for different groups on the basis of different factors.

Plotting flow charts with greater technical precision is a large part of the work of any designer or engineer. Again from the example of a study of the aerodynamic properties of a wing design, the charts which follow illustrate how a technical writer uses graphics to support and expand the verbal discussion in the text of a document. Words alone and mathematical demonstrations alone cannot summarize the response of the wing to various angles of attack as well as a drawing or a graph; all must work in concert to communicate.

Graphics in Technical Writing

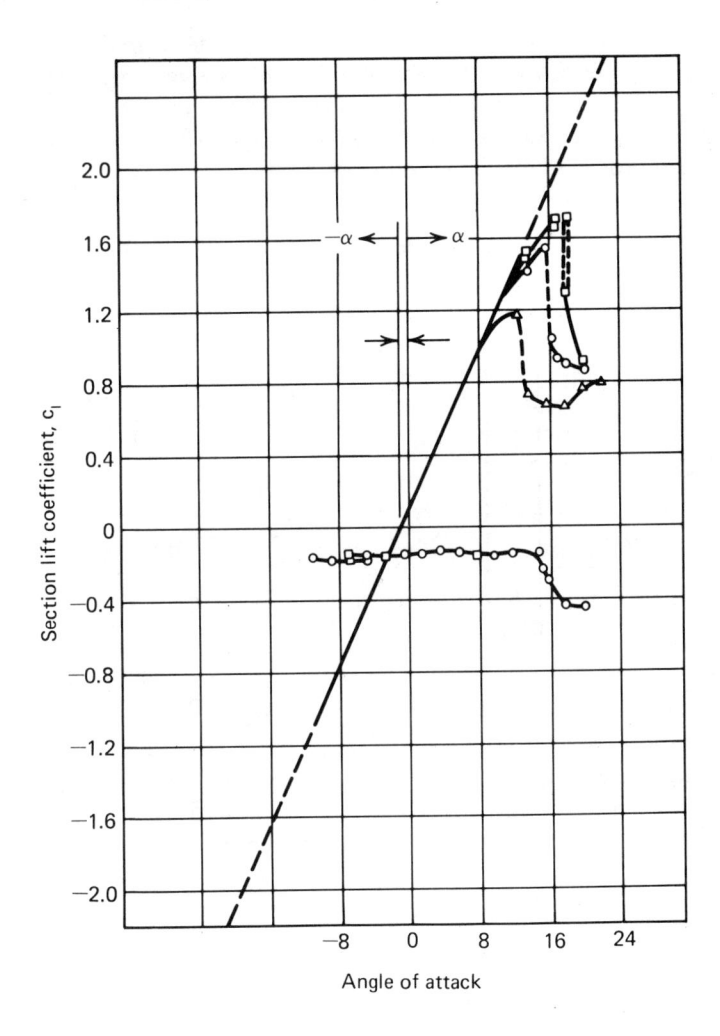

NOTE: Take particular note that this graph is another representation of the drawing marked Figure 17.9 which also described angles of attack for the wing.

Not simply in the work of illustrating individual items in a report, but also in showing larger aspects of planning, graphics plays a very important part. The three charts in succession below illustrate how a project is set up on a planning schedule, how the tasks are represented in a flow chart, and how needs for materials are projected through the use of a chart:

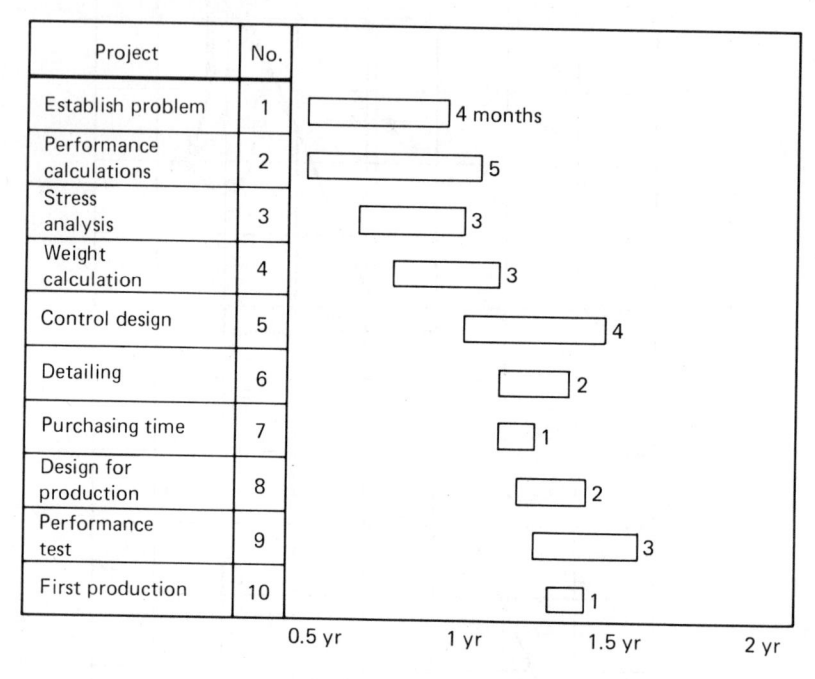

Project	No.				
Establish problem	1	▭ 4 months			
Performance calculations	2	▭ 5			
Stress analysis	3	▭ 3			
Weight calculation	4	▭ 3			
Control design	5	▭ 4			
Detailing	6	▭ 2			
Purchasing time	7	▭ 1			
Design for production	8	▭ 2			
Performance test	9	▭ 3			
First production	10	▭ 1			

0.5 yr   1 yr   1.5 yr   2 yr

Product Design Plan

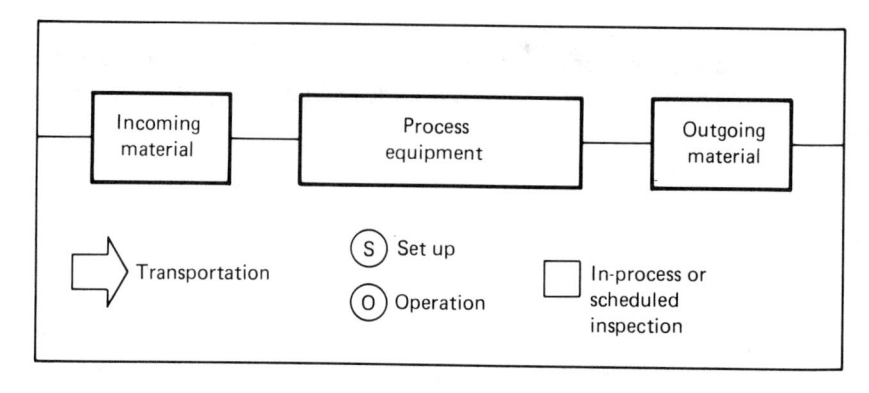

Incoming material → Process equipment → Outgoing material

⇨ Transportation

Ⓢ Set up
Ⓞ Operation

▭ In-process or scheduled inspection

Flowchart.

Tool-machine load	No.	Sept. 26	Oct. 1	Oct. 2	Nov. 3
Screws	43				
Hammers	5				
Wrenches	20				
Paint (gal.)	30				
Pliers	4				
Bolts	82				
Nails	56				
Screwdriver	3				

Schedule Design.

These graphics represent the precision of technology at a very high level, but they are merely alternative methods of formatting how we think in rhetorical patterns. One form of communication or another is used, appropriately to the situation; each embodies the same principle.

In addition to rhetorical strategies for using graphics, one of the most difficult decisions for technical writers is *when to place data in an appendix rather than insert it in the body of a report.* Make that judgment on this basis: 1) place in the text a figure needed to *advance* the reader's immediate understanding; 2) append a figure needed only to *support* a statement in the text or used to show in detail how you *arrived at* that statement. The answer lies within the reader's process of understanding, not in the mechanics of writing and illustration.

Figure in Text: Example 21.1

Records of electric consumption show that a measure of conservation is possible under existing conditions. Comparing the consumption of KWH during the summer months of 1988 and 1989 illustrates this point:

	1988	1989
June	117,000	127,500
July	129,000	105,000
August	109,000	105,000
September	102,000	85,000

> This reduction, however, is not enough to achieve the desired level of conservation. Moreover, weather can influence these figures in any given month.

How

1. An axiom in technical writing is: never express mathematically what can be expressed verbally. In this case, the figures speak for themselves; had they followed a complicated formula, you would have wanted to explain the process verbally, then illustrate it with the working-out of the formula in the appendix.
2. Make an analogy between how you speak and how you write. Don't you frequently illustrate an oral explanation with numbers and facts? Putting a table or an illustration in the text serves the same purpose and has the same logical flow.
3. Take care that your transitions lead the reader into the illustration, and explain how the figure relates to the explanation you are making. Don't assume that the reader will be able to create for himself the desired association and achieve the understanding you intend.

Why

1. In the communications chain in report writing, a client may verbally describe a problem to an engineer. The engineer's work translates that problem into technical and mathematical terms. The process of report writing translates the problem back into words at the other end of the process. Along that continuum, not everything *can* be reduced to words (e.g. tables of computer data) and not everything *should* be reduced to words (e.g. the table of KWH consumed). So in making your decision about when to illustrate, be aware of the interplay of media and the effectiveness of each in the communications process.
2. Audience analysis is enormously important here. What is the audience's level of expertise? What is the reader's need and ability to understand? The answers to these questions are much different for a teachnical reader and a non-technical reader, and the differences tells in making a decision about using figures.
3. You may do a brilliant job of engineering, and then undersell it by writing 10 pages of report and appending 100 pages of data—in effect, leaving the reader to do your work of making a report.

Appended Figures: Example 21.2

> These operational schedules were used by the programs to simulate the operation of the building and its energy consuming equipment. These schedules are shown on page 108 as a part of the computer printouts in Appendix E. The paragraphs which follow supply a narrative description of where and how these schedules were input to produce that data

How

1. Once you make your judgment about appending a figure, the mechanics of writing is mainly a matter of adequate transitions. Such transitions let the reader know how such figures fit logically in the report as a whole and how to use the data you have supplied.
2. Use transitions in writing to make the use of figures easy to understand:
   a. These summaries are shown in Appendix A, on page 5 of each program printout.
   b. ... according to the management chart. (Fig. 3, Appendix B.)
   c. The flow chart (Fig. 12) on the following page illustrates the comparative levels of use for 1978.

Why

1. Logically, your task is to blend the technical and the verbal to make the combination readable and understandable.
2. Visually, the judicious use of figures within the text can give needed emphasis to a point expressed by an illustration.
3. Technically, it is your obligation to the reader to make illustrations easy to understand as they relate to the written text. Good graphics become an invaluable part of the communications process here.
4. Physically, make it easy for the reader to refer to the appendices if you clearly indicate in the text where an appendix is located in the report, and number the appendices clearly. Good graphics is also important here. In a word, if you cite appended material, make it quick and easy for the reader to *find* the damn thing.

Finally in our discussion of graphics, every writer of any technical document should always be aware that every page of technical prose is also a graphic design which illustrates the logical relationship among the ideas presented.

Page Design: Example 21.3

> ... before entering the boilers. This is not only very costly, but it also puts an undue strain upon the boilers.
>
> 3.1.2. <u>Boiler Intake</u>: The boilers should receive water that is oxygen-free, and low in hardness content. Oxygen under elevated temperatures causes corrosion within the boilers and piping systems. Hardness causes scale to build up in the boilers and makes necessary the costly removal of the scale, or even boiler-tube replacement. Therefore, the injection of raw water from the city mains into the boilers as make-up must be held to a minimum.
>
> 3.1.3. <u>Air Dryer Intake</u>: The existing air dryers also use compressed air to regenerate the desiccant chamber. Air is passed through the desiccant during regeneration to carry away the moisture, and then exhausted into the atmosphere. This process creates two problems:
>
>   a. The total air exhausted is 3% of the total capacity of the air dryers, or 16,500 cubic feet per minute.
>
>   b. Because this air must be supplied by the air compressors, it is a part of the cost of providing compressed air.

How

1. Every page of a report should be broken.
2. Divide each topic in a chapter to a second or third level; then number, indent, and underline as you see illustrated above.
3. For quick and easy reference, the outline is a key to each of these parts.

Why

1. Technical matter is difficult to read. For that reason, a reader needs these so-called "external" and "mechanical" techniques of organization.
2. Technical writers and technical readers are likely to have greater visual aptitude than verbal aptitude. The design of a page with clearly identified parts and adequate white space thus works for both reader and writer.
3. Unlike a book which is read through serially and laid aside, a report is likely to be referred to, read in parts, and physically handled more than a book. Broken pages create, by analogy to computers, a random access memory.
4. How well you break a page depends upon how good your initial outline is. Using a good outline as a guide, a writer in a hurry can write the parts of a

report despite interruptions. The process of writing becomes more mechanical and less a matter of establishing an association of ideas. These external devices which break the page do much of the work of creating transition, and they can cover a multitude of writing sins. If you can think logically and display that logic in how you break a page, that process reduces the demand on your writing capacity by at least 50%.

5. *If you are in a desperate hurry,* make a detailed outline, break each page into small parts, and let the process of writing be little more than writing a "bloated outline."

Manuscript Specifications

See Appendix A, pages 476, 480.

chapter 22

Outlines

The scope of a sustained writing project makes an outline absolutely basic to managing the mass of data, for a writer's mind cannot store and retrieve sufficient data in manageable form to reel off a long report or scientific paper. Just as a blueprint or a set of shop drawings serves as a guide to the construction of a building or the manufacture of a machine, an outline serves as a set of plans for a longer manuscript.

THE THINKING PROCESS

The first critical process whereby a writer thinks about a subject for a long manuscript is different from the logical process whereby he writes about research for a reader. Thinking, even about the most technical problem, is somewhat random and certainly associative—even if the components of a technical problem may be highly objective and experiments may develop in an extremely logical and systematic way. New ideas and innovative solutions derive from an associative process of synthesizing old ideas and moving from the already known to the new idea or application. This process of seeing relationships among old ideas in order to develop new ideas is the same for a student researching a library paper and for the Nobel Prize Laureate devising a cure for a disease. Only the scope and complexity differ; the process is the same.

The major purpose here in describing the communications process is to make clear the difference between the work of the technical researcher and the work of the technical writer. The mental activity of the researcher is somewhat associative; that is, one insight and one sample of data suggests how to look for other relationships. The mental activity of the writer is more deliberate and more directed to a reader; the task is to create a clear, logical progression of facts and ideas for the reader, not to retrace the randomly creative and investigative workings of the mind.

Technical communication very often fails when the writer gives the reader a "total mind dump" by simply retracing all that he did and thought on a project; instead, he should translate the work process into a clear and logical technical manuscript.

TECHNIQUES OF ARRANGING DATA

A writer who must impose a logical communications order upon a body of data has many techniques at her disposal. One of the simplest cookbook formulas derives from Aristotle. When you must write an outline, apply three questions to the body of data: 1) what are the parts? 2) what is the order of the parts? 3) why are the parts in this order? Being able to answer those questions enables you to break a large subject into manageable components and to arrange the data as a set of logical relationships.

Exercise 22.1

To test that method, apply those 3 questions to some concern on your campus:

- parking
- registration
- housing

What are the basic components of each problem? What is the order of the components: a progression of events through time for a historical process? a series of actors in cause-and-effect relationships? a group of factors which have a logical relationship to each other? a comparison and contrast, a classification in terms of a larger problem, or a division in terms of a series of interrelated problems?

Learning to think in answer to such questions is a way of determining the best arrangement of data in a document.

RELATIONSHIP OF PARTS

Once the parts of a problem are identified, there are four major ways to determine the relationship of the parts: historic process, causal process, comparison and contrast, and classification and division. These may be further divided into *linear* and *vertical* sequence. In the chart which illustrates linear thought processes, letters which represent major factors are capitalized and underlined; minor factors are represented by lower-case letters. An entire line shows the interrelationship of factors:

Types of Process	Subject	Treatment	Sequence of MAJOR and (minor) events
historical	events in a sequential order	record, distinguish between major and minor events	$A\underline{BCD}(ef)G_{hi}\underline{J}(lkn) = \underline{N}$
mechanical	operation of parts in sequence	observe, record, explain, interpret	$A\underline{B}CD\,df\,(ghi)\underline{JK} = \underline{L}$
physical	operation of physical laws	explain, relate, interpret	$A\underline{BC}\begin{smallmatrix}d\\ \\e\end{smallmatrix}fg = \underline{H}$
cause and effect	interrelationships in sequence	explain, interpret, extrapolate	(a?)—$\underline{A}$ $\quad$ g(??) (b?)—$\underline{BC}$ $\underline{D} = \underline{E}$—(F?) h(?) $\quad\quad\quad\quad\quad\quad\quad\quad\quad\quad$ i(???)
logical	development of related ideas in a planned sequence	plan, arrange, and state for specific effect	$a+b+c=d$ $\quad$ $\begin{array}{l}A=B\\B=C\\C=A\end{array}$

LINEAR FORMS

Historical Sequence

In a *historical sequence* of events, the main purpose is to describe what happened in chronological order: a, b, c, d, e, f, g. An inexperienced writer will

Linear Forms

do just that: describe what happened, step by step, with no effort to interpret the relative significance or the interrelationships of the steps. This is the mind dump: first this happened and then this and then this and then this—with no attempt to explain and interpret for the reader that step A was more important than step B and that D represented a critical turning point. In communicating, simply reporting what happened, when, is seldom sufficient; an important part of a writer's job is to interpret the sequence of events as they occur and in their relationship to each other.

In the sequence of events which led to the construction of a new dormitory on your campus, some of those events would be more important than others. The treatment of events a, b, c, d, e, f, and g would probably take this form:

a B C (d) e FG

This configuration shows some events as less important than others and (d) as a side issue which nevertheless took some part in the events leading to the final construction FG.

Causal Sequence

In a *causal sequence* of events which leads from the causes to the final effects, the relationships of cause and effect require clear distinction:

- single cause—single effect
- multiple cause—single effect
- single cause—multiple effect
- multiple cause—multiple effect

Dealing with causal sequences makes the linear arrangement of details more complex than simply relating a sequence of historic events and their relative significance.

In the explanation of why the decision was made to construct a new dormitory on your campus, the data might take this configuration:

a   e
(minor causes) led to C (a major effect) and (minor effects)
b   f

Exercise 22.2

Using the sequence a b c d e f g, reorder these factors of cause and effect into a number of different configurations, depending upon whether the steps in the process are single cause/single effect, multiple cause/single effect, single cause/multiple effect, or multiple cause/multiple effect. Reporting data to a reader in a logical way also means describing the relationship of the parts in a logical way.

Physical and Mechanical Process as Causal Sequence

The operation of an experiment or of a machine very frequently requires the technical writer to arrange data in configurations of causal sequence. (The operation of the internal combustion engine is a good general example of physical and mechanical sequence.) Moreover, the technical writer frequently must describe such sequences to a multiple readership, some of whom understand the technical aspects of the process (engineers and technicians) and some of whom do not (financial and executive officers with decision-making responsibility in technical projects).

The writer's task here is to treat the causal sequence on three different levels: the *technical* workings of the sequence itself (for instance, the technical operation of an internal combustion engine); the physical or mechanical *principle* involved (e.g., compression, combustion, and exhaustion of gasses, as principles of operation which a well-informed and intelligent lay person can readily understand); and the *implications* of that causal sequence (in terms of time, cost pay-back, and other such concerns of management). It may not always be sufficient, then, simply to communicate technical information in a logical way; beyond that may be the necessity to reach a wider, non-technical readership which has a very practical need to understand the sequence in more general terms.

In fact, three longer forms of technical writing reflect the three levels of writing about a sequence:

a *lab report* has as its major purpose demonstrating a problem and its technical solution

a *professional article* not only describes the problem and how it was solved technically, but also deals with theories, previous research, and applications of the solution.

a *technical recommendations report* describes a problem and how it was solved technically, the physical and mechanical principles in operation, and also the more far-reaching implications of the time and money involved in applying the results of research and investigation.

In increments determined by the nature of the subject, the needs of the audience, and the purpose in writing, the complexity of describing events in a historic or a causal sequence grows exponentially. At base, however, we are still dealing with the method which also produces the 500-word theme: what is the problem? (the thesis stated in paragraph 1); what are the parts of the problem?

(the division of the thesis into three parts for the purposes of the 500-word theme); what is the order of the parts? (the development of the three statements which derive from the thesis).

The technical writer who develops linear patterns of arrangement has a major advantage: the organizational pattern mirrors the pattern of what actually happened. The a, b, c, and d steps in an outline follow the sequence of events; the writer's major task is to interpret what happened. In a linear sequence, the thinker perceives the order of events; in vertical forms, the thinker must impose a new order upon the data.

VERTICAL FORMS

Vertical forms call upon the writer to arrange the data in logical configurations which may not reflect what happened in sequence. Here the writer must create a new configuration of the data. A writer may have to compare and contrast two different causal sequences and discuss how one is more efficient than another. Or a writer may have to consider the many factors—economic, historical, demographic—which call for the reconstruction of a state prison system. In such instances, the mass of data does not contain within it a basic pattern of organization which gives the writer a way of imposing a logical order upon his subject.

There are two principal vertical forms for arranging such data: comparison and contrast, and classification and division. Either of these methods can be used to arrange data into logical, vertical patterns. Wherever two sets of factors must be considered, or when one must choose among several alternatives, these vertical patterns come into play. They establish a logical relationship between components in data which are not necessarily historic or causal in their relationship. Bar charts give a good example of how the vertical forms work.

Comparison and Contrast

Perhaps a good way to explain how the various patterns work with words and ideas is the following example. In the professional football season, the schedule of games has a historic sequence. Commentators comment endlessly about the causal relationship in the series of games before the play-off games and the Super Bowl. When we approach Super Bowl Sunday, commentators shift into comparison and contrast on a broad scale: every factor about the two opposing teams which can be compared and contrasted is subjected to that pattern of organization: coaches, offense, offensive lines, quarterbacks, defense, past records, and on and on.

An essential factor in the various studies of the two teams is the *basis* for comparison and contrast. A commentator does not compare the coaches of one team with the offensive line of another, or the quarterback of one team with the offensive backs of another. The basis is like a common denominator which sorts out directly comparable or contrasting factors.

If a sportswriter outlined his presentation on the two opposing quarterbacks, he might arrange the data in this way:

Alternating pattern:	Opposing pattern:
1. passing yardage	1. QB X
1.1. QB X	1.1 passing yardage
1.2 QB Y	1.2 passing percentage
	1.3 running yardage
2. passing percentage	2. QB Y
2.1 QB X	2.1 passing yardage
2.2 QB Y	2.2 passing percentage
	2.3 running yardage
3. running yardage	
3.1 QB X	
3.2 QB Y	

Probably many factors in the writer's experience played a part in his analysis of the quarterbacks, but in organizing these data for a reading audience, he used a systematic arrangement. He used his judgment to select a topic, he decided to include a selection of details about the topic, and he used a particularly effective order for communicating and interpreting those details.

Classification and Division

If our sports writer wanted to compare and contrast the quarterbacks in the present Super Bowl to quarterbacks of past Super Bowls, he might classify and divide an entire group of players in this way:

1. drop-back quarterbacks
    1.1. QB X
    1.2 QB Y
    1.3 QB Z
2. scrambling quarterbacks
    2.1 QB A

Vertical Forms 417

    2.2 QB B
    2.3 QB C
3. combinations
    3.1 QB Y
    3.2 QB C

    In the process of discussing the development of styles and techniques among quarterbacks in the recent history of professional football, the sports writer would have an orderly and systematic way of communicating these data. In the actual process of writing his notes and outline, he may have thought first of Joe Montana as a younger quarterback; then his mind may recall Bart Starr, who retired to coaching, or Jack Kemp, who became a politician. His final story, however, would not record the random associations of this thinking process; instead, he would communicate his thoughts in orderly fashion.

    Those illustrations of how the sportswriter's mind operates are not essentially different from how any mind operates in arranging data in logical form. Aristotle's thoughts followed that process, Sir Isaac Newton's mind followed that process, and that is, in fact, how people in western culture tend to process data which they then communicate.

Exercise 22.3

    As practice in the orderly operation of thought, rapidly sort out the parts of the following subject:

A. the players on a football team
B. the players on a baseball team on the basis of
    (1) their positions on the field
    (2) their positions in the batting order.
C. foods you eat in an average day on the basis of
    (1) the basic food groups
    (2) calories
    (3) your preference
    (4) price.
D. automobiles on the basis of
    (1) size (sub-compact, compact, . . .)
    (2) number of cylinders
    (3) combustion capacity
    (4) body style
    (5) fuel economy.

E. the players in a symphony orchestra on the basis of
    (1) the instruments they play
    (2) where they sit on stage.

Exercise 22.4

Classify and divide the reasons for improving some condition on your college campus: parking, registration, or housing, perhaps.

A. classify these problems under general headings such as economic, logistic, demographic, geographic.
B. divide each larger area into its component parts. For example, economic factors may be divided into costs to the student and costs to the administration. Those costs may be further divided into direct and indirect costs to the student and direct and indirect costs to the administration.

Selecting an Organizational Pattern

To write a preliminary outline, first determine the overall organizational pattern of the data:

- what are the parts?
- what is the logical order of the parts?

Then, determine whether you will arrange the data in linear form:

- historical sequence
- causal sequence

or in vertical form:

- discussion of two major components (with subdivisions)
- discussion of three or more major components (with subdivisions)

If the overall pattern which emerges is historic, concentrate upon describing the sequence; add a logical dimension by discussing the relative importance of the events in sequence.

If the overall pattern which emerges is causal (more frequent in topics having to do with physical and mechanical processes), then determine the arrangement of a single and multiple causes and effects, allow for the principles involved in the operation of each step in the causal sequence, and allow for the implications of the causal sequence.

If the overall pattern which emerges is a logical relationship between two major factors, select a basis for comparison and contrast and outline accordingly.

If the overall pattern which emerges shows a group of factors which are logically but not sequentially related, use classification and division. This approach usually applies if there are three or more major factors to consider in treating the subject.

Classification and division is among the most difficult patterns of thinking and, at the same time, among the most widely useful. Of the four types of arrangements mentioned above, classification and division requires dealing with the most factors and in the most ways.

A good review of the process of classification is to outline this chapter. We began by saying that there are two basic ways to arrange data in a piece of sustained writing: linear and vertical. Each of those is further divided: linear forms into historic sequence and causal sequence; vertical forms into comparison-and-contrast and classification-and-division. The comparison-and-contrast method can be still further divided, into the alternating pattern and the opposing pattern. Each of these methods is like a software program for arranging the same data in several different ways to serve different readers.

Exercise 22.5

The outlines on the next several pages were taken from student papers. In studying them, note the use of historic and causal sequences, and the use of comparison-and-contrast and classification-and-division. Study them carefully to see how the *Table of Contents* in a sustained writing project becomes a writing guide for the writer, a reading guide for the reader, and a reference guide for the user of technical and scientific writing.

Outline 22.5.1

FACTORS IN OPERATING A STEREO TAPE RECORDER

TABLE OF CONTENTS

1. Understanding Tape Recorders ....
    1.1. Four Main Parts of Recorders ....
        1.1.1. Transport Mechanism ....
        1.1.2. Heads ....
        1.1.3. Electronics ....
        1.1.4. Controls ....
    1.2. Modulation Noise
2. Tape ....
    2.1. Materials in Tape ....
    2.2. Thickness and Width ....
    2.3. Properties Which Should Be Controlled ....
        2.3.1. Cupping ....
        2.3.2. Curl ....
        2.3.3. Poor Edges ....
        2.3.4. Coating Quality ....
        2.3.5. Sticking ....
        2.3.6. Stretching ....
        2.3.7. Print-through ....
        2.3.8. Noise ....
    2.4. Tape Speed ....
    2.5. Tape Tracks ....
3. Types of Tape Recorders ....
    3.1. Reel-to-Reel ....
    3.2. Cassette ....

3.3. Eight-Track ....
3.4. Comparison ....
4. Comparison of Disc and Tape ....
5. Appendix ....
6. Bibliography ....

Outline 22.5.2

USE OF TIRE CASINGS FOR EROSION CONTROL

TABLE OF CONTENTS

1. Introduction ....
2. Discussion ....
   2.1. Theory of Experiments ....
   2.2. Evaluation for Lift Force ....
   2.3. Evaluation for Drag Force ....
   2.4. Predictions of Lift and Drag Forces ....
   2.5. Predictions of Sliding Friction Forces ....
3. Experiments ....
   3.1. Lift and Drag Experiments ....
      3.1.1. Procedure ....
      3.1.2. Results ....
   3.2. Sliding Resistance ....
      3.2.1. Procedure ....
      3.2.2. Results ....
4. Interpretation ....
5. Conclusions and Recommendations ....
6. Appendices ....
7. Glossary of Terms ....

Outline 22.5.3

NUCLEAR POWER REACTOR SAFETY

TABLE OF CONTENTS

1. Introduction ....
2. Nuclear Power Reactor Safety ....
    2.1. Nuclear Power Compared to Other Energy Sources ....
        2.1.1. Coal ....
        2.1.2. Oil ....
        2.1.3. Solar Energy ....
        2.1.4. Wind Energy ....
    2.2. Regulation of Nuclear Facilities ....
        2.2.1. Medical Regulation ....
        2.2.2. Military Regulation ....
        2.2.3. Power-plant Licensing ....
    2.3. Operating a Nuclear Reactor and the E.P.A. ....
    2.4. Conclusion ....
        2.4.1. Human Safety ....
        2.4.2. Environmental Safety ....
3. Footnotes ....
4. Bibliography ....
5. Appendix ....

Outline 22.5.4

ZIRCONIUM CORROSION AND ZIRCONIUM-CORROSION PREVENTION IN THE NUCLEAR INDUSTRY

TABLE OF CONTENTS

1. Executive Summary ....
2. Introduction ....
   - 2.1. Subject ....
   - 2.2. Purpose ....
   - 2.3. Scope ....
   - 2.4. Method of Development ....
3. General Background ....
   - 3.1. Zirconium in the Nuclear Industry ....
   - 3.2. Zirconium Environment ....
4. Corrosion Discussion ....
   - 4.1. General Information ....
   - 4.2. Zirconium Protective Film ....
   - 4.3. Radiation Effects ....
   - 4.4. Chemical Effects ....
5. Corrosion Protection ....
   - 5.1. General Information ....
   - 5.2. Alloying ....
   - 5.3. Coating ....
6. The Nickel/Zirconia Coating ....
   - 6.1. Basic Principles ....
   - 6.2. Coating Evaluation ....
7. Summary and Conclusions ....
8. Glossary ....

Outline 22.5.5

AN EXAMINATION OF AN EDUCATIONAL DATA BASE MANAGEMENT SYSTEM

TABLE OF CONTENTS

1. Introduction
2. Advantages of Automating
3. Logical Schema
   - 3.1. Administrative-academic
     - 3.1.1. Housing
     - 3.1.2. Employee
     - 3.1.3. Teacher
     - 3.1.4. Finance
     - 3.1.5. Academic History
     - 3.1.6. Courses
     - 3.1.7. P.O. Box
     - 3.1.8. Department
     - 3.1.9. Student
4. Physical Representation
   - 4.1. Housing
   - 4.2. Employee
   - 4.3. Teacher
   - 4.4. Finance
   - 4.5. Academic History
   - 4.6. Courses
   - 4.7. P.O. Box
   - 4.8. Department
   - 4.9. Student

>   4.10. Research ....
> 5. Summary and Conclusions
> 5. Summary and Conclusions ....
>   Footnotes ....
>   Bibliography ....
>   Appendix 1 ....
>   Appendix 2 ....
>   Appendix 3 ....

Outline 22.5.6

> # THE DESIGN OF A HANGING LIGHT FIXTURE
> # FOR A HOTEL LOUNGE AREA
>
> ## TABLE OF CONTENTS
>
> 1. Executive Summary ....
> 2. Introduction ....
> 3. Establishment of Environment ....
>   3.1 Environment ....
>   3.2 Mood ....
> 4. Research Stage ....
>   4.1 Contemporary Work ....
>   4.2 Materials ....
>   4.3 Types of Lighting ....
>   4.4 Creating Visual Interest Within the Fixture ....
>   4.5 Research Conclusions ....
> 5. Concept Stage ....
> 6. Mechanics Stage ....

7. Prototype Stage ....
8. Illustrations ....
   8.1 Concepts ....
   8.2 Mechanics ....
10. Bibliography ....
11. Appendix ....

Outline 22.5.7

EIGHT-PHASE TEAM PLANNING AND DESIGN PROCESS
FOR A PRIMARY HEALTH-CARE FACILITY

TABLE OF CONTENTS

1. Executive Summary 1
2. Introduction 2
   2.1. Subject 2
   2.2. Purpose 2
   2.3. Scope 2
   2.4. Method of Development 3
3. Planning Factors 4
   3.1. Establish a Planning Team 4
      3.1.1. Selecting an Architect 4
      3.1.2. Comparing Alternatives 5
      3.1.3. Construction Systems and Methods 5
   3.2. Economic Considerations 6
      3.2.1. Cost of Renovation vs. Cost of
             New Construction 6
      3.2.2. Project Costs and Budget Guidelines 7

4. Process 7
    4.1. Process Background 7
    4.2. Predesign and Program Evaluation 8
        4.2.1. Open Process 8
        4.2.2. Project Funding and Methodology 9
        4.2.3. Design-construction Systems 10
    4.3. Scheduling 10
        4.3.1. Network Analysis 10
        4.3.2. Network Diagram and Bar Chart 10
    4.4. Site Analysis 11
    4.5. Schematic Design 12
        4.5.1. Cost Estimates 13
        4.5.2. Schematic Documents 13
    4.6. Design Development 13
        4.6.1. Design Modifications 13
        4.6.2. Design Approval 14
    4.7. Construction Documents 14
    4.8. Bidding and/or Negotiations 15
5. Conclusions and Recommendations 15
6. Appendix 1 16
7. Appendix 2 26
8. Bibliography 30

INTENSIVE READING

Learning how to observe how others write enables you to use one of the best guides to writing well. The exercise which follows will be particularly useful for those who plan to submit a manuscript to a publication. Such writing requires conforming to the exact specifications of the given publication, so knowing how to read it helps one to write an article to order.

Reading Exercise 22.1

Read a major article in *Scientific American*, giving careful attention to the points listed below.

Introductory Techniques

1. Notice the scope of the first two or three paragraphs. How does the writer give a background for the study to place this specific topic in perspective? Are there references to other researchers and other publications? With dates? With locations? In a word, the subject of this article does not exist in a vacuum. The writer first creates a general context for this specific subject of research. How is that accomplished?
2. Does the article then define terms so that readers will understand the writer's definition of basic terms, and, in particular, so that a general reader is fully prepared by the definitions to read the remainder of the article?
3. At some point in the first page of the printed article, does the writer specify the purpose of the article? Does the writer establish the scope of the article by explaining what the article will, and will not, cover?
4. Does the writer then map out the major points in the article in a paragraph which lets the reader know what to expect? Do you see the writer narrowing the topic to a very specific scope?

Transitional Techniques

1. Look at the next two pages of the article. Do you see major sub-headings in bold-face type? Does this overview give you direction for reading?
2. Select a full page without illustrations and study the transition techniques which lead you through the article:
   a. Look in the left-hand column for key words related to the subject. Circle each key word and all forms of the word: *burn, flame, ignite, ignition, combustion, detonation*.
   b. Look in the middle column for names and pronouns: *we, he, she, they, their, our, that, which, this, these, such, anyone, everyone*.
   c. Look in the right-hand column for other transitional devices: *however, therefore, moreover, because, since, when, then, such, and, but, yet, so, first, second, next, last*.

Reference Techniques

1. Note how the writer refers to names: "*name* reported the results of a study performed at *place* on *date*."
2. Note how the writer refers to illustrations "(Fig. 3)," and note the description lines which accompany illustrations.

Concluding Techniques

1. Note how the writer makes all the details in the article add up to a general statement of results or significance.
2. Does the writer point toward further study which must be done on this subject?
3. Note the tone of the last two paragraphs. Compare that with the tone of the first two paragraphs. Compare them with the tone of the body of the article.

Reading Exercise 22.2

Read an article in a journal of research in your academic major. Repeat all the steps outlined in exercise 22.1 and add the following steps:

1. Carefully note the use of reference techniques in the article. How often does the writer refer to a source? How often does a writer quote directly? How long are the quotations?
2. What is the manner of use of mathematical calculations? How does the writer explain them verbally? How are they placed on the page?
3. What are the techniques for introducing sources? (*e.g.* "In their 1970 study Rea and Holmes ...")
4. In the end notes, what variety of sources do you see? Books? Articles? Manuscripts? Government publications? Accounts of interviews?

Reading Exercise 22.3

Search your library for transcripts of research papers, theses, or dissertations in your academic major. Repeat exercises 22.1 and 22.2.

chapter 23

Technical Persuasion

Along with words, sentences, paragraphs, and outlines, logic is a basic component in the full range of communications skills. As other components of writing skill must be adapted to technical writing, traditional concerns of logic come to bear upon technical writing in a specifically defined sense.

To prove a point in a logical, systematic way is a very compatible and familiar *method* to most students in technologically oriented curriculums. To use words and ideas as the *medium* for proof, however, may seem unfamiliar and unreliable. As with other forms and methods within this text, the important point to remember is this: while the verbal mode may be unfamiliar, the thought process required for problem-solving is exactly the same as in math, science, and technology.

Just as there are certain rules or protocols whereby a chemist may make an experiment under controlled conditions, and just as there are certain prescribed methods a researcher may follow in taking data samples, there are certain rules which traditionally apply to argument.

These rules of argument are traditionally established because, at one time in history, those who sought answers had only words to rely upon. The ancient Greek thinkers believed that rational discourse of the type we see in formal argument could solve any problem. Ancient philosophers speculated about the universe; contemporary scientists gather empirical data. In the eighteenth century, Leibnitz and Voltaire discoursed upon whether this is indeed the "best of all possible worlds"; contemporary planners resort to a "decision tree." Napoleon and Robert E. Lee were geniuses at the deployment of men and materiel; contemporary planners have elaborate project-control techniques, such as the Critical Path Method (CPM) and the Shortest Path Method (SPM).

Even though we have far transcended words and ideas as the primary medium for our thoughts in making decisions, we nevertheless need some knowledge of how the mind works in making arguments, and of how minds operate upon each other in creating persuasive effects.

When someone writes a proposal which causes a client to contract for tens of millions of dollars in a large project, that person is using persuasion of a very high order. When a writer must write a technological assessment of the impact a corrections facility will likely have upon a small rural community, that also is persuasion of a high order. When a special-interest group acts against a road or a dam or a project to derive oil from shale, the most argumentative methods are employed in deciding the most technological issues. When an irate client threatens court action, that also is a situation which calls for skilled argument and persuasion.

In each of these situations, technology may inscribe the last word on the issue, but words constitute the medium whereby issues in our society are discussed. To know the proper and effective techniques of argument and persuasion creates an advantage in situations, and, even more important, it avoids a disadvantage when the technical merits of your case are superior.

This chapter will go on to discuss ways to use technical matter in persuasion, and ways to refute persuasive techniques unsupported by the technical merits of the case they advance. By performing only those two tasks, this chapter limits its scope to the concerns of technical communicators rather than to offer a comprehensive guide to logic.

WAYS TO PERSUADE

Words and ideas may be the primary modes of persuasion, but technical documents must nevertheless be underpinned with facts, figures, and graphics. Within the verbal matrix of a proposal, a report, or a letter, there are two major ways of persuading a reader to act on technical matters: argument by exposition, and argument based on authority.

Argument by Exposition

As this text points out in several sections, a clear explanation of a technical principle or problem has a greater persuasive force than almost any other technique available to the engineer or technologist with integrity.

To give an immediate example: my heating and air-conditioning contractor is a neighborhood businessman who checks my furnace each fall. He recently persuaded me to contract with him for central air-conditioning, rather than with another firm, by giving a thorough technical explanation. He explained that another unit I was about to buy contained aluminum tubing rather than copper tubing, and copper is a more durable material for an air-conditioning unit. The most convincing point, however, was his discussion of coolants. He said that, within the next few years, coolants will be improved and will greatly increase the achievable cooling efficiency; in the unit he will install, he can later exchange the coolant and thus offer me a potential cost savings. His technical explanation convinced me to spend more for his product and service than for another.

Very often, it is the willingness and the ability to explain that carries the greatest persuasive force when dealing with technical material. To make clear how a process operates, to be able to use analogies and figurative language to make the process clear to the hearer, and to make the complex plain—these are the most important persuasive skills of the technical communicator. These skills rest with basic skills in writing and rhetoric, however, rather than with more sophisticated techniques of argument and logic.

Argument Based on Authority

Arguments based on authority fall into two classes: an appeal to other authorities, and establishing oneself as an authority.

The Appeal to Other Authority

Often, to cite and identify an authority in a technical specialty adds weight to a technical point a communicator is trying to make. "According to Dr. X, professor of chemical engineering at YZU" lends an additional authority to a statement a junior engineer may make. "The latest figures from the Labor Department show that ..." is another way to appeal to an established or published authority to buttress an argument.

Citing authorities is as much a matter of technique as it is one of content, for as a student's experience increases, he develops a sense of how and when to cite authorities. Close observation of how textbooks and other printed sources use the appeal to authority is the best way to develop that technique of persuasion.

A very important note about the appeal to authority is that both the communicator and the reader or listener should agree on the validity of the authority. Three people with different views would rapidly reach an impasse in discussing a subject if they could not reach an agreement about the reliability of the authorities whom each cited.

Therefore, audience analysis and a study of the nature of the authority you cite are the best tests for an authority. An authority which has the strength of law is followed by a regulation under a law as a test of authority. Codes and specifications carry the weight of authority. Published works carry authority as do established and well-known experts in a given field of expertise. In all these instances, an appeal to authority strengthens the statements made by a single individual.

The Appeal to One's Own Authority

A second way to appeal to authority is to establish oneself as an authority. "Ethical proof" is another term for this technique whereby a person establishes herself as a reputable professional specialist with a proven record of success. Identification with a particular school or company, or as a member of professional organizations within your specialty, has the effect of lending added weight to your statements about technical matters.

There are special warnings about appeals to authority, however. That one is an authority in the field of medicine may not necessarily make the physician's political statements more authoritative. That one is a noted chemist may not make statements about nuclear engineering more valid. Therefore, make certain that a noted authority is an authority in the specialty under discussion and not someone who is lending the weight of his reputation in another field to an issue outside his expertise.

The basic skills involved in using an appeal to authority in argument and persuasion are reading and scholarship. Furthermore, one who has become widely and deeply knowledgeable in an academic specialty has had the opportunity to develop judgment about how to use this technique of citing authorities.

This list of ways to persuade in technical communication is a short one, for the technical communicator does not employ the entire range of the techniques enjoyed by companies who market consumer goods. The advertising and selling of professional services is considerably more conservative in tone and style. The chief difference, however, is the technical content of the products and services in engineering and technology. A quick way to see such a distinction is to read the ads in a general-interest magazine and then to read the ads in a specialty magazine like *The Consulting Engineer*. Claims about generators cannot be nearly so extravagant as claims about men's cologne—and that restriction in tone and content illustrates the constraints within which technical persuasion operates.

WAYS TO REFUTE PERSUASIVE TECHNIQUES

Engineers and technologists have basic methods for testing designs and structures for their feasibility and durability. Those who work with computers have programs which will test for flaws or errors in a given program. In the same way, there are basic methods for testing what you read and hear for the soundess of the argument's persuasive force.

Any persuasive statement, whether it be technological, political, economic, or theological, should be tested for the cause-and-effect connections within it. The section of the text which deals with cause-and-effect paragraphs outlines the basic patterns for cause and effect in writing. Adding persuasive force to cause and effect increases the demands upon that way of thinking and makes careful testing necessary.

Flaws in Cause-and-Effect Statements

post hoc ergo propter hoc ("After this; therefore, because of this")

This flaw confuses a relationship in time with a causal relationship. A most familiar example of the *post hoc* flaw is to attribute an accident to its being Friday the 13th. In fact, most superstitions involve a confusion between the juxtaposition of events in time and their causal connections.

In technical reporting or persuasion, testing for cause becomes more crucial than determining whether to be careful on Friday the 13th. For instance, when a suspended walkway collapsed in a hotel during a tea dance, there was an immediate search for cause: was it the number of people on the walkway? was it the vibration of couples dancing on the walkway? was it a flaw in the design? All those questions had a technological answer in that, after the design of the suspended walkway was changed, that design flaw caused the walkway to fall and kill many people in the process. The expression of cause and effect thus becomes crucial in any report which follows, and in any litigation which attempts to persuade a court to pay damages.

Oversimplification (to ascribe a single cause where there may be a multiplicity of causes)

To oversimplify in search of an answer is another flaw in statements which seek to prove. In Atlanta, a group of fans believe that professional sports teams in the city lose games because nearby Stone Mountain emits unfavorable radiation which affects their performances. Doubtless there are many causes for the football and baseball teams' losing seasons, but it is an oversimplification to ascribe a single cause to that phenomenon.

In technical situations where one must make a statement of cause, it is the mark of thorough research to explore a range of possible causes, to rule out those which are not operable, and to qualify statements of single or absolute cause. In making a report, a civil engineer may have considered, tested, and ruled out a number of possibilities before recommending one course of action. To report the range of possible causes and the method of determining a single cause is more nearly a sign of thoroughness and strength than of weakness. Therefore, when only one cause is advanced for an effect, look further and test that statement for possible flaws.

Unreasonable Extrapolation

It is often possible and necessary to make predictions by extrapolating from an existing body of data. When the prediction goes beyond reasonable limits, however, the extrapolation is called unreasonable. Engineers may predict with reasonable certainty that a nuclear power plant on a geological flaw will not be damaged by earthquakes, but they cannot assert that such an accident will *never* happen. Nor can those who oppose such a structure predict with absolute certainty that a nuclear power plant on that site will inevitably cause death and destruction.

In the sections in the text on words and verb selection, there is advice about the careful choice of words in making predictions: *shall* is almost always contractual in specifying that certain actions or materials are to be a part of a project; *will* carries a greater certainty than *can*, *might*, or *may*. In predicting all future effects, E/Ts must remain within the realm of reasonable limits rather than make unreasonable extrapolations from existing data about what will occur in the future.

Call to Perfection

Practicing E/Ts soon learn that compromise is necessary in most designs. A flaw in argument, then, can be the call to perfection, or disallowing a design because it does not meet one minor criterion.

In specific terms, anyone who shops for a car realizes that some compromises have to be made: does the buyer want the comfort of air-conditioning and power options, at the price of reduced engine efficiency? is the buyer willing to surrender some comforts for greater economy? are magazine reports that this very classy sports car is unstable in cornering sufficiently discouraging that a buyer will forgo the cachet of owning it?

Particularly in discussions where decisions must be made, test for the call to perfection as a flaw in argument.

All technical communication has so frequently to do with cause and effect that knowing the basic design of cause-and-effect statements and the basic flaws in cause-and-effect design can be a great help to communicators and an even greater help to those who must test communications.

Flaws in Language

Engineers and technologists are accustomed to calculations, graphics, and calibrated instruments which have very fine tolerances. Language as an equally precise tool for communication of technological findings must be very carefully chosen if it is to bear the weight of argument and persuasion. Much language on general subjects relates the word to an idea. Technical writing relates the word to an object; therefore most technical statements relate closely to hard data. To bid for a government contract by asserting that one's company is "made up of true patriots whose only wish is to serve this great nation of ours" is insufficient to prove that one's company can bring in the project on time and within budget. However fine the sentiments, words must be supported with facts or a technical writer's statement may reify (substitute the idea of a thing for the actual object). Even a sales document must offer satisfactory proof. Thus the nature of technical subjects places strict controls upon the precisions of language.

Exercise

1. Read the editorial pages of a newspaper to find persuasive discussions of technical subjects: ecology, nuclear energy, or any current issue which has a technological basis. Study the language for emotionally charged words and appeals to feeling. Take careful note of how authorities are cited to add strength to statements.
2. Select a product like an automobile or some electronic device, and look for ads in a wide variety of magazines. Note in particular how an ad for the same product will contain one kind of persuasive appeal in one magazine and another kind in a different magazine.
3. Take note of bumper stickers which express opinions on technical issues (NO NUKES, SOLAR POWER). Create short bumper-sticker statements which reflect a technical view: SOME NUKES, WHEN AND WHERE FEASIBLE or SOLAR POWER WITH SUFFICIENT AVERAGE DAILY SUNLIGHT. What do such attempts tell you about the nature of specifically technical persuasion?
4. Order copies of advertising brochures and annual reports from major technical firms. Make a careful study of the use of language, graphics, etc.

chapter 24

Researching and Writing a Library Paper

SUSTAINED WRITING

Any long-term writing project can be called "sustained writing." That term refers to the length of the manuscript, but it also refers to the method a writer uses to organize, manage, and sustain a system for producing a manuscript which may be hundreds of pages long and which may require years of work to produce.

Fortunately, a student writer's education builds toward sustained writing in increments. First there is the 500-word theme as a standard unit in learning to write. Then there is a lab report, a case study, or a research paper for a history class. For a student who enters graduate school, there is a progression of increments from a series of lab reports, to research papers, to a master's thesis, and further to a doctoral dissertation. Practicing professionals in engineering and technology usually do two major types of sustained writing: the technical recommendations report and the professional article. Writing a long paper in a writing course thus becomes an important stage in building skills for any career.

The First Step in Sustained Writing

Taken together, your analysis of the subject, the reader, and your purpose in writing become the first step in any sustained writing project.

What You Already Know

If you are in a technical writing class, you know that you can write a 500-word theme. The 500-word theme has probably been taught as well in fraternity and sorority houses as in classrooms: a brother or sister will take the freshman aside and say, "This is how you write a theme: state a thesis in paragraph 1, break the thesis into three parts, devote a paragraph to developing each part of the thesis with examples, and write a conclusion." The 500-word theme, then, is a part of the lore of college life.

The work of sustained writing can be viewed as a work of wider scope than a 500-word theme. Whether the manuscript is a doctoral dissertation or a 500-page environmental impact study, the work has the same features which Aristotle noted centuries ago: a beginning, a middle, and an end; a division of parts into some logical relationship; and support of general statements with examples and details. As the chart below illustrates, a standard thought pattern operates:

LOGICAL PATTERNS OF ORGANIZATION

Aristotle	Math Problem	Lab Experiment	500-Word Theme
beginning	state equation	state problem	state thesis and division of parts
middle	demonstrate the working out of the equation	method of investigation to derive results	develop, discuss, illustrate the parts with data
end	answer	results	summary and conclusion

The writing skills which various long forms require are exactly the same: appropriate diction, effective sentences, unified and coherent paragraphs. The difference lies in how well you can plan and how well you can manage—or sustain—a work of greater scope and complexity. Many of these skills are analytical and management skills which lie completely outside the actual writing of a manuscript.

How the Scope of Sustained Writing Differs

A sustained writing project differs from a 500-word theme in that you must manage the project differently simply because of its larger scope. To illustrate that contention, consider your brain as analogous to a computer. If there are about 5 bytes per word, a 500-word theme calls for a storage capacity of 2500 bytes. You have comparatively little difficulty in storing the 2500 bytes in your mind and accessing them in logical sequence as you record your thoughts in a 500-word theme.

Sustained writing calls for a far greater capacity to store, retrieve, and arrange data. A 20-page paper calls for a capacity of about 200 words (1000 bytes) per page, or 20,000 bytes of memory altogether. A writer cannot reel off that many pages serially, even in the most desperate effort to submit a finished paper at 8 a.m. tomorrow. Therefore, you must have some more orderly and systematic means of gathering and managing a larger body of data than a 500-word theme requires.

How Your Relationship to a Subject Differs

Writing in an academic situation is primarily *re*active. That is, an authority figure who knows more than you know about a subject assigns a topic for writing, and you react by writing what you know and think. A part of the burden of proof is upon you to provide data, but a very large part of the burden of proof is upon the professor as judge of your competence.

The further you progress in your education, the more the burden of proof shifts to you as the writer. When you are a freshman in chemistry writing a lab report, your professor probably reads hundreds of such reports. When you have your first lab in aerospace engineering, thirty-four other students may also be writing lab reports on boundary layer theory. Later on, however, your topic may not be fully understandable even to your professor; that, in fact, is the very reason why you have chosen the topic for research.

As a practicing professional, you write a scientific paper because your colleagues do not know what you have discovered. As an E/T on the job, you write a technical recommendations report for a non-technical client precisely *because* the reader does not know the subject. *You* are the person who brings all the knowledge to the situation.

This different relationship to your subject places upon you the responsibility for collecting and interpreting data in a systematic, well-managed way and for reporting that data responsibly. The more you know, the more you become the authority and the less you react to someone else in authority. In a more mature role, the ability to do sustained writing is one of the principal indications of your competence.

How Your Treatment of Subject Differs

Selection of detail is another factor which changes as you do sustained writing. The more you learn, the more important it becomes for you to select from all that you know precisely what a reader in a given situation requires. In an academic situation, you are often reactive in a situation which requires only that you respond to a professor's questions. The reader in that situation is chiefly responsible for determining whether you have made an appropriate selection of data on the subject, and you receive a grade accordingly. One of the chief differences which characterizes sustained writing projects is that you as the researcher/writer must make appropriate decisions about what to say and what not to say about a subject. Responsibility for managing your body of knowledge on a given subject lies with you as the writer. The further you progress in your education and career, the more true that statement becomes.

How Your Relationship to a Reader Differs

In academic writing, you write in response to a single professor who knows more than you know about a topic. It is also the professor's major responsibility to assign you a topic and, in that way, assume some responsibility for the direction your writing will take. As you develop academically and as you graduate to professional writing, you must take more and more of the responsibility of deciding what to write for a given audience. The professional who writes must exercise considerable sensitivity and imagination to analyze a reader's needs and to meet those needs by writing with accuracy, thoroughness, clarity, and integrity.

How the Purpose Differs

A student in an academic situation most often deals in a reactive way in fulfilling a pre-determined purpose for writing. As sustained writing becomes more complex in subject, scope, and relationship to the reader, your task becomes more complex in determining the purpose of what you write. The age-old cry of the student to the professor, "But you didn't explain the assignment so that I could understand it," is not a sufficient answer to a graduate committee who has read your thesis, to the editor of a scientific journal, or to a paying client who has asked for a technical recommendations report. The writer must bear the burdens of determining the purpose of a sustained writing project and of fulfilling that purpose.

GUIDE FOR SELECTING A TOPIC

Engineers and technologists most often look to the results of laboratory experiments or field observations to form the data base for sustained writing projects. However, not every student who is learning to write long forms has a research project to serve as the data base for a long paper. That circumstance makes a library the logical data source for a research paper, and a resourceful student can learn much about sustained writing on such a project. Choosing the library as a resource presents these special concerns:

A. The great mass of material in a library makes it absolutely essential to set up strict parameters:
   - a carefully selected and narrowed topic; and
   - a specified length and purpose.

   Most technical projects have inherent limitations of subject which, in turn, limit the scope of the paper. A student who enters a library sets out to isolate a 20-page paper from the huge body of knowledge there.

B. After a topic is determined and narrowed the problem becomes a sequence of decisions:
   1. where to look for material;
   2. what material to look for;
   3. what to scan and what to read intensively;
   4. what notes to select and how to record them.

   In lab work or field work, the project itself helps the researcher to make those determinations. With an entire library at his disposal, a student writer must learn how to make judgments which manage a writing project within a narrowly defined range. The operative word in that statement, however, is *manage*; for writing a research paper is—first, last, and always—a management task. These management concerns are basically the same for a writer producing the first research paper and for the scientist preparing a paper for a scholarly journal.

   First, however, be certain that you know something about research papers. Do not set out blindly to write a library-based paper until you have first established clearly in your mind the major features of the finished product you must produce. Your assigned paper may be shorter, less complex as a subject, and less complex in the way you write about it; but your task is still to replicate as closely as possible a professional-quality paper. Your use of library sources for a data base rather than laboratory or field research is the only significant difference between your work and the work of more advanced researchers.

What to Write About

Few academic exercises can seem so artificial and so arbitrary as having to learn how to write a library research paper. Students commonly complain,

"But I'm only a sophomore and I don't know anything about my major yet!" In a sense, this is a valid complaint; having to write the results of research usually occurs much later in an academic career, with a major senior project, an undergraduate honors thesis, or a paper for an upper-division course in a major. For the purposes of a writing course, however, there are many ways to select and narrow a suitable topic.

Select a Topic Which Interests You

Many areas of your life require systematic research. Very probably, you do not make a major purchase of an appliance or a vehicle without first doing a thorough search of literature related to your purchase. Consider, then, writing a paper on some technical aspect of a product you may one day purchase: a vehicle, electronic equipment, or any one of the whole range of consumer goods potentially available to you.

When the American sub-compacts were first introduced to the market, one student wrote a comparative study of two American models compared to the most popular European models. That fulfilled the essential purposes of a research paper: to read a variety of sources in the library, to take notes, to draw together the data he gathered and form his own conclusions in a paper.

The following shows how the student outlined his paper.

A Comparative Study of Three Sub-Compact Cars

TABLE OF CONTENTS

1. Manufacturer and type
   - 1.1. Chevette
   - 1.2. Horizon
   - 1.3. Rabbit
2. General specifications
   - 2.1. Options
     - 2.1.1. Chevette
     - 2.1.2. Horizon
     - 2.1.3. Rabbit
   - 2.2. Engine
     - 2.2.1. Chevette

2.2.2. Horizon
2.2.3. Rabbit
2.3. Drive train
    2.3.1. Chevette
    2.3.2. Horizon
    2.3.3. Rabbit
2.4. Dimensions and capacities
    2.4.1. Chevette
    2.4.2. Horizon
    2.4.3. Rabbit
2.5. Suspension
    2.5.1. Chevette
    2.5.2. Horizon
    2.5.3. Rabbit
2.6. Steering
    2.6.1. Chevette
    2.6.2. Rabbit
2.7. Brakes
    2.7.1. Chevette
    2.7.2. Horizon
    2.7.3. Rabbit
2.8. Wheels and Tires
    2.8.1. Chevette
    2.8.2. Horizon
    2.8.3. Rabbit
3. Performance
  3.1. Acceleration
    3.1.1. Chevette
    3.1.2. Horizon

      3.1.3. Rabbit
  3.2. Top Speed
      3.2.1. Chevette
      3.2.2. Horizon
      3.2.3. Rabbit
  3.3. Braking
      3.3.1. Chevette
      3.3.2. Horizon
  3.4. Fuel Economy
      3.4.1. Chevette
      3.4.2. Horizon
      3.4.3. Rabbit
  3.5. Handling
      3.5.1. Chevette
      3.5.2. Horizon
      3.5.3. Rabbit
  3.6. Interior Sound Level
      3.6.1. Chevette
      3.6.2. Horizon
      3.6.3. Rabbit
4. Price
  4.1. Chevette
  4.2. Horizon
  4.3. Rabbit
5. Driving impressions
  5.1. Chevette
  5.2. Horizon
  5.3. Rabbit
6. Analysis

> 7. Conclusion
>   7.1. Chevette
>   7.2. Horizon
>   7.3. Rabbit
> 8. Bibliography
> 9. Appendix

Recycle Existing Data

A major failing of undergraduate library papers is this: too often a student will select a topic about which a number of books and/or encyclopedia articles have already been written. That defeats the purpose of research for the researcher sets out to pull data together into new associations, not to repeat what has already been researched and reported.

A library is a wealth of data which you may recycle for your own purposes. For example, an interest in the nutritional content of baby foods, especially the percentage of carbohydrates, provided a student with a topic. Her paper traced the changes in the content of baby foods and the consumer pressures which had brought about that change.

To research the paper, she consulted *The Readers' Guide to Periodical Literature*, books on nutrition, and the labels of three popular brands of baby food on the grocery store shelves. More important, she replicated the process of research: a general search of published sources in books and magazines and an application of what she discovered there in field research to gather data. Then she used the data she gathered to devise her own purpose: to trace the recent history of changing nutrition through several sources which established points along that continuum. Finally, she derived her own conclusion about the quality of nutrition in baby foods. Her outline shows how she presented the subject.

Nutritional Quality of Commercial Baby Foods, 1973–1977

TABLE OF CONTENTS

1. Executive Summary
2. Introduction
    2.1 Subject
    2.2 Purpose
    2.3 Scope
    2.4 Plan of Development
3. Baby Foods Before 1973
    3.1 Added Ingredients
        3.1.1. Explanation of Additives
        3.1.2. Levels of Each Additive
    3.2 Health Problems
        3.2.1. Sugar
        3.2.2. Salt
        3.2.3. Modified Starches
        3.2.4. Other Additives
4. Current Baby Food
    4.1 Discontinued Additives
    4.2 Nutritional Quality of Current Commercial Foods
5. Conclusion and Recommendations
6. Appendices
    6.1 Nutritional Quality of 1973 Commercial Foods
    6.2 Nutritional Quality of Current Commercial Foods
7. Bibliography

Apply the Results of an Existing Research Project

It is a very common occurrence that the state of the art in any field advances when existing methods of research are applied to new projects. A student used an existing research model to explore the nutritional value of the breakfasts which students selected in the college cafeteria. He considered that the percentage of students who selected a soda and two jelly donuts for breakfast was worthy of further study. He was able to find that breakfast programs for elementary school children are supported by government publications on nutritional requirements. There is additional research on the correlation between nutrition and scholastic performance.

To produce a valid research paper, the student first read the literature on school nutrition; he learned a method of collecting data which he could apply to his own research; and he then applied that established research technique to a series of observations of breakfast trays in the college cafeteria. He did what many researchers do on a much more complex level: he took an established method of investigation and an existing body of data; then he applied that to a specific project of his own.

Apply Your Knowledge of an Academic Subject to a Life Problem

Any student who says, "But I'm only a sophomore and I don't *know* anything about my major!" has probably not stopped to apply what he knows to existing situations in the day to day world around him. A student derived a research paper topic from a daily problem for which she had devised an innovative solution. She lived in the suburbs a considerable distance from the college campus; that meant she had to be a shrewd observer of traffic patterns and how they were affected by weather, accidents, time, placement of traffic signals, and many other factors. She also knew the rudiments of systems design in industrial and systems engineering. Combining her academic knowledge with her daily observations about traffic, she was able to write an acceptable research paper on how to select the best route to school through morning commuter traffic. She was able to make a synthesis of her informal observations and the rudiments of academic knowledge and thus produce a report of the results of research, as shown by the outline and Executive Summary below.

Alternate Commuter Routes From College Park to Georgia Tech

TABLE OF CONTENTS

1. Executive Summary
2. Introduction
3. Conditions Affecting Route Choice
    - 3.1. Time of Day
    - 3.2. Traffic Flow
    - 3.3. Weather
4. I75-85 Route
    - 4.1. Route Description
        - 4.1.1. Roads Used
        - 4.1.2. Condition of Route
        - 4.1.3. Traffic Lights and Speed Limit
    - 4.2. Conditional Route Completion Times
        - 4.2.1. Time of Day
        - 4.2.2. Traffic Flow
        - 4.2.3. Weather
    - 4.3. Ideal Usage of I75-85 Route
5. Georgia Avenue Route
    - 5.1. Route Description
        - 5.1.1. Roads Used
        - 5.1.2. Condition of Route
        - 5.1.3. Traffic Lights and Speed Limit
    - 5.2. Conditional Route Completion Times
        - 5.2.1. Time of Day
        - 5.2.2. Traffic Flow
        - 5.2.3. Weather
    - 5.3. Ideal Usage of the Georgia Avenue Route

6. Stewart Avenue Route
   6.1. Route Description
      6.1.1. Roads Used
      6.1.2. Condition of Route
      6.1.3. Traffic Lights and Speed Limit
   6.2. Conditional Route Completion Times
      6.2.1. Time of Day
      6.2.2. Traffic Flow
      6.2.3. Weather
   6.3. Ideal Usage of Stewart Avenue Route
7. Conclusion and Recommendations
8. Appendix

---

1. EXECUTIVE SUMMARY

This report presents a study of three alternate routes (I75-85 route, Georgia Avenue route, and Stewart Avenue route) for commuter travel from College Park to Georgia Tech. It gives the commuter the knowledge he needs to make a logical route choice by evaluating each route according to physical condition and the effect of certain external factors (time of day, traffic flow, and weather). The results are reflected in this set of commuter guidelines which have been established to yield the best route completion time under any given circumstances:

1. If I75-85 becomes congested between the starting point and the Lakewood Freeway exit, it is best to choose the Stewart Avenue route.

2. If I75 becomes congested between the I75-85 junction and the Georgia Avenue West exit, it is best to choose the Georgia Avenue route.

> 3. If I75 is clear through the Georgia Avenue West exit and there is no indication of problems ahead, it is best to take the I75-85 route.

Remarks

Modern research methods began in just such everyday ways—when Sir Francis Bacon hypothesized that ice would preserve meat, he went outside to pack a chicken in snow. Unfortunately he caught his death of cold, but innumerable scientists and researchers in every discipline have followed in his path to observe, to collect data, to draw conclusions, and to report the results of the study. To be an educated engineer or technologist, you must enter at some level, however simple and humble, that distinguished company of researchers and report-writers who have gone before you. Selecting a topic for a library-based paper is one place to begin.

GUIDE FOR COLLECTING DATA

Once you have selected a topic, you must then manage the process of finding information.

What You Should Know About Reference Sources

To develop skills in your major you should know the following:

a. the major reference sources in your academic major: dictionaries (there are dictionaries of terms in many fields such as electrical, civil, and mechanical engineering), encyclopedias (the *Encyclopedia of the Social Sciences*, for instance, contains a wealth of material, including extended definitions of key terms, for the disciplines within that area), standard works of history and theory (a multi-volume history of mathematics is an example of such a work).

b. standard textbooks in your field. Apart from the texts for your courses, there is usually a group of well-established and widely-known textbooks; know them and their authors.

c. the professional organizations in your field and their journals of research. The proceedings of conferences of such organizations also provide valuable

resources of research. Find out whether student memberships are available in the major professional organizations.
d. the "flagship journals" in your field—that is, the major prestigious publications of research in your field: who edits them, when they are published, and where to find the indexes to the articles they include;
e. collections of abstracts in your field, both published and available through computer search; and
f. computer sources for research in your field.

Taken together, these items will give you a comprehensive outline of the research sources in your field; in any technological or scientific field, the continual advance of research will make you quickly obsolete unless you know how to keep informed.

How to Manage Library Sources

The system for locating, gathering, and managing data from printed sources presents a range of specific problems. The steps in locating data in a library move from the most general to the most specific. Once you have selected and limited a topic (a separate selection task for each long form), further research tasks occur, as shown below.

How to Find Library Sources

A key word from your topic should lead you either to a library card catalogue or to a computer search for publications on your topic.

The computer search will provide you with a bibliography and abstracts so that you can determine what sources seem potentially useful to you. A hand search of the library card catalogue will give you an immediate sense of the material available to you, as will various indexes to periodical literature and bibliographies of specific publications or specific disciplines.

If this is your first venture into a library to do research, do not—repeat, do *not*—begin by searching out a reference librarian to say piteously, "I have this paper to write for a class, and I wonder if you could help me find..." Reference librarians are angels of mercy without whom research could not take place, but begin here to be resourceful and self-reliant about doing research. Only when you have done your own general site preparation and begun to break ground for your project should you request help for more specific and specialized references.

How to Begin Using Library Sources

First, read for background to provide yourself a context for the specific topic you will develop as a paper. As you do general reading in encyclopedias, textbooks, and general books on a given subject, do not expect all of that reading to be directly useful to the actual construction of the paper you write. Much of that work is simply site preparation for the actual structure you will build later; or much of that general reading is a subsurface foundation which does not appear to the reader of your paper even though you provide your limited topic with a solid foundation of information and understanding.

As your general subject begins to take shape in your mind, you will want to begin more intensive reading to collect notes. The reading you do at this stage should begin to feed directly into your manuscript outline and later into the manuscript itself. Here you have begun to assemble the materials for constructing the manuscript.

With a lab report or with a technical report, the active research process usually stops here. You have at this point a sufficient understanding of the background and basic technology of your project to proceed from the library to the laboratory or the field. Evidence of your reading will appear in your lab report or your technical report as a general reference bibliography rather than as quotations, citations with end notes, or references to sources in your text. In other words, secondary printed sources are not the content of a lab report or technical report. For a library-based paper, however, notes for quotations, end-note citations, and references in the text are the basic material of the manuscript.

How to Manage and Trace Notes for a Library Paper

As you begin intensive reading to gather notes for your paper, you must begin to make decisions about note-taking.

a. As you read each book or article, insert slips of paper to mark the pages which contain data you need.
b. Stop here to include this publication in your bibliography. Make a 3 × 5 bibliography card which includes: author (last name first for alphabetizing later on); second author (first name first); title (underline the title of a book; for a part of a larger publication, such as a chapter or an article, use quotation marks; place, date, and name of publisher; volume number if applicable; number of the issue if it is a periodical; editor's name if applicable. To save time later on, prepare each bibliography card as it will be typed on the bibliography pages of your finished paper. If you do that, preparing your bibliography becomes a matter of alphabetizing your cards and transferring the citations directly to the final paper.
c. Copy the pages you want to use on a copying machine (the expense is nominal compared to the time spent taking notes by hand), but make *sure*

you note the author and a short title on *each* photocopied page to create a link to your bibliography card. Mark the portion of the page(s) which you want to use in your paper.

d. As you begin to plan an outline, set up a file for each major section of your paper in a manila folder. This filing system helps to judge how much information you have in order to complete the details of the paper.

As the topic becomes clearer in your thinking, and as you break the major headings down into the second and third levels of your outline, the filing system helps you to determine what you will say in your paper and also what you still need to find in order to provide thorough coverage of the subject.

If your paper includes field observations and laboratory work as well, you will want to keep a detailed field notebook, rough sketches of the finished drawings you may include in your paper, and both tables of data and mathematical calculations to be included. Keep these notes also in a thorough, systematic way; do not spend undue time on making your notebook picture-perfect, but be neat enough so that you reduce to a minimum the work required to transfer material from your notebook to your paper.

e. The next major management task in dealing with notes for a paper is to make sure you can trace any citation from your rough draft back to your bibliography so that you can make an accurate end note. Do this by citing author, short title, and page number within the text of your rough draft (the data you include is determined by the reference form you use):

"This kind of preparation was repeated in both settled and isolated places all over the world, for the 1769 transit would produce 138 observations from 63 positions (Wolf, 1959)."

If you include your bibliographical data in your rough draft as you go, you can always trace a quotation of a citation back to the exact source. Thus writing end notes becomes easily managed. There is no more characteristic and useful feature of thorough, accurate research writing than end notes. It is the note of the careful scholar, connecting one work to all the works which went before it, that makes the research system function. *Source,* then, becomes a very important word in research, for even an undergraduate library paper must be connected to its source with citations to be a part of the living system of scholarship.

The steps in note-taking relate to your management skills and your decision-making skills more than to your writing skills. If you do not perform these tasks carefully and well, however, you may be assured that you will not do research writing well, either.

How to Incorporate Notes in a Library Paper

There are two marks of an inexperienced research writer: long quoted passages in the text of the paper, and end notes which cite one source three times, another source four times consecutively, and a third author several times in sequence with page numbers for the citations in close proximity (4, 5, 7, 9, 12–17). This is the mark of unprocessed data dumped into a paper; it is more the sign of cutting and pasting than of using source material to form a synthesis of new ideas.

When to use end notes and when to quote or summarize are always questions of judgment. The ability to make such judgments comes with experience, but we can narrow the range of questions and set up parameters for making such judgments. Use these tests in deciding about citations:

a. Data which are readily available in several sources (for instance, that a transit of Venus occurred in 1769) need not be cited in a footnote. Such information is sometimes called "encyclopedic information" to indicate that it is widely available. This is particularly true of any dictionary definition you may use, for any subject that appears in alphabetical order hardly needs an end note to lead another researcher to the source.

b. Another test is whether a dictionary or encyclopedia entry is signed by an author. In such a case, an end-note citation is appropriate, for the source represents the work of a named expert in the field.

c. Three or four words in sequence is more than you should quote directly from a source without citing the source in an end note. (Hint: use a far greather proportion of summaries and paraphrases than direct quotations.)

Conclusion

This list of methods for managing data cannot include an answer to every question you may have about research. Much depends upon your own resourcefulness; whatever else we may say about writing a long paper, *management* is the key concept. Not just the management of ideas, but also the management of time and other resources, and even the management of the production aspects of assembling a long paper, are among the surest tests of competence in any student or professional.

END NOTES AND BIBLIOGRAPHY

One of the most exacting tasks for any writer is producing accurate and consistent end note and bibliography documentation for a long paper. Technically educated students should bring to this task the same understanding of the demand for complete accuracy as issued in engineering drawings or measurement in a chemical solution. To achieve such technical accuracy in research documentation, it is first necessary to understand four principles:

1. The demand for technical accuracy is based on the assumption that any documented paper will be typeset or photocopied. Therefore, the copy a writer produces is *exactly* what will be printed. As with ⅛ inch on a shop drawing or 2 feet on a blueprint, any variation from the standard makes a *measurable* difference.
2. Correct form for an individual end note or bibliographical citation is not something a researcher memorizes. Even the most experienced research writer should always use a model form as a kind of template for an individual citation.
3. There is no one right way to do end note and bibliography entries; forms vary from discipline to discipline or publication to publication. *Consistent* use of *one method* therefore becomes essential.
4. The form of an individual citation also varies with the same kind of material the writer is documenting:

    a book with one author

    a book with two or more authors

    a book or pamphlet with no author listed

    an article in a journal

    a signed article in an encyclopedia

    an unsigned article in an encyclopedia

    a government report or pamphlet

    an unpublished thesis or dissertation

    a map or chart

    a manuscript or photocopy of a manuscript

    an interview

    conference proceedings

    transactions of a professional society

In applying these four principles on end notes and bibliography, a researcher should first decide upon one form to use, should be consistent in using that form, and should identify each type of material to be cited.

In dealing with an individual end note or bibliography citation, it is best to break it down into four components:

1. the work's author(s): whether the full name or initials are used, whether the last name appears first, the order of the second or third author's name in the reference, or whether there is an editor or compiler.
2. the work's title: whether it is a part of a larger work, a single book, a volume in a series, or a second, third, or later edition.
3. the work's publisher: whether it is published by an identified university or commercial publisher in a specific geographical location, or whether it is part of a periodical identified by day, month, year, volume, and number.
4. the page number(s): whether the form calls for a specific page reference, total pages in an article or part of a longer work, or whether the form calls for the total number of pages in a book.

End note and bibliography form will vary in the process of changing end notes into bibliography; in addition, different types of material will mean that each of those components will vary slightly to accommodate the different kinds of data needed to identify a source. Finally, the different overall methods of documentation will mean differences in each of those four components. Despite such variations, however, each citation has exactly the same purpose: to lead the reader back to the exact source of the data used.

Two further concerns relate to every end note or bibliographical citation: spacing between each element in a citation and punctuation within the citation. In following any model, a researcher should pay close attention to the position of commas, periods, quotation marks, italics (underlining), parentheses, colons, upper (capitalized) and lower case letters, upper and lower case Roman numerals, p. and pp. as abbreviations for page(s)—or the absence of such punctuation in a given form or in an individual citation.

The two major divisions in the variations among documentation methods are made along the lines of discipline: humanities and sciences. In the following examples, the same article is documented according to the two different methods. The two methods are displayed below to make some of the distinctions clear and specific.

Humanities: book with a single author

    end note

    [9]Harry Woolf, *The Transits of Venus* (Princeton: Princeton University Press, 1959), p. 280.

    bibliography

    Woolf, Harry. The Transits of Venus. Princeton: Princeton University Press, 1959.

Sciences: book with a single author

internal note

(Woolf, 1959)

bibliography

Woolf, Harry, 1959. The transits of Venus. Princeton University Press, Princeton, 258 p.

Humanities: journal article with a single author

end note

[10]Robert R. Rea, " 'Graveyard for Britons,' West Florida, 1763-1781," *Florida Historical Quarterly* 47 (1969), 358.

bibliography

Rea, Robert R. " 'Graveyard for Britons,' West Florida, 1763–1781." *Florida Historical Quarterly* 47 (1969), 345–364.

Sciences: journal article with a single author

internal note

(Rea, 1969)

bibliography

Rea, Robert R. 1969. Graveyard for Britons, West Florida 1763–1781," Fla. Hist. Quat. 47:345–364.

Exercises

A. Apply to those examples of documentation the discussion of the four components of an individual citation presented earlier in the chapter: 1. names, 2. titles, 3. publication data, 4. page reference. Also apply the discussion of spacing between each element and punctuation of each element and between elements.

B. Turn to the examples of professional articles from which those documentation examples were taken. Note the use of names, titles, publication data, page reference, spacing, and punctuation in each citation. In addition, identify the following types of citations:

work with more than one author

two works by the same author in a bibliography

a note which discusses the source in the citation

a copy of a manuscript

C. To gain a better understanding of the flow of research citations as other sources combine to make a new research work, study the end notes below.

How well do you think the writer has processed the data from other sources? Do these end notes make it appear that the author has simply done a "cut and paste job" from other sources?

[1] Joan Bennett, "An Aspect of Seventeenth-Century Prose," *Review of English Studies*, 17, No. 67 (July, 1941), 282.

[2] Bennett, p. 282.

[3] Bennett, p. 284.

[4] Bennett, pp. 286–288.

[5] George Williamson, *The Senecan Amble* (Chicago: The University of Chicago Press, 1951), p. 275.

[6] Williamson, p. 278.

[7] Williamson, pp. 279–283.

[8] Williamson, p. 284.

Now study the second list of end notes and answer these questions: 1. What is the relationship of notes 1 and 4 and notes 7 and 8? Do you see that ideas or quotations are used from the sources, but not necessarily in the same order that they appear in the published work? Do you see that separate items of data are being used to suit the present author's purpose? 2. Does a citation for p. 64 seem more precise to you than a citation for pp. 108–109? Do these, in turn, seem more precise than pp. 34–42? If you were a researcher who needed to check those sources, which would be most helpful and seem most accurate to you?

[1] Joan Bennett, "An Aspect of Seventeenth-Century Prose," *Review of English Studies*, 17, No. 67 (July, 1941), 282.

[2] George Williamson, *The Senecan Amble* (Chicago: The University of Chicago Press, 1951), p. 275.

[3] Dorothy Stimson, *Scientists and Amateurs* (New York: Henry Shuman, 1948), p. 64.

[4] Bennett, p. 281.

[5] Stimson, pp. 108–109.

[6] Silvio A. Bedini, *Thinkers and Tinkerers* (New York: Charles Scribner's Sons, 1975), p. 73.

[7] Raymond Phineas Stearns, *Science and the British American Colonies* (Urbana: University of Illinois Press, 1970), p. 105.

[8] Stearns, pp. 115–116.

[9] Kenneth B. Murdock, *Literature and Theology in Colonial New England* (Cambridge: Harvard University Press, 1949), pp. 34–42.

This discussion of forms and techniques in documentation does not answer every question about what to do with a specific type of material in a specific form. However, this treatment of documentation should enable anyone who uses this text book to select a form and follow that. Many handbooks are available for research purposes; among these, *A Manual of Style* published by the University of Chicago Press is perhaps the most complete, useful, and widely used. The end notes and bibliographies in this text offer basic patterns for most common references; using a manual of style to document other types of material according to the same method is an immediate practical solution to the needs of most students.

As a final word for students, if documentation forms seem arbitrary, they are. Although the chief virtue of doing them in a certain way lies in using one form consistently and to the letter, there are certain technological reasons why a particular form is used. The replacement of end notes for footnotes is perhaps the best example: to print footnotes at the bottom of a page has become prohibitively expensive for publishers and hence the shift to end notes at the end of a chapter or at the end of a book. Using one continuous line to indicate italics is also technically dictated by the continuous underlining feature on printing and word processing equipment. In many instances, however, an upper or lower case letter, a comma or a period, or a space or the absence of a space is simply an arbitrary standard to be exactly replicated. While such standards may have nothing to do with writing skill, they nevertheless indicate how well a writer works as a professional who writes.

part five

Appendices

Appendix A: Manuscript Specifications
Appendix B: Evaluation Checklists

appendix A

Manuscript Specifications

The manuscript specifications in this text are based on the assumption that a student will type assignments for class or closely supervise a typist or word processor. Whichever option a student chooses, the ability to produce the final draft of a manuscript is a separate skill.

Preparing manuscripts to specification involves the following skills:

1. management of time to allow sufficient time for preparation of a finished draft once the work of writing is complete.
2. supervision of others' work: if someone else is the typist, to give complete and easy-to-follow directions and to allow time for the work without unduly burdening the person who must prepare the typescript.
3. understanding all the technical features of a manuscript—spacing, selection of type, features of the page design, etc.—so that only the work of typing is delegated and not the full responsibility to handle all specifications of manuscript.

Although typing is a skill which has suffered a decline because of recent political thought, both male and female students will greatly enhance their effectiveness as writers and as computer users by learning basic typing skills. In the enormously expensive work of preparing technical documents, not knowing how to type and not knowing how to supervise the production of a final draft add greatly to the costs involved.

MEMORANDUMS

Design

```
                        CORPORATE ENGLISH

March 27, 1990


MEMORANDUM
─────────
TO:   George P. Burdell
FM:   D. R. Prof, Ph.D.   DRP
RE:   Manuscript Design and Production
─────────────

1. Purpose: The purpose of this memorandum is to respond to a
   statement you frequently make in writing class: "But my
   secretary will know how to do all that."  This memorandum
   describes how changes in business and technical writing have made
   your statement obsolete.

2. Office Automation: Word processing equipment, expanded capacity
   for producing graphics, and a whole host of other equipment can
   help you to produce clear, effective documents.  Along with those
   changes, however, are other changes which will influence how you
   write:

      2.1. The concentration of office equipment and personnel in a
      central location will replace the secretary nearby who can confer
      with you about what you write.  Whoever types your manuscript will
      be likely to copy exactly what you put on the page.

      2.2. Opportunities for women have removed the ceiling for that
      generation of women who could not advance beyond the executive
      secretary level.  Clerical personnel, while still indispensable,
      will more nearly work directly under your supervision in your
      immediate office setting.

      2.3. You may even find yourself having to produce first drafts
      on word processing equipment so that an editor or skilled clerical
      person can make revisions at another terminal.

3. Conclusion: While the work of writing will become easier in
   some ways, in all likelihood your responsibilities as a professional
   who writes will increase.  Designing and writing a memorandum like
   this one will be a process which begins and ends with you.




        Post Office Box 5267    Atlanta, Georgia 30307    (404) 525-6225
```

Memorandums

Specifications

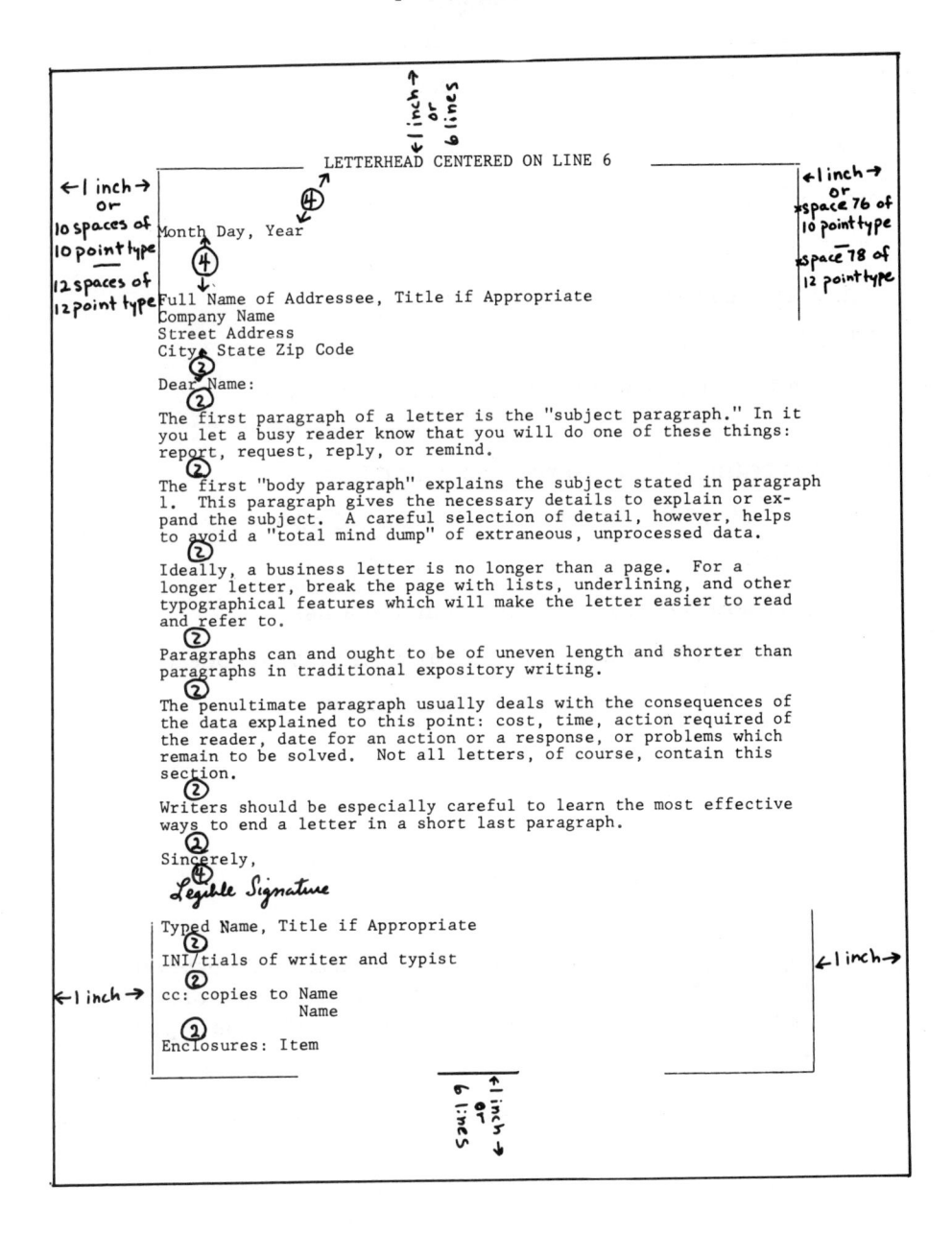

```
                    ←1 inch→
                    or
                    6 lines
              LETTERHEAD CENTERED ON LINE 6
←1 inch→                                              ←1 inch→
or              ④                                     or
10 spaces of  Month Day, Year                         *space 76 of
10 point type                                          10 point type
              ④
12 spaces of                                          *space 78 of
12 point type Full Name of Addressee, Title if Appropriate  12 point type
              Company Name
              Street Address
              City, State Zip Code
              ②
              Dear Name:
              ②
              The first paragraph of a letter is the "subject paragraph." In it
              you let a busy reader know that you will do one of these things:
              report, request, reply, or remind.
              ②
              The first "body paragraph" explains the subject stated in paragraph
              1.  This paragraph gives the necessary details to explain or ex-
              pand the subject.  A careful selection of detail, however, helps
              to avoid a "total mind dump" of extraneous, unprocessed data.
              ②
              Ideally, a business letter is no longer than a page.  For a
              longer letter, break the page with lists, underlining, and other
              typographical features which will make the letter easier to read
              and refer to.
              ②
              Paragraphs can and ought to be of uneven length and shorter than
              paragraphs in traditional expository writing.
              ②
              The penultimate paragraph usually deals with the consequences of
              the data explained to this point: cost, time, action required of
              the reader, date for an action or a response, or problems which
              remain to be solved.  Not all letters, of course, contain this
              section.
              ②
              Writers should be especially careful to learn the most effective
              ways to end a letter in a short last paragraph.
              ②
              Sincerely,
              ④
              Legible Signature

              Typed Name, Title if Appropriate
              ②
              INI/tials of writer and typist
              ②                                              ←1 inch→
←1 inch→      cc: copies to Name
                             Name
              ②
              Enclosures: Item

                                    ←1 inch→
                                    or
                                    6 lines
```

Design

```
                         Letterhead

date 4 lines beneath letterhead logo

MEMORANDUM 4 lines below date

TO: 2 lines below

FM: 2 lines below

RE: 2 lines below

_____   a line of 10 spaces 2 lines below the heading


The text begins 4 lines below the dividing line.

In any text which is left-hand justified, double space between
paragraphs.

        1. Indent 5 spaces to display items in a list.

            a. indent an additional 3 spaces if the list has subheadings.

            b. Double space between items in a list.


                      OTHER FEATURES


    left margin 10--------------------------------------------------72 right
margin

    many companies prefer 10 point type, that is, 10 bytes to
    the inch rather than 12; and most now avoid the use of elite
    type faces. Do not use an Italic type face for business pur-
    poses.
```

Spacing for page 2 and succeeding pages of a Memorandum:

Georgia P. Burdell
April 26, 1990
page 2

The second and succeeding pages of a letter or a memorandum should include reference to the recipient of the letter or memorandum, the date, and the page number. The reference is an essential aid to the work of copying, collating, and filing large volumes of correspondence.

LETTERS

The specifications listed and discussed assume that the student writer will either type himself or supervise a typist in preparing manuscripts for class. Practices vary from company to company; these specifications are designed to work for amateur typists.

Overall Page Design

CORPORATE ENGLISH

March 20, 1990

Mr. George P. Burdell
1234 Battery Place
Charleston, Utah 18603

Dear Mr. Burdell:

In response to your question about whether engineering technologists really have to write letters, I am writing to explain some basic principles of this important skill.

Letter writing is an essential form of communication in all professional activities. Letters not only communicate to one reader who then replies, but they also make up a large system of communications on any given project. Generally, what you write in a letter also becomes a part of a legal record.

Letters are also a sales tool. For professionals who deal in complicated technical matters, that sales appeal is a subtle art which develops only with long practice.

For all those reasons, then, you will greatly benefit in your career if you learn to write good letters.

Two factors which are especially important to technical professionals are these:

   1. Effective visual design is important because your readers are likely to have a high degree of visual aptitude.

   2. Learning to produce a great volume of letters of high quality is also important -- just as a golfer or field goal kicker develops a "grooved swing" to increase the percentage of success.

This letter should give you a good idea of what a routine letter sounds like -- and looks like. If you do have further questions, please consult the letters chapter in your text.

Sincerely,

Mack T. Thompson

Mack T. Thompson

MTT/sbg

Post Office Box 5267   Atlanta, Georgia 30307   (404) 525-6225

Specifications

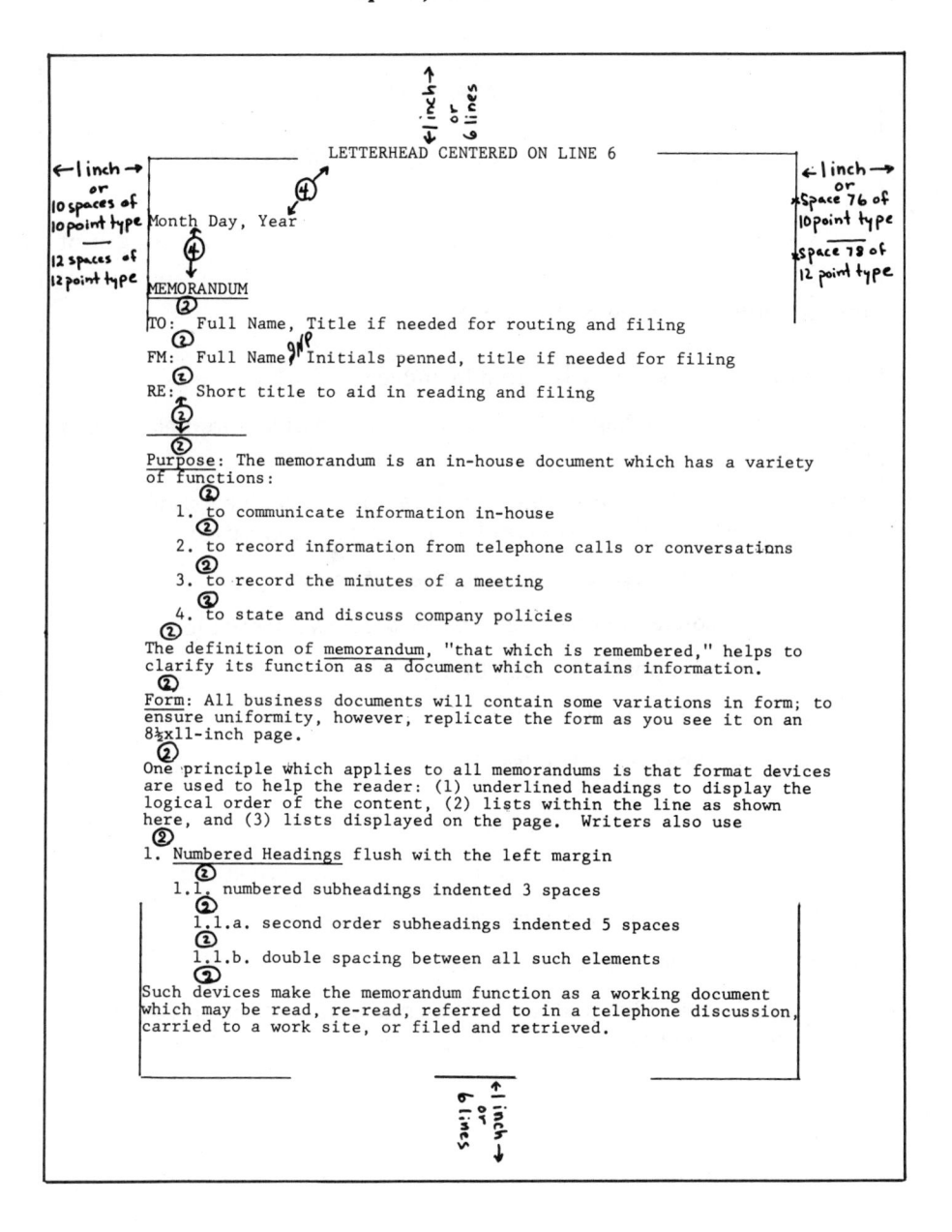

```
                    ←1 inch→
                      or
                    6 lines
            ↕
            LETTERHEAD CENTERED ON LINE 6
←1 inch→                                              ←1 inch→
   or                                                    or
10 spaces of                                          Space 76 of
10 point type  Month Day, Year                        10 point type
               ↕ ④
12 spaces of                                          space 78 of
12 point type  MEMORANDUM                             12 point type
               ↕ ②
               TO:  Full Name, Title if needed for routing and filing
                 ②
               FM:  Full Name ⤺ Initials penned, title if needed for filing
                 ②                 gnp
               RE:  Short title to aid in reading and filing
                 ②
                 ↕
                 ②
               Purpose: The memorandum is an in-house document which has a variety
               of functions:
                 ②
                    1. to communicate information in-house
                       ②
                    2. to record information from telephone calls or conversations
                       ②
                    3. to record the minutes of a meeting
                       ②
                    4. to state and discuss company policies
                 ②
               The definition of memorandum, "that which is remembered," helps to
               clarify its function as a document which contains information.
                 ②
               Form: All business documents will contain some variations in form; to
               ensure uniformity, however, replicate the form as you see it on an
               8½x11-inch page.
                 ②
               One principle which applies to all memorandums is that format devices
               are used to help the reader: (1) underlined headings to display the
               logical order of the content, (2) lists within the line as shown
               here, and (3) lists displayed on the page. Writers also use
                 ②
                 1. Numbered Headings flush with the left margin
                       ②
                    1.1. numbered subheadings indented 3 spaces
                          ②
                       1.1.a. second order subheadings indented 5 spaces
                          ②
                       1.1.b. double spacing between all such elements
                 ②
               Such devices make the memorandum function as a working document
               which may be read, re-read, referred to in a telephone discussion,
               carried to a work site, or filed and retrieved.




                                    ↕
                                  ←1 inch→
                                     or
                                   6 lines
                                     ↓
```

Overall Page Design

LETTERHEAD

(usually 1 inch from the top edge of the page)

*date 4 or 8 lines below the letterhead, depending on the length of letter

inside address 4 lines below the date

salutation 2 lines below the inside address

The text of a letter begins 2 lines below the salutation. Letters are single spaced.

Between paragraphs, letters are double-spaced to give a visual indication of the division of paragraphs. If a list appears within a letter,

    1. Indent 5 spaces to display the list on the page

        a. indent an additional 3 spaces to show subdivisions in a list

        b. double space between items in a list

last line of the last paragraph of a letter.

Sincerely, 2 lines below the last line

typed name 4 lines below Sincerely

*INI/tials of writer and typist 2 lines below the typed name

*cc: indicates copies 2 lines below the initials

*Enclosure: 2 lines below the copies

  *Adding 2 to 4 additional lines to the placement of the date, initials, or other elements can enhance the vertical design of a very short letter.

> OTHER FEATURES
>
> left margin 10---72 right margin
>
> many companies prefer 10 point type, that is, 10 bytes to the inch rather than 12, and most now avoid the use of elite type faces. Do not use an Italic type face for business purposes.

Spacing on Page 2 of a Letter

```
GEORGETTE P. BURDELL
MARCH 20, 1990
PAGE 2


THE SECOND PAGE OF A LETTER OR A MEMORANDUM SHOULD INCLUDE THE
NAME OF THE RECIPIENT, THE DATE, AND PAGE NUMBER.  IF YOU FIND
YOURSELF GOING BEYOND 3 PAGES, YOU SHOULD CONSIDER RE-WRITING
THE LETTER OR MEMORANDUM AS A SHORT MEMORANDUM REPORT WITH A
LETTER OF TRANSMITTAL.
```

> Georgia P. Burdell
> April 26, 1990
> page 2
>
> The second and succeeding pages of a letter or a memorandum should include reference to the recipient of the letter or memorandum, the date, and the page number. The reference is an essential aid to the work of copying, collating, and filing large volumes of correspondence.
>
> Always carry at least 2½ lines to a second page of a letter. To include only one line with the complimentary close and signature lines is not only poor design, but also an enormous waste of labor and materials.

INI/tials

The writer's initials in all caps separated by a diagonal from the typist's initials is useful in tracing the work of clerical personnel in a company. Some companies also add a second set of initials INI/TIA/ls to show that a junior member of the staff wrote the letter for a senior manager's signature. This also is a valuable aid for tracing the work of letter writing.

cc

In the era before copying machines, this meant "carbon copies." The multiple use of a single letter for the information and files of other concerned parties in a transaction makes the copies line a standard feature of many letters. For instance, a project manager on a large project might copy a letter to his immediate superior, to the chief engineer, to the architect, and to the contractor. It is considered discourteous to send a copy of a letter without recording the names on the cc line.

Enclosure

Business letters very often contain enclosed material in addition to the text of the letter itself. In this instance, to indicate that there are enclosures is valuable for record keeping and reading purposes. It is also an effective technique to make a list of enclosures to motivate and guide the reader's attention to other material sent with the letter.

Avoid

> GEORGE P. BURDELL
> MARCH 27, 1990
> PAGE 2
>
> PLEASE LET ME KNOW.
>
> SINCERELY,
>
> *Georgia Brown*
>
> GEORGIA S. BROWN
>
> CARRYING OVER THE LAST LINE, COMPLIMENTARY CLOSE, AND SIGNATURE IS ESPECIALLY POOR DESIGN, IN ADDITION TO BEING TIME AND COST INEFFICIENT.

> 3. SPACING MANUSCRIPT: DO NOT BEGIN THE FIRST LINE OF A PARAGRAPH OR A MAJOR SECTION AT THE BOTTOM OF A PAGE, OR

> 12
>
> BEGIN A PAGE WITH THE REMAINDER OF ONE LINE.
>
> 4. NEXT MAJOR HEADING: A SINGLE LINE AT THE TOP OF THE PAGE CREATES A DISPROPORTIONATE DESIGN.

ARRANGEMENT OF PARTS: FORMAL TECHNICAL DOCUMENTS

	Serial or Continuous Texts	*Division of Text into Chapters*	*Professional Article*
BEGINNING	*title page	*title page with Abstract	*title as prescribed by publication, may include Abstract
	*SUMMARY as a separate heading in a serial or continuous text	*EXECUTIVE SUMMARY entered as a separate chapter with chapter heading	*introductory paragraphs as a logical or rhetorical division, not as a physical division of the pages
	*INTRODUCTION as a separate heading in continuous text	*TABLE OF CONTENTS entered as a separate chapter	
		*LIST OF FIGURES as a chapter	
		*INTRODUCTION entered as a separate chapter	
MIDDLE	*STATEMENT OF THE PROBLEM	*major logical divisions of the subject as separate numbered CHAPTERS	*logical partition signalled by transitional words and phrases
	*APPARATUS		
	*PROCEDURE Each as a separate heading in the continuous text	*each page designed with underlined and numbered subheadings which correlate with the Table of Contents	*underlined subheads and numbered lists prescribed by the typography of the publication
END	*SUMMARY AND DISCUSSION as a separate heading	*SUMMARY as a separate chapter in very long reports, or may be combined with	*concluding paragraphs as a logical or rhetorical division, not as a physical division of the manuscript
	*CONCLUSION as a separate heading	*CONCLUSIONS AND RECOMMENDATIONS entered as a separate numbered chapter	*possible list of conclusions, suggestions for further research, or as determined by publication typography and style
	*RECOMMENDATIONS separate		
	*REFERENCES as a separate numbered chapter	*REFERENCES as a separate numbered chapter	*END NOTES on a separate page as prescribed by the publication
	*APPENDIX as a separate chapter with numbered subsections and captions for figures	*APPENDIX as a separate chapter with numbered subsections and captions for figures	*BIBLIOGRAPHY on a separate page as prescribed by the publication.

> 1. CHAPTER HEADING CENTERED ON LINE 12
>
> The first page of a chapter contains a title centered on line 12. The first paragraph begins on line 16 and is indented 5 spaces. The text of any long-term writing project is double spaced. The specifications page for longer papers contains further information about margins and spacing.

TYPIST'S CHECKLIST FOR PRODUCING A LONG-TERM WRITING PROJECT

Any concern with effective graphics and page design must be extended to clear and accurate typing of a manuscript. The checklist which follows is a step-by-step checklist for actually typing the manuscript.

Before Beginning to Type

1. Is the typewriter in good working order?
2. Is the type face a 10 point, or a 12 point which can be easily read and reproduced?
3. Is the type clean? (check a, o, p, e, g especially for blurred impressions.)
4. Is the ribbon new or sufficiently dark to give a clear, easily reproduced impression through to the end of the manuscript?
5. Are the margins set for 10 and 72 (16 and 72 if the paper is to be punched and bound in a thesis binder)?
6. Is the paper a good quality bond, not transparent, and can it be easily fed into a copy machine?

Beginning the Text

1. Begin on the 12th line from the top edge of the page.
2. Center the title of the first section of the text.
3. Type title of the section in ALL CAPS.
4. Skip 4 lines.
5. Indent 5 spaces from the left margin and begin typing the first line of text. (Indent each paragraph 5 spaces unless directed otherwise.)
6. Be *sure* that the typewriter is set to double space.

Ending Page 1 of the Text

1. Leave 6 to 12 spaces at the bottom of the page.
2. Do not begin a section or a paragraph as the last line of a page.

Beginning Page 2 of the Text

1. 4 lines below the top edge of the page and 6 spaces from the right hand edge, type a 2.
2. Skip 2 lines and begin typing the text.
3. Do not carry over one line or a part of a line as the first line on page 2.

Break Each Page

As is discussed throughout this text, make sure that the page design features format devices such as lists, underlining, and indentations. This rule applies to *every* page of a technical document. If you are using a word processor, accessing parts of the manuscript for revision is easier if the parts are clearly formatted. Indent lists 8 spaces from the left.

Allow Space to Add Graphics

Measure the illustrations you want to add to the text and be sure to allow ample space to include them. Also leave space for typed, numbered captions.

Beginning a New Section in a Continuous Text

1. In a continuous text, when you come to the end of SUMMARY, skip 4 lines.
2. Write the heading of the next section, centered, ALL CAPS: INTRODUCTION.
3. Skip 4 lines.
4. Indent 5 spaces to indicate a paragraph break.

Beginning a New Chapter in a Longer Text

1. As with page 1, Chapter 1, begin on the 12th line from the top edge of the page.
2. Center the title of the chapter and type in ALL CAPS.
3. Skip 4 spaces.
4. Indent 5 spaces and begin typing the first line of the text.
5. Note that the first page of a new chapter has no page number.

Citing Reference Sources in a Text

1. If you use numbers to refer to endnotes, raise the numeral one-half space at the end of the line after the end punctuation.[6] Do not place the number within the sentence.
2. If you use scientific style (Rea and Holmes, 1990), you may insert such references at any point within the text, and within the end punctuation if the citation occurs at the end of the sentence: (Tiger, 1990).
3. A quotation from a secondary source which is more than 2½ or 3 lines should be "displayed":

    Indent 10 spaces from the left margin, indent 5 spaces from the right margin. Single-space, and attempt to right justify the text. Do not use quotation marks but place a numeric or scientific end-note reference at the end of the passage.[7]

    Then resume typing double-spaced from margin to margin, having allowed two spaces before and after the displayed quote.

Listing End Notes

1. End notes should be included at the end of the entire manuscript.
2. As with Chapter 1, begin on the 12th line from the top edge of the page.

3. Center and type in ALL CAPS: END NOTES as a separate, numbered chapter.
4. Skip 4 lines.
5. Indent 5 spaces, [1]place the numeral ½ space above the line (or 1. for scientific form).
6. [2]for the second and succeeding notes, follow the same procedure, single-spacing the lines of the note and double-spacing between notes.
7. Page 1 of the End Note chapter does not have a page number.
8. The second and succeeding pages of end notes are numbered sequentially with the remainder of the text; i.e., if the last page of text is 42, the first page of the endnote chapter is 43, and the first numbered page of end notes, the second page, is 44.

Listing Bibliography

1. As with Page 1 of Chapter 1, begin on the 12th line from the top edge of the page.
2. Center the title and type it in ALL CAPS: BIBLIOGRAPHY as a numbered chapter.
3. Skip 4 lines.
4. Begin typing *at the left margin*.
5. Using alphabetical order, write the last name of the author first.
6. Single-space the lines in an individual bibliographical reference.
7. Indent 5 spaces for the second and succeeding lines of a bibliographical reference.
8. Double-space between bibliographical references.
9. Note that the first page of bibliography has no page number, but the second and succeeding pages of bibliography are numbered sequentially with the remainder of the entire text.

Adding Appendices

1. Page 1 of the Appendix(es) may be a single page with APPENDICES written in ALL CAPS on the 12th line from the top edge of the page.
2. Appendix I, Appendix II, and so on are usually included at the bottom of the page, at least 1 inch (or 6 lines) from the bottom edge of the page.
3. Figure 1, Figure 2, and so on are also included at the bottom of the illustration with a descriptive caption.

4. Except for the first page of the Appendix section, on which the number is not shown, number the pages sequentially with the remainder of the entire text.

Writing the Table of Contents for a Manuscript with Chapters

1. As with Page 1 of Chapter 1, begin on the 12th line from the top edge of the page.
2. Center the title and write in ALL CAPS: TABLE OF CONTENTS.
3. Skip 4 lines and begin at the left margin.
4. Write the number and title of Chapter 1.
5. Write a leader .... to the right hand portion of the page, 14 or 16 spaces from the righthand edge of the page.
6. Write the page number for Chapter 1 at the end of the leader ...... 1.
7. Continue in that manner with each major chapter.
8. For subheadings in chapters, indent 5 spaces for the first subhead, 3 spaces for the second level of division, and do not carry the division beyond three levels: using scientific style for numbering,
    1. Chapter in Caps and Lower Case
        1.1. Second Level Division in Caps and Lower Case
            1.1.a. third level division in lower case except for capitalized words (proper nouns)
9. At page 10 in the sequence of numbers in the Table of Contents, shorten the leader by one period to make the right hand margin of the numbers even:
    9
    10
10. At Chapter 10 in the Table of Contents, back space one space beyond the left margin to align the numbers:
    9.
    10.
11. End the page with a 1 to 1½ inch margin at the bottom.

Typing Page 2 of the Table of Contents

1. Page 2 of the Table of Contents is numbered.
2. Begin on the 6th line from the top edge of the page at the left margin.

Listing Appendices or Lists of Figures

1. Begin on the 12th line from the top edge of the page.
2. Center and type in ALL CAPS: APPENDICES or Center and type in Caps and Lower Case: List of Figures
3. List the titles of Appendices with leaders and page numbers OR list the captions on Figures with leaders and page numbers
4. The page which lists the APPENDICES or the List of Figures has no page number shown.

Adding Graphics to the Typed Draft

1. Turn through the manuscript, adding graphics in the order in which they are to appear.
2. Trim the edges of the graphics you will insert so that the edges are clean and straight.
3. Either glue or tape the graphics in the spaces provided.
4. Make it appear that the graphics are a part of the page:
   - by using glue, or
   - by making a very high quality bond copy of the pages which contain graphics.

   Make certain that any graphics you add will feed into a copy machine without catching and crumpling in the feed mechanism.

Typing the Title Page

Type the title page according to the specifications at the end of this appendix.

Binding the Paper in a Thesis Binder

1. Align one cover of the thesis binder with the left edge of the paper, and lightly circle with pencil where holes should be punched to fit within the cover. Leave a sufficient margin on the left to make certain that the copy will be clearly visible.
2. Punch holes at the top and bottom of a few pages.

3. Use 2 or 3 of the punched pages to indicate where to punch the next batch of pages and continue until all the pages are punched.
4. Insert the metal spine device in the back cover of the binder, then place a few pages at a time onto the flexible metal clamps, beginning with the last pages of the paper.
5. Place the front cover on the flexible metal strips, add the front plate of the spine device, bend the flexible metal strips and secure them with the sliding metal loops along the front plate.

Adding a Cover Label

The thesis binder should have a small rectangular indentation on the front cover for a label.

1. Slightly moisten one corner of a label and attach it to a $4 \times 6''$ card or to a sheet of typing paper.
2. Gingerly insert the page or card with the label into the typewriter.
3. Centered at the top of the label write the exact title of your paper IN ALL CAPS.
4. Center your name in Caps and Lower Case beneath the title.
5. Beneath your name include some identification as to class and hour to facilitate the handling and grading of the papers.

ONE FINAL NOTE

In an era of accessible and inexpensive copying services, always retain a copy of any manuscript for your files.

THE TITLE PAGE

Example A.1

```
                    space 42 * is center




           TITLE OF PAPER: USUALLY TEN OR FEWER WORDS      (line 12)




                              Abstract                    (line 18)


           Knowing when and why to write an abstract       (line 21)
           and how abstracts are used is the best
           basis for the writing process itself.  Do
           the abstract last, even after writing the
           Executive Summary.  Then you should be able
           to reduce your subject to 100 - 150 words.
           A researcher who reads your abstract should
           be able to know whether your paper relates
           to her research.  Reading abstracts is most
           often the first step in collecting secondary
           sources for a research project and are thus
           a valuable research tool.







                         George P. Burdell              (line 46)
                            English 2023                (line 48)
                  Introduction to Technical Writing     (line 49)
                           January 19, 1990             (line 51)
```

Example A.2

```
                  space 42 * is center




         EXPERIMENTAL VERIFICATION      (line 12)
                  of the                (line 14)
         INTEGRAL MOMENTUM EQUATION     (line 16)







                    by                  (line 30)
              Georgia P. Burdell        (line 32)




                   for                  (line 40)
               AE3001/ENGL3023          (line 42)




              October 16, 1990          (line 50)
         SCHOOL OF AEROSPACE ENGINEERING (line 52)
         GEORGIA INSTITUTE OF TECHNOLOGY (line 54)
```

REPORT PAGE SPECIFICATIONS

SPACE 42 ↓ IS CENTER

PAGE NUMBERS ON LINE 4 AT SPACE 78

PAGES WITHIN A REPORT BEGIN ON LINE 6

CHAPTER HEADINGS ARE CAPITALIZED AND CENTERED ON LINE 12

THE FIRST PARAGRAPH OF A CHAPTER BEGINS ON LINE 16, THE FIRST LINE IS INDENTED 5 SPACES, AND THE TEXT IS DOUBLE SPACED.

SPACE 10 CREATES A 1-INCH MARGIN ON THE LEFT IN 10 POINT TYPE

SPACE 15 CREATES A 1½-INCH MARGIN FOR A BOUND REPORT

A 1-INCH RIGHT MARGIN IN 10 POINT TYPE IS AT SPACE 76

SPACES 77-80 CREATE A BUFFER TO AVOID EXCESSIVE HYPHENATION

PUNCH HERE FOR THESIS BINDER

EACH PAGE MUST HAVE A 1-INCH MARGIN AT THE BOTTOM (6 LINES)

MEMORANDUM DRAFT

BURDELL
Engl. 2023
MWF 9

March 27, 1990

MEMORANDUM

TO: D.R. PROF, Ph.D.

FM: GEORGE P. BURDELL *GPB*

RE: Ms. SPECS. FOR IN-CLASS ASSIGNMENT

PURPOSE: The purpose of this memorandum is to demonstrate how an 8½-11-inch sheet of ruled paper can be well designed for an in-class assignment or in preparing copy for a typist.

SPACING: The heading should be proportionally spaced. Even though typing specs call for 4 vertical spaces between the date and memorandum heading, 2 spaces create a better design on ruled paper. Typed memorandums are single spaced; double spacing handwritten copy is an aid to the reader, the writer can add words, and there is space for revisions or grader's marks. *RUN ON*

PEN AND INK: A pen with a medium point and black or blue ink work best for written assignments. Other technical tasks may require pencil and very small characters, but technical writing must be clearly visible to a reader, a grader, or a typist.

LETTER DRAFT

THOMPSON
Engl 2023
MWF 9

March 20, 1990

Mr. George P. Burdell
165 Battery Way
Charleston, Georgia 31061

Dear Mr. Burdell:

You will have to write many letters during the successful career which awaits you. The first letters you write may be on ruled paper in a writing class. At other times you may prepare a handwritten draft for a typist. This letter shows what a handwritten letter should look like.

Note that one vertical space on ruled paper is much wider than one space on a typewriter. Exercise good judgment about design, therefore, instead of using the same number of spaces in typing specs.

As with a typed letter, vary the length of paragraphs in your letter. Avoid long, dense blocks of copy.

Use a pen with black or blue ink and, even if you do not have a beautiful cursive style, try to make your writing neat and legible. If you have learned lettering in a drafting class, use that style even for a finished letter for mailing.

While this letter represents minimum standards, you can produce functional copy for a class or a typist in this way.

Sincerely,

Mack Thompson

230 Lucy Lane
Dallas, Texas 81085
(914) 555-6666

REPORT DRAFT

ROUGH DRAFT DESIGN

22

1.1. **PAPER**: Theme paper with lines every ½ inch is preferable for a handwritten draft of a report. Two types of paper to avoid are as follows:

1.1.a. Unlined paper tempts a writer to crowd too much on the page.

1.1.b. Legal sized paper can be difficult to handle and especially difficult to place on a typing stand.

1.2. **HANDWRITING**: If your handwriting is very large, be aware that your typed copy may contain many paragraphs which are too short.

1.3. **WRITING INSTRUMENT**: If you prepare your draft in pencil, use a dark lead. Remember that the typist will be reading your copy at slightly more than arm's length.

1.4. **DIRECTIONS TO TYPIST**: Write special directions about spacing in red and make all such marks to help the typist.

appendix B

Using Evaluation Checklists

The evaluation checklist is useful in several ways:

1. as a sheet to accompany an assignment and therefore act as an aid to the writer who may need a list of reminders about the major points to check before submitting the assigned document;
2. as a basis for grading papers;
3. as a basis for peer evaluations done by classmates;
4. as a form used by senior clerical staff who must bear some of the responsibility for the form and style of letters, memorandums, and reports;
5. as a form used by managers to spot-check files of employee writing and to make an evaluation of job performance in communications.

A feature which makes the checklist especially useful is that it gives both writer and evaluator a standard to refer to. Discussions of form, therefore, tend not to dissolve into what one senior secretary calls "a war of wills" between people who must work together to produce technical documents. Matters of content always create the necessity for the writer and the evaluator to exercise judgments, but the checklist handles much of the work not related to serious and substantive professional judgments about the text itself.

MEMORANDUM CHECKLIST

                            Reader_____   Writer_____

1. Are the pages easy to see, easy to handle? − 1 2 3 4 5 +
   Comment:
2. Has the writer followed the prescribed form? − 1 2 3 4 5 +
   __capitals __spacing __titles __names __initials __other __RE:
   Comment:
3. Is the memo divided into logical parts? − 1 2 3 4 5 +
   Can you glance down the page and get a good idea of what the memo covers?
   Comment:
4. Are individual paragraphs clearly written and easy to follow?
                        − 1 2 3 4 5 +
   __unity __neither too long nor too short __logical transitions
   Comment:
5. Are there awkward sentences which you had to read more than once? Underline any part which is difficult to read. − 1 2 3 4 5 +
   Comment:
6. Is the vocabulary precise and are definitions clear? − 1 2 3 4 5 +
   __colloquial __vague __wordy __telegraphic __unidiomatic
   Comment:
7. Is the punctuation an adequate guide to reading? − 1 2 3 4 5 +
   __commas __semicolons __colons __dashes __" " __( )
   Comment:
8. If the memo called for a conclusion or an interpretation of data, did that clearly add up to an answer? − 1 2 3 4 5 +
   __logical __appropriate to what had come before
   Comment:

In the space below, cite the weakest and the strongest points of this memo.

Weakest:

Strongest:

LETTERS CHECKLIST

Reader _____ Writer _____

Format: Has the writer followed the prescribed format on these items?:

__ date __ inside address __ salutation

__ complimentary close __ signature __ typed/written name

other _____ , _____ , _____ , _____ , _____

Rating: − 1 2 3 4 5 +

Comment:

Convention: Has the writer followed the conventions used by knowledgeable people of good taste?

first paragraph	__ language	last paragraph	__ tone
	__ length		__ motivation
	__ for the reader		__ easy answer
			__ sales and PR

Rating: − 1 2 3 4 5 +

Comment:

Overall Construction: Has the writer presented the content of the letter in complete and readable form?

___ All necessary information required by this type of letter.

___ Appropriate use of language for this audience and occasion.

___ A combination of sentences and punctuation which "read themselves:" that is, the reader does not have to puzzle over a sentence and read it more than once to get the meaning.

___ Are the paragraphs logically arranged; that is, do the parts appear in an order which aids the reader's thinking or does the reader have to do the work of making sense of the letter?

___ If appropriate for this kind of letter, has the writer used lists, underlining, and other typographical devices to make the letter physically easy to handle and visually easy to read and see?

Rating: − 1 2 3 4 5 +

Comment:

EVALUATION CHECKLIST: SHORT TECHNICAL REPORTS

Reader _____ Writer _____

1. Is the report visually easy to see and physically easy to handle? Rate your first impression of the report: − 1 2 3 4 5 +
2. Does the heading of the report tell you immediately __ the subject, __ the author, __ the audience, __ the date? Is this an adequate record in addition to being a communication? − 1 2 3 4 5 +
3. At first glance, does the report reflect a clear and well-proportioned design on the page: __ major headings, __ subheadings, __ underlining, __ numbering, __ indentation? Are format devices __ adequate, __ too many, __ too few? − 1 2 3 4 5 + Does a quick review of the page give you an idea of the logical arrangement of parts?
4. Is the content of the report clear and relatively easy to understand? __ clear definitions, __ processes clearly described and interpreted, __ subject sufficiently limited for thorough treatment? After reading the report do you feel that you understand the subject even if you are not expert in this field? − 1 2 3 4 5 +
5. Are the illustrations in the report clearly and neatly produced: __ clear directions in the text for finding the graphic illustrations, __ clear captions which identify the illustrations, __ a sufficient balance of verbal text and graphic illustration? − 1 2 3 4 5 +
6. Are the end note and bibliographical references written to specification? Are there references in the text which make clear the distinction between the writer's ideas and material from other sources? If you wanted to locate a reference quickly and easily, do the references in the report facilitate that? − 1 2 3 4 5 +
7. Is the writing in the report clear: __ are there sentences which you must re-read in order to understand them (if so, underline them), __ is there too much use of passive voice, __ are there too many *and*'s and *but*'s (circle those you consider ineffective), __ is the vocabulary appropriate to the subject and audience (circle any words you consider better suited to conversation than to reporting), __ is the point of view consistent (circle *I* or *you, me* or *my*, and similar pronouns), __ is the mood consistent or does the writer shift into the *you* point of view as though giving directions, __ are there choppy sentences? − 1 2 3 4 5 +
8. Is the manuscript neat and easy to read: __ are the pages crowded, or do you get a sense of sufficient space, __ are there errors in typing, __ is the

type face clear and clean, __ is the type dark enough for easy reading, __ type of sufficient size to assure ease of reading, __ is the paper of good quality rather than thin and transparent? − 1 2 3 4 5 +

Discuss the STRONGEST point about this report:

Discuss the WEAKEST point about this report:

Assume that you are a manager for whom the writer wrote this report; write the writer a letter of evaluation and advice about writing reports.

EVALUATION CHECKLIST: LAB REPORTS

Reader 1 _____   Writer _____
Reader 2 _____

1. Describe your first overall impression of the Lab Report. Is it neat, easy to handle, easy to see?

   Reader 1:

   Reader 2: I agree and/or disagree with Reader 1 because

2. Does the Summary tell you enough about the contents of the report that you would not have to read further? If not, what is left out?
   Reader 1:

   Reader 2: I agree and/or disagree with Reader 1 because

3. Does the paper include a list of symbols? Are these clearly written and would this page be easy to use as a reference?
   Reader 1:

   Reader 2: I agree and/or disagree with Reader 1 because

4. Does the Introduction give you a clear idea what the Report is about and what direction the discussion will take?
   Reader 1:

   Reader 2: I agree and/or disagree with Reader 1 because

Evaluation Checklist: Lab Reports

5. Is the Analysis clearly presented? Are there clear transitions to mathematical illustrations and are the references throughout the paper a clear and convenient means of finding appendixes?
   Reader 1:

   Reader 2: I agree and/or disagree with Reader 1 because

6. Are the explanations and descriptions in the Apparatus and Procedures sections clear? If you had not also done the work in the lab would you understand them with no difficulty?
   Reader 1:

   Reader 2: I agree and/or disagree with Reader 1 because

7. Is the Results and Discussion section clear? Does the writer show an understanding of the principles involved in rocess writing as well as a grasp of the technical aspects of the process itself?
   Reader 1:

   Reader 2: I agree and/or disagree with Reader 1 because

8. Does the Conclusion really "add up" to a logical and sufficient answer to the problem explored in the lab?
   Reader 1:

   Reader 2: I agree and/or disagree with Reader 1 because

9. How about the writing in the Lab Report?
   Reader 1: Did you find __ sentences which you had to read more than once __ punctuation which slowed your reading __ use of conversational words __ use of the first person (*I, me, my*) __ use of the second person (*you*) __ consistent forms in numbers as described in the 3023 memo __ which were too long __ paragraphs which were too short?

   Reader 2: I agree and/or disagree with Reader 1: my overall critique of the style and mechanics of this paper is

10. Such forms as this are used more and more by industry for evaluative purposes. Eventually you may have to use a similar form in making a decision about whether to hire, promote, or give a raise to a writer. Overall, what would be your judgment of this report in such a situation?
    Reader 1:   __positive   __negative
    Reader 2:   __positive   __negative

PROPOSAL EVALUATION CHECKLIST

Reader_____ Writer_____

1. Is the proposal visually easy to see and physically easy to handle? Would those factors influence your decision to accept it? − 1 2 3 4 5 +

2. Does the heading of the proposal tell you immediately __ the subject, __ the audience, __ the date? Is the heading an adequate record in addition to being a communication? − 1 2 3 4 5 +

3. Does the page design enable you to glance down the page and determine the logical arrangement of parts: __ major headings, __ subheadings, __ underlining, __ numbering, __ indentation? Are these design features adequate, __ too many, __ too few? − 1 2 3 4 5 +

4. Is the statement of the problem or purpose clearly understandable? __ Are terms clearly defined? __ If no, list the terms you found difficult:

   Do you gain a clear idea of how this specific problem relates to a larger conect? __
   Is it clear to you how the writer plans to solve the problem? __
   − 1 2 3 4 5 +

5. Do you find elements of technical persuasion as they are defined in the Writers' Handbook: __ clear and persuasive explanations, __ ethical proof which inspires your confidence in the writer, __ reasonable cause-and-effect connections, __ clear and objective language? Note any weaknesses in those elements of persuasion. − 1 2 3 4 5 +

6. Is the statement of fees reasonably made? __ Do you understand the reason for the costs as they are listed? __
   Does the schedule seem reasonable? __
   Do you consider the description of the personnel adequate? __ − 1 2 3 4 5 +

7. If you were a manager, would you contract for the services and problem-solving capabilities contained in this proposal? __ On the back of this page, fully discuss why or why not in terms of the text discussion of proposals and technical persuasion.

Long Report Evaluation

LONG REPORT EVALUATION

Grade _____   Writer _____   Reader _____

Using 1 as the lowest mark and 5 as the highest, rate the report on each of the questions below. On each question, make a comment which justifies your rating.

1. Is the report in a binder which makes it attractive and easy to handle?
   − 1 2 3 4 5 +
   Comment:

2. Does the title page follow the prescribed form from the text?
   − 1 2 3 4 5 +
   Comment:

3. Is the Table of Contents a clear guide to the report, both as a logical overview and as a useful reference guide to access the report at random?
   − 1 2 3 4 5 +
   Comment:

4. Does the introduction contain an adequate guide to subject, purpose, and scope, and a preview of the major parts the report will include?
   − 1 2 3 4 5 +
   Comment:

5. Does the body of the report contain a clear and logical pattern of development; are there numbered and underlined subheadings keyed to the Table of Contents? − 1 2 3 4 5 +
   Comment:

6. Is the report understandable, even if you are not expert in this field? Are the definitions clear? Are the parts clearly explained and is the relationship of the parts logical? − 1 2 3 4 5 +
   Comment:

7. Is there a suitable balance between illustrations within the text or the report and in the Appendix(es)? Do the illustrations aid your understanding; are they clearly identified; do you receive clear directions to finding an appendix? − 1 2 3 4 5 +
   Comment:

8. Are there quotations from other sources? Are they so long that the writer seems to be borrowing heavily from another source? Does the writer refer to authors, titles, and previous research on the topic to add authority to the statements in the paper? − 1 2 3 4 5 +
Comment:

9. Now read the Conclusions and Recommendations. Do they follow logically from the report? − 1 2 3 4 5 +
Comment:

10. Now read the Executive Summary. Is it an adequate and effective overview? − 1 2 3 4 5 +
Comment:

11. Now read the Abstract. Is it a 100-150 word condensation which would aid you as a researcher who needed to know the contents of this report? − 1 2 3 4 5 +
Comment:

12. Rate the writing of this report. Did you find sentences which were so long that you had to re-read them for understanding? ___ Did you find words better suited to conversation than to writing? ___ Did you find the use of first person (*I, me, my*) ___ or second person (*you*) ___ pronouns? Was the punctuation placed in a way that it impeded your reading? ___
− 1 2 3 4 5 +
Comment:

13. Managers are frequently required to read the work of their employees and make evaluations of job performance. If you were a manager reading this report, how would you rate the worker in your firm?
___positive
___negative

14. In the space below discuss the strongest and weakest points of this report:
STRONGEST

WEAKEST

Index

This index is compiled to function, first, for an audience which will read *Technical Writing: A Practical Approach* for the purpose of studying technical writing theory, and second, for the user who needs a quick reference guide to writing and revision.

READERS' INDEX FOR STUDYING TECHNICAL WRITING THEORY

Pages 15–20 in Chapter 2 of this text define *writing* in the traditional terms of *invention, arrangement, style, grammar,* and *mechanics* and applies those basic components to the practice of technical writing. Pages 21–27 in Chapter 3 of the text expand upon the principle of *arrangement* to discuss how technical writing is formal. Chapter 4 discusses the technical writer's relationship to the audience, the critical factor that makes technical writing different from general academic writing.

This section of the Index arranges all topics discussed in the book under the major headings used to discuss writing. Under *arrangement* are listed not only the traditional rhetorical patterns, but also all the major technical forms and the standard parts for each form. A student, for example, who wants to learn more about the Summary as a standard feature of technical writing thus has a guide to everything the book has to say about that aspect of technical writing.

This arrangement also continues the definition of technical writing by grouping major elements under *invention* and *style*. One who wants to understand more about how imagination and invention function for the technical writer, as distinct from the creative writer, can thus give a systematic reading to the headings under *invention*.

Style, along with the word *image*, is a word often used when describing a particular "*style* of management" or concern about projecting an appropriate *image* in business situations. Taken together, all of the individual bits of advice about using names and titles, sales effectiveness in technical professions, and responding to problem situations add up to advice about an appropriate management style for engineering and technical professionals. And writing is, of course, one of the major ways that an individual or a company forms and projects a defined image.

Sales elements in technical writing cannot be distinguished from the more traditional concerns with Argument and Persuasion, nor can technical writing function apart from carefully designed and managed systems. Points in the book which relate to those subjects are collected in this Index under the headings *persuasion* and *systems*.

This part of the Index becomes, then, not only a reference guide to what is in the book, but also a serious attempt to make the book useful for teaching pre-professionals about writing as an essential component in a long list of professional skills.

ARRANGEMENT, BASIC WRITING, 16, 252, 411
   CAUSE AND EFFECT, 18
      flaws in, 434
      graphics, 399
      linear outline, 413
      paragraphs, 391–393
      persuasive, 169, 392, 434
      process, 169, 391–392
      short technical report, 168
      style in, 392, 393
      technical, 391
      technical recommendations report, 35
   CLASSIFICATION AND DIVISION, 18
      basis in, 380, 416
      description, 387
      exceptions within, 381
      graphics, 389–399
      incomplete, 381
      paragraphs, 380–383
      vertical outline, 415–416
   COMPARISON AND CONTRAST, 17

ARRANGEMENT, BASIC WRITING (*Contd.*)
      alternating method, 399, 416
      basis in, 380, 415
      graphics, 399
      opposing method, 399, 416
      vertical outline, 415–416
   DEFINITION, 17
      analogy, 380
      connotation, 385–386
      graphics, 397
      by negation, 386
      non-technical, 385–386
      paragraphs, 383–386
      punctuation in, 385
      short technical report, 167
      stipulative, 386
      synonyms, 384–385
      technical, 255–256, 384–385
      of unfamiliar terms, 384
   DESCRIPTION, 17
      building, 387
      informal report, 119

Index

ARRANGEMENT, BASIC WRITING (*Contd.*)
   paragraphs, 386–389
   proposal, 235
   site, 119, 164, 235, 254, 256, 388
   space, 146, 164
   technical recommendations report, 254
DIRECTIONS
   letters, 149
   memorandums, 59
   short technical report, 170
DIVISION, 18
   cook book method, 411
   in cause and effect, 391, 393
   of groups, 382
   of ideas, 383
   of objects, 381
   paragraphs, 380–383
   in process, 389
      CHART, 412
   technical presentation, 317
   vertical outline, 315–318
LOGICAL ORDER, 16, 31
   CHART, 438
   correspondence systems, 146
   informal report, 112
   memorandum, 48–49
   proposal, 232, 234
   technical recommendations report, 262
PARAGRAPHS
   banks, 379
   in letters, 80
   rhetorical patterns
      definition, 383–386
      description, 386–389
      cause and effect, 391–393
      classification and division, 380–383
      process, 389–390
   transitional words in, 428
PROCESS, 18
   CHART, 412
   graphics, 402–403
   historical, 412
   implications in, 167, 390
   linear outline, 412–413
   logical, 412
   mechanical, 412
   paragraphs, 389–390
   physical, 412
   principle of, 167, 379–388
   proposal, 232
   relationship to cause and effect, 169, 391–392
   relationship of parts in, 412
   short technical report, 166
   technical recommendations report, 256–257

ARRANGEMENT, TECHNICAL WRITING, 21–27
   Abstract, Appendix A, 482–485

ARRANGE., TECHNICAL WRITING (*Contd.*)
   Apparatus, lab report, 213
   Appendix, 34
      lab report, 222
      technical recommendations report, 258
      when to use, 405
   Authorization, 33
      informal report, 119
      proposal, 237
   beginning a chapter, 267
   beginning a letter, 75–76
      how *not* to begin, 359
   CASE STUDY, 183–197
   cost analysis
      technical recommendations report, 237, 255
   Conclusions
      lab report, 220
      professional article, 428
      short technical report, 168, 171
      technical presentation, 319
      technical recommendations report, 269
CONSTRUCTION REVIEW, 13–14, 112, 119, 236
   Design, 22
   Appendix A, 463–487
   for audience, 31, 34
   CHART, Appendix A, 474
   employment letters, 123
   informal report, 117
   letters, 70
   memorandums, 48
   proposal, 11, 230
   short technical report, 165, 166
   professional article, 328
   RESUME, 128–129
   for ease of physical handling, 48, 59, 233, 234, 262
   for sales appeal, 11, 233, 234
   for visual effect, 48, 49, 233, 262
   as *working* documents, 81, 262
Discussion
   correspondence system, 146
   lab report, 219, 221
   short technical report, 168
   technical recommendations report, 254
example, in graphics, 397
Executive Summary, 260, 263
exordium, 165, 266
Fees, statement of, 233
Field Procedures, 166
Figures
   how to append, 212, 405, 480
   how to cite in text, 170, 405, 476
GRAPHICS, 12, 13, 34, 394–409
   in appendices, 407, 418
   classification and division, 398
   comparison and contrast, 407
   lab report, 221

ARRANGE., TECHNICAL WRITING (*Contd.*)
   mathematical illustrations, 218, 406
   partition, 395
   process, 402
   professional article, 327
   technical presentation, 317
   technical recommendations report, 258
   how to cite in text, 405, 476
GROUP PRESENTATION, 320–321
INFORMAL REPORTS, 111–119
Introduction
   group presentation, 320
   lab report, 212
   professional article, 428
   short technical report, 165–166
   technical presentation, 317
   technical recommendations report, 259, 264
Investigation, proposed, 231
Investigative Procedure
   informal report, 119
   short technical report, 165, 170
LABORATORY REPORT, 198–228
LETTERS, 69–110
   correspondence system, 144–153
   employment, 121–144
   routine situations (*see* STYLE)
   problem situations (*see* STYLE)
   mathematical illustrations, 218, 406
MEETING MINUTES, 59, 64–65
MEMORANDUMS, 47–68
Personnel
   how to cite, 146, 234
Problem, Statement of
   informal report, 112
   lab report, 212
   memorandum, 66, 67
   short technical report, 170
   technical recommendations report, 254, 255
Procedure
   field, 236
   investigative, 165
   laboratory, 166
   lab report, 215
PROGRESS REPORT, 52–53
Project Information
   proposal, 231, 236
   short technical report, 164
PROPOSAL, 229–250
   informal, 145
Recommendations, 33
   for further study, 170
   short technical report, 168
   technical presentation, 319
   technical recommendations report, 259, 269–270
References
   bibliography and end note form, 455–457

ARRANGE., TECHNICAL WRITING (*Contd.*)
   in lab report, 221
   professional article, 429, 455–457
   sources, 450
   techniques, 429, 455–457
      CHART, 160–162
      Manuscript specifications, Appendix A, 474, 477–479
Results
   in lab report, 221
   professional article, 327
REPORTS
   informal, 111–115
   LABORATORY, 198–228
   SHORT TECHNICAL, 163–182
   TECHNICAL RECOMMENDATIONS, 251–270
RESUME, 129
Summary
   group presentation, 321
   informal report, 118
   lab report, 211
   short technical report, 172, 177
   technical presentation, 317
   technical recommendations report, 259, 270
SUMMARY EXAMPLES
   CASE STUDY, 184–187
   LABORATORY REPORT (*See parts* under LAB REPORT in Writers' Index)
   LETTER
      formal technical, 81
      routine business, 79
      routine technical, 80
   MEMORANDUM, 58–59
   PROPOSAL, 231–233
   SHORT TECHNICAL REPORT, 164–171
   TECHNICAL RECOMMENDATIONS REPORT (*See parts* under TRR in Writers' Index)
Symbols
   lab report, 211
Table of Contents
   manuscript specifications Appendix A, 479
   technical recommendations report, 261
TECHNICAL PRESENTATIONS, 12, 312–321
total mind dump, 120
TRANSMITTAL LETTER
   in correspondence system, 145
   informal report, 115, 117–118
   proposal, 237
   short technical report, 171
   technical recommendations report, 237

INVENTION, 15, 252
AUDIENCE, 28–37
   CHART, 158
   correspondence system, 146
   employment letter, 122

Index 501

INVENTION (*Contd.*)
  graphics, 406
  lab report, 199
  letters, 31–32
  memorandum, 31, 59
  non-technical, 32, 146, 221
  persuasion, 432–433
  professional article, 323, 326
  proposal, 32, 232
  research project, 440
  short technical report, 318, 321
  technical recommendations report, 33, 253–254, 260
PURPOSE
  CHART, 158–159
  employment letters, 122
  lab report, 198
  letters, 75, 76
  memorandums, 59
  professional article, 324, 328
  short technical report, 165, 170
  sustained writing, 440
  technical recommendations report, 260, 266
SCOPE
  CHART, 158
  construction review, 119
  short technical report, 165
  sustained writing, 435, 438
  technical recommendations report, 260, 266
SOURCES
  CHART, 160–162
SUBJECT
  CHART, 158
  lab report, 198
  professional article, 324, 325, 329
  proposal, 230
  short technical report, 164
  sustained writing, 439–440
  technical presentation, 317
  technical recommendations report, 260
SUSTAINED WRITING, 437–440
TOPIC SELECTION, 441
  CHART, 158–459

PERSUASION, 430–436
ARGUMENT
  based on authority, 432
  by exposition, 237, 257, 431–432
  fallacies in argument
    call to perfection, 435
    oversimplification, 434
    post hoc, 434
    reification, 436
    unreasonable extrapolation, 435
  flaws in argument
    cause and effect, 434
    language, 436

PERSUASION (*Contd.*)
ETHICS, 38–44
PERSUASION, 34, 430–436
  cause and effect, 392
  correspondence system, 145
  group presentation, 321
  memorandum, 59
  motivating readers, 79, 95, 98, 149
  letters, 75, 77, 79, 80, 99, 100
  professional article, 327
  proposal, 237
  technical presentation, 318
  technical recommendations report, 257–258
Political Issues, professional article, 324
REFUTATION, 433

STYLE, BASIC WRITING
analogy, 19, 32, 74, 256, 390
creative writing, 10, 16, 19, 39
definition, 167–168, 255–256, 382–385
description, 387
cause and effect, 169, 392, 393
lab report, 119
letters, 74
memorandums, 49, 49
process, 167, 390
professional article, 327, 330
short technical report, 171
SENTENCES, 369–378
WORD USAGE
  causation, 362, 392
  colloquial, 356
  correct, 358
  "correct rule", 71
  current business, 358
  precise technical, 361
  pronouns, 358
  quantities, 361–362
  unidiomatic, 357
  verbs, 363–364, 435

STYLE, BUSINESS AND MANAGEMENT
CHARTS
    conventions in letters, 83
    situations in letters, 83
CONVENTIONS, 22
  employment letters, 123
  memorandums, 48
  letters, 70–74
  Names
    citing in a text, 429
    company name, 81, 127
    dropping a name, 127
    first names, 73
    full name, 72, 73
    record keeping, 72
    signature, 73
  Titles
    academic, 89

STYLE, *(Contd.)*
   gentlemen, 26, 72, 82, 85, 97
   ladies and gentlemen, 98
   medical, 87
   Miss, 72
   Mrs., 72
   Ms., 72
   official, 52, 62, 64, 72, 73
   professional, 80
   religious, 92
dictation, 356
fear of speaking, 313–314
fear of being too repetitious, 267
motivating readers, 79, 95, 98, 149
SALES, 34
   beginning a letter, 76
   correspondence systems, 145, 146
   ending a letter, 76–77
   how *not* to end, 360
   informal report, 111
   letters, 75, 77, 79, 80, 99, 100
   persuasion, 432, 433
   proposals, 11, 230, 233, 234
   technical presentation, 321
   technical recommendations report, 253, 257
PROBLEM SITUATIONS
   apologies, 150
   bad news, 150
   business disagreement, 113
   changes, 150
   compromises, 149
   cost overruns, 150
   delays, 150
   explanations, 96, 112, 150–151
   motivating readers, 79, 95, 98, 149
   perspectives, 151
ROUTINE SITUATIONS
   administrative, 65, 67
   directive, 61–62, 97–98, 149
   informal reports, 146–146
   informative, 60, 87–88, 89–91, 92–93, 99–100, 145–147
   personal contacts, 112–113
   requests, 85–86, 94–96
   reporting, 47–48, 95–97
   telephone conversations, 147–148
   transmitting documents, 115, 171, 237, 239
TECHNIQUES, 41

STYLE, *(Contd.)*
   letters, 74
   memorandums, 48

SYSTEMS IN TECHNICAL WRITING, 10–14
   CHART, 2
Correspondence Log
   employment letters, 123
   large technical project, 148
   professional article, 329
CORRESPONDENCE SYSTEMS, 121–153
Critical Path Method, 459
field book, 253
First Draft of a Technical Recommendations Report, 253–260
Functions of Technical Writing, 5–9
Group Presentations, 12, 320–321
managing time
   employment letters, 123
   large technical project, 147–148
   technical presentation, 316, 319
Manuscript Production and Specifications Appendix A, 463–487
note taking techniques, 452–453
physical setting for technical presentation, 351
publishing a professional article, 328
record keeping
   correspondence system, 145–146
   letters, 69–70, 71
   memorandums, 58
Shortest Path Method, 431
System for Sustained Writing, 437–440
System for Planning a Technical Presentation, 312–321
   setting, 315
   situation, 12
   timing, 316, 319
   visual aids, 317
      bar chart, 399
      chalk board, 316
      exploded drawing, 12, 395–396
      flip chart, 316
      flow chart, 404
      mathematical illustrations, 218, 406
      overhead projector, 318
      pie chart, 398
      slides, 12
      transparency, 318
Work Ethic, 38–44

WRITERS' INDEX FOR WRITING AND REVISION

The actual practice of technical writing leaves little time to stop and ponder over whether to write a memorandum or a letter, whether to use *Mrs.* or *Ms.*, and just what an Executive Summary looks like and where it goes in a report that is already past the due date for being delivered out the back door to the waiting audience.

This part of the Index is designed as a quick reference for the writer at work. Two traditional components of the writing process, *grammar* and *mechanics*, are reserved to this section of the Index to aid both the writer in need of a quick answer and the student whose graded paper presents the next, most essential step in learning to write—revision.

The student who has read about the Summary as a feature of technical writing will see in this section of the Index the same heading listed under *Memorandum, Informal Report, Technical Recommendations Report,* and *Case Study* as a guide for writing a specific summary in a specific document to the specifications of that particular form and, beyond that, to the *specifications* of that kind of document as it is *designed* on the page, for the *audience*, to fulfill the *purpose* of that form as it deals with the *subject*, in a given *situation*. Hence the definition of technical writing as "English for *special* purposes."

The general order of items under each form is (1) audience, (2) the checklist in Appendix B which covers all major features of a given form, (3) conventions which govern the form, (4) manuscript specifications as they appear in Appendix A, (5) a parts catalogue both as a guide to planning and a quick reference to answer questions during writing and preparing final copy, (6) references to the summary charts on purpose, scope, and subject to clarify the function and form of a given kind of document, and (7) the Summary Example for each of the shorter forms as an overall guide.

The rationale for this part of the Index is prescriptive—to reduce the number of instances where a student says of a graded paper, "But you didn't *tell* me!" or "But I didn't know what you *wanted*!" If the Index is prescriptive in that sense, it is humane in another; for in some instances the most serious study and practice of writing may occur between midnight and 8 a.m. on the morning before a paper is due—when the Library is closed, and no known expert would welcome a telephone call at 3 a.m. Listed below should be some workable answers for writers in that dark night of the soul.

ARRANGEMENT, FORM, AND STYLE IN WRITING AND REVISION

CASE STUDY 183–197
manuscript specifications CHART, Appendix A, 474
parts
    Analysis, 189
    Alternate Solutions, 195
    Competitive Dimension, 193
    Current Trends, 191
    Industrial Analysis, 185
    Major Issues, 188
    Policies, 196
    Strategies, 196
    Summary of Analysis, 187
page design, 408–409, 476
purpose, 183
    sources and citations techniques, Appendix A, 455–457
    CHART, 160–162
SUMMARY EXAMPLE, 184–187

INFORMAL REPORT, 111–120, 143–145
audience, 158–159
checklists
    letter, Appendix B, 490
    memorandum, Appendix B, 489
manuscript specifications, Appendix A, 464–465, 477, 468–469, 471
page design, 408–409
parts, logical organization, 111–112
purpose, 112–113
scope, 111
situation, 112–113
style, 112–113
subject, 112–113

LABORATORY REPORT, 198–228
audience, 199
    CHART, 158–159
checklist, Appendix B, 492
manuscript specifications, Appendix A, 474
page design, 408–409, 476
parts
    Abstract, Appendix A, 482
    Apparatus, 213
    Appendix, 222
    Conclusion, 220
    Discussion, 219
    Graphics, 221
    Introduction, 212
    Problem Statement, 212
    References, 221, 455–457
    Results, 219
    Title Page, Appendix A, 482–483
purpose, 199
    CHART, 158–159
scope, CHART, 158–159
sources and citations techniques, CHART, 160–162, 455–457, 477–478

ARRANGE., FORM AND STYLE (*Contd.*)
style, 198
subject, 198
    CHART, 158–159

LETTERS, 69–110
audience, 31
checklist, Appendix B, 490
conventions
    CHART, 83–84
    discussion, 70–74
manuscript specifications, Appendix A, 468–473
mistakes to avoid, 473
parts
    address, 72
    attention line, 81, 97–98
    beginning a letter, 75–76
        how *not* to begin, 359
        cc (copies), 472
    complimentary close, 72
    date, 71
    enclosures, 472, 93
    ending a letter, 76–77
        how *not* to end, 360
    names (*see* STYLE, conventions)
    page 2 of a letter, 473
    reference line, 80, 89
    salutation, 72
    signature, 73
    titles (*see* STYLE, conventions)
    typed name, 73
    typist's initials, 472
    writer's initials, 472
purpose, 69
    CHART, 83–84
scope (length), 111
situation CHART, 83–84
style (*see* CONVENTION)
SUMMARY EXAMPLE
    formal technical, 80
    routine business, 79
    routine technical, 80
visual design, 48, 49

MEMORANDUMS, 47–68
audience, 31
checklist, Appendix B, 489
convention, 48
mistakes to avoid, 473
manuscript specifications, Appendix A, 464–465, 477
page design, 408–409, 489
parts
    heading, 58
    names (*see* STYLE, convention)
    organization
        cook book method, 165
        logical, 112
    reference line, 58

Index

ARRANGE., FORM AND STYLE (*Contd.*)
titles (*see* STYLE, conventions)
purpose, 47
situation
    when to write a memo, 70, 111–113
    administrative, 65
    directive, 61–62
    explaining a problem, 112
    informative, 60
    meeting minutes, 64
    reporting an action, 60
style, 49
SUMMARY EXAMPLE, 58–59
visual design, 48, 59
writer's initials, 58, 464, 465

PROFESSIONAL ARTICLE, 322–352
audience, 326
    CHART, 158–159
bibliography, 455–457, 477–478
checklist
    informal, 238, 330
endnotes, 455–457, 477–478
manuscript specifications, CHART, Appendix A, 474
note taking techniques, 452
parts
    logical order, 418–419
    outlines, 330
    tell 'em method, 259, 411
purpose, CHART, 158–159
scope, 327
    CHART, 158–159
situation, 328–329
sources and citations techniques, CHART, 160–160, 455–456, 477–478
subject, 325
    CHART, 158–159

PROPOSAL, 229–249
audience, 32, 230
    CHART, 158–159
checklist, Appendix B, 494
manuscript specifications, 11, 230
    CHART, Appendix A, 474, 475–481
parts
    Authorization, 237
    Fees, Statement of, 233
    Job Number, 231
    Personnel, listing, 233
    Project Information, 231
    Schedules, 233
    Scope of Services, 231, 236
    Site Description, 235
    Transmittal Letter, 237
page design, 233, 408–409, 476
purpose, 327
    CHART, 158–159
scope, CHART 158–159
situation, project, 231

ARRANGE., FORM AND STYLE (*Contd.*)
sales, 230
sources and citations techniques, CHART, 160–162, 455–457, 477–478
style, 234
SUMMARY EXAMPLE, 231–233
subject, 230, 231
    CHART, 158–159

SHORT TECHNICAL REPORT, 163–182
audience, 33, 170
    CHART, 158–159
checklist, Appendix B, 491
manuscript specifications, Appendix A, CHART, 474
page design, 166, 408–409, 476
parts
    Conclusion, 168
    discussion, 168
    further studies, 170
    graphics, 394–409, 170
    introduction, 165–166
    Job Number, 164
    Problem, Statement of, 170
    Procedures
        field, 166
        laboratory, 166
    Project Information, 164
    Recommendations, 168
    Site Description, 164
    Transmittal Letter, 171
purpose, 164
    CHART, 158–159
scope, 165
    CHART, 158–159
situation, project information, 164
SUMMARY EXAMPLE, 164–171

TECHNICAL RECOMMENDATIONS REPORT, 251–311
audience, 33, 258–254
    CHART, 158–159
checklist, Appendix B, 495
manuscript specifications, Appendix A, CHART, 474, 475–481
page design, 408–409, 476
parts
    Abstract, Appendix A, 482
    Appendix, 258, 407, 478
    Beginning a Chapter, 267, 477
    Cost Analysis, 258
    Conclusion, 269
    Executive Summary, 263
    Discussion, 254
        logical organization, 418–419
    Figures in the text, 405, 476
    Graphics
        bar chart, 399
        exploded drawing, 395–396
        flow chart, 404

ARRANGE., FORM AND STYLE (*Contd.*)
    lists, 98
        pie chart, 398
    Recommendations, 269
    Table of Contents, 261–262, 479
    Title Page, Appendix A, 482–483
    Transmittal Letter, 237
purpose, 440
    CHART, 158–159
scope, CHART, 158–159
sources and citations techniques, CHART, 160–162, 455–457, 477–478
subject, 439, 442–447
    CHART, 158–159

GRAMMAR AND MECHANICS IN WRITING AND REVISION
abbreviations, 384
acronyms, 384
Arabic numerals, 262, 361
awkward sentence, 378
BASIC WRITING SKILLS, 36, 49, 78
colloquial usage, 356
colloquial verbs, 364
colon, 375
    salutation in letter, 80
commas, 374–375
    complimentary close in letter, 80
    items in series, 385
    unnecessary, 378
current business usage, 359
conjunctive adverbs, 372
CONVENTIONS (*see* STYLE)
coordinating conjunctions, 371
    faulty use, 377
"correct rules", 29, 71
correct usage, 358
correlative conjunctions, 372
dangling modifier, 377
dash, 375
flaws in sentences, 376
    absent semicolon, 378
    awkward sentence, 378
    faulty coordination, 377
    loose adjective clause, 377
    faulty parallelism, 377
    misused semicolon, 378
    run-on, 377
    unnecessary comma, 378
faulty parallelism in sentences, 377
Gentlemen, 72
Initials
    typist's, 472
    writer's, 464, 472
Jargon, 327
listing items in series, 98
loose adjective clause, 377
Ms., 72
names (*see* CONVENTIONS)

GRAMMAR AND MECHANICS (*Contd.*)
PARAGRAPHS, 379–393 (*see* ARRANGEMENT)
    transitions, 428
passive voice, 363, 377
precise technical usage, 361
PRONOUNS
    he, she, 359
    I, me, my, 358
    myself, 358
    who, whom, 358
    yourself, 358
PUNCTUATION
    in bibliography, 456
    colon, 375
        salutation, 79
    comma, 374, 378, 385
        complimentary close, 79
    dash, 375
    in definition, 384–385
    in end notes, 456
    hyphen, 385
    italics,
        words, 384
        titles, 456
    leaders, 262
    quotation marks
        words, 385
        titles, 456
    semicolon, 375, 378
run-on sentence, 377
semicolon, 372, 375
    absent, 378
    misused, 378
    punctuated items in series, 375
SENTENCES
    awkward sentence, 378
    conjunctive adverbs with semicolon, 372
        misuse, 378
    coordinating conjunctions, 371
        faulty use, 377
    correlative conjunctions, 372
    dangling modifier, 377
    flaws in sentence structure
        absent semicolon, 378
        awkward sentence, 378
        faulty coordination, 377
        loose adjective clause, 377
        faulty parallelism, 377
        misused semicolon, 378
        run-on, 377
        unnecessary comma, 378
        faulty parallelism, 377
    loose adjective clause, 377
    Multiple Components, listing items in series, 98
    passive voice, 363, 377
    run-on sentence, 377
    transitional words, 167, 387, 390, 392, 393

Index

GRAMMAR AND MECHANICS (*Contd.*)
    conjunctive adverbs, 372
    coordinating conjunctions, 371
    correlative conjunctions, 372
    punctuation, 374–375
    testing for sentence flaws
        length, 376
        and-but; but-so, 376
        and-which; but-which; but-because, 376
        which clause; *because* clause, 376, 377
        *bo*ing, *bo*ing effect, 376, run-on, 377, 390
        O'Shea effect, 376
STANDARD ENGLISH, 74, 355
total mind dump, 120
TRANSITIONAL WORDS AND PHRASES, 428
    cause and effect, 392, 393
    description, 387
    process, 390, 167
    techniques for use
        citing authority, 432
        citing a name, 429
        referring to an appendix, 212, 405
        referring to a figure in the text, 170, 405
        referring to a reference source, 477
        in a technical presentation, 318, 320
unusual words, 168
    when to define, 384, 168
    when to punctuate, 384
WORDS, 355–363
    colloquial, 356
    correct, 358
    current business, usage, 359
in technical communication, 430–431
precise technical usage, 361
pronouns
    he, she, 359
    I, me, my, 358
    myself, 358
    who, whom, 358
    yourself, 358
    relationship to cause, 392
        as, 362, 392
        because, 362, 392
        due to, 362
    relationships to quantities
        amount/number, 362
        few/less, 362
        over/more than, 362

GRAMMAR AND MECHANICS (*Contd.*)
    plus/in addition to, 362
    unidiomatic usage, 357
verbs
    colloquial, 364
    passive voice, 363, 377
    strong verbs, 363–364
    verb forms, 363
        can, 363
        may, 363, 435
        might, 363, 435
        ought, 363
        shall (contractional), 363
        should, 363
        will, 230, 363, 435

EXERCISES FOR FURTHER STUDY
Bibliography and Endnotes, 458
Letters, Checklist, Appendix B, 490
    employment letters, 137–144
    letter of technical information, 101–110
Memorandums, Checklist, Appendix B, 489
    basic form and organization, 49–58
Reports
    lab report, Checklist, Appendix B, 492–493
    professional article, informal checklist, 329
        intensive reading, 428–429
    proposal, Checklist Appendix B, 494
    short technical report, Checklist Appendix B, 491–492
    technical recommendations report, Checklist, Appendix B, 495–496

DEVELOPING WRITING SKILLS
Outlines
    thinking, 411, 413, 417, 418, 419
    reading, 428–429
    writing (*See* Thinking and Reading Exercises)
Paragraphs
    reading, 49–50, 393, 428
    writing, 49–50, 393
Persuasion, 436
Sentences
    imitation, 372
    reading, 428–429
    writing, 372–373
Words
    Listening, 364
    Reading, 49–50, 366, 393, 428
    Writing, 356, 436

A62TCCS2395